**BEHAVIOR OF MARINE
ANIMALS • VOLUME 6**
Current Perspectives in Research

SHOREBIRDS

Migration and Foraging Behavior

BEHAVIOR OF MARINE ANIMALS

Current Perspectives in Research

Series Editors: Howard E. Winn and Bori L. Olla

**BEHAVIOR OF MARINE
ANIMALS • VOLUME 6**
Current Perspectives in Research

SHOREBIRDS

Migration and Foraging Behavior

Edited by
Joanna Burger
Rutgers University
Piscataway, New Jersey

and

Bori L. Olla
Oregon State University
Newport, Oregon
and Northwest and Alaska Fisheries Center
Seattle, Washington

PLENUM PRESS • NEW YORK AND LONDON

Library of Congress Cataloging in Publication Data

Winn, Howard Elliott, 1926–
  Behavior of marine animals; current perspectives in research.

  Vol. 4 edited by J. Burger, B. L. Olla, and H. E. Winn; v. 6, by J. Burger and B. L. Olla.
  Includes bibliographies.
  Contents: v. 1. Invertebrates. — v. 2. Vertebrates. — [etc.] — v. 6. Shorebirds, migration and
foraging behavior.
  1. Marine fauna — Behavior — Collected works. I. Olla, Bori L., joint author. II. Burger,
Joanna, joint author. III. Title.
QL121.W5   591.5′2636   79-167675
ISBN 0-306-41591-7 (v. 6)

© 1984 Plenum Press, New York
A Division of Plenum Publishing Corporation
233 Spring Street, New York, N.Y. 10013

Printed in the United States of America

CONTRIBUTORS

Joanna Burger Department of Biology and Bureau of Biological Research, Rutgers University, Piscataway, New Jersey 08854

Peter R. Evans Department of Zoology, University of Durham, Durham DH1 3LE, England

J. D. Goss-Custard The Institute of Terrestrial Ecology, Furzebrook Research Station, Wareham, Dorset BH20 5AS, England

R. I. G. Morrison Canadian Wildlife Service, Ottawa, Ontario, Canada K1A 0E7

J. P. Myers Academy of Natural Sciences, Philadelphia, Pennsylvania 19103; Bodega Marine Laboratory, University of California, Bodega Bay, California 94923; and Biology Department, University of Pennsylvania, Philadelphia, Pennsylvania 19104

Michael W. Pienkowski Department of Zoology, University of Durham, Durham DH1 3LE, England

Gillian M. Puttick Percy Fitzpatrick Institute of African Ornithology, University of Cape Town, Rondebosch 7700, South Africa. *Present address:* Museum of Comparative Zoology, Harvard University, Cambridge, Massachusetts 02138

INTRODUCTION

Among birds, shorebirds provide some of the more unique opportunities to examine basic problems in behavior, ecology, and evolution. This is in large measure due to the diversity, both behaviorally and ecologically, of a group closely related taxonomically and distributed throughout the world. The overall aim of these two volumes is to provide a representative selection of current research being conducted on shorebird behavior and ecology.

Traditionally, marine birds have included those species that breed in large colonies on offshore islands along coasts (see Volume 4 of this series). Although shorebirds have generally not been considered within this group, the fact that almost 40% of the species breed along coasts and more than 60% often or always spend the nonbreeding season in coastal habitats more than justifies their inclusion as marine birds (at least those species that totally or partially depend upon the marine environment). Their inclusion markedly increases species diversity in marine birds since shorebirds add about 217 species to the 280 that are traditionally thought of as marine.

Over the last twenty years there has been a veritable explosion in studies on shorebirds. Prior to this, there were few studies of breeding behavior and only cursory examinations of their migratory patterns. The increased interest in shorebirds was not the least caused by the increased awareness of the diversity and plasticity of their social systems. More pragmatically, the increased accessibility to researchers of Arctic breeding grounds as well as increase in the availability of fixed winged planes and helicopters for censusing populations and monitoring movements during the nonbreeding season also did much to foster the research effort.

Advances in shorebirds research have been of vital interest to those studying other animal groups, especially in the area of foraging behavior, with many of the paradigms and theories developed for optimal foraging theory being readily applicable to, and testable in shorebird species. Unlike the traditional marine species, their prey (primarily invertebrates) can be readily sampled both qualitatively and quantitatively. Further, since they forage primarily in estuaries it is possible to follow individuals in the nonbreeding season. The variability imposed by tidal regimes only make the system more valuable for field testing foraging theory.

Future studies on shorebirds will continue to center on examining breeding behavior and ecology, migration, and foraging behavior. However, more emphasis will be placed on examining the entire life cycle of particular species, comparing the success of populations that differ in breeding sites, breeding and nonbreeding habitats, and migration routes. Finally, the mechanisms that control behavioral and ecological plasticity in shorebirds should provide some of the more interesting avenues for future research.

J. Burger
B. Olla

PREFACE

Most chapters in these two volumes on the behavior of shorebirds include a balance between literature review, original data and synthesis. The research approaches of the authors range from in-depth study of a single species to a comparative one involving groups of species that are taxonomically or ecologically related.

The subject has been divided into two volumes, one dealing with breeding behavior and populations, and the second with migration and foraging behavior. The first two chapters of Volume 5 serve as an introduction to the two volumes and address the issues of shorebird classification and whether shorebirds are marine animals. Most of the remaining chapters in Volume 5 concern aspects of breeding behavior including breeding site fidelity, polyandrous mating systems, communication, parental behavior, and antipredator behavior. Two other aspects discussed in Volume 5 are population dynamics and conservation of shorebirds.

Volume 6 covers behavior aspects of shorebirds during the non-breeding season and includes abiotic factors affecting migration, migratory behavior in the Western Palearctic and New World, foraging and activity patterns, food supplies, and spacing patterns of nonbreeding migrants. Taken together these two volumes provide paradigms for the further study of organisms that move freely between marine and terrestrial ecosystems, as well as summarizing current research with a particular group of birds. Some other important areas of shorebirds behavior (general mating systems, sexual dimorphism) have not been included because other reviews of these topics exist.

In recent times man's encroachment has been particularly strong on the beaches and shorelines of marine habitats used extensively by shorebirds. A volume elucidating aspects of their breeding, migration, and wintering behavior and ecology is particularly useful to managers and conservationists involved with coastal planning and protection of marine avifaunas. Their accessibility for study in breeding and nonbreeding habitats permits the identification of those factors which play a role in controlling population levels, a prerequisite for the application of rational management plans. For ornithologists, ecologists and behaviorists the two volumes provide a review of the concepts concerning shorebird behavior and the elucidation of the basic mechanisms affecting behavior.

We are especially grateful to Michael Gochfeld, Betty Green, Jill Grover, Brook Lauro, Cindy Paszkowski and Gary Braun for their help in indexing, and to the authors of Chapter 1 in Volume 5 for providing a shorebird classification that could be used as a basis for the other chapters in the two volumes.

J. Burger
B. Olla

CONTENTS

Chapter 3
Migration Systems of Some New World Shorebirds

R. I. G. Morrison

Chapter 4
Foraging and Activity Patterns in Wintering Shorebirds

Gillian M. Puttick

Chapter 5
Intake Rates and Food Supply in Migrating and Wintering Shorebirds

J. D. Goss-Custard

Chapter 6
Spacing Behavior of Nonbreeding Shorebirds

J. P. Myers

Chapter 1

ABIOTIC FACTORS AFFECTING MIGRANT SHOREBIRDS

Joanna Burger

Department of Biology and Bureau of Biological Research
Rutgers University
Piscataway, New Jersey 08854

I. INTRODUCTION

Abiotic factors affect the distribution and behavior of all marine organisms at all times of the year. The timing and location of reproductive activities are determined by seasonal and weather-related patterns. Similarly, sedentary and migratory life history strategies of marine animals evolved in response to seasonality and environmental constraints.

Shorebirds have evolved both sedentary and migratory life history patterns. Whereas only half of the Charadrii species are migratory, 85% of the Scolopaci migrate (Burger, Vol. 5 of this series). Migratory shorebirds congregate on inland marshes and lakes, and along coastal bays and estuaries, often feeding on mudflats, marshes, and tidal sloughs (see Morrison, this volume). Recently, several studies have documented the numbers of migrating shorebirds (Pitelka, 1979; Prater, 1981), often concentrating on habitat utilization (Storer, 1951; Recher, 1966; Thomas and Dartnall, 1971; Robertson and Dennison, 1979) and population dynamics (Hartwick and Blaylock, 1979; Isleib, 1979; Prater, 1979; Page *et al.*, 1979). The importance of abiotic variables (such as time of day or year, tides, salinty, weather-related factors) has been mentioned in passing by several authors, but only recently have investigators concentrated on finding out how particular abiotic factors affect shorebirds.

In shorebirds, abiotic factors have shaped the evolution of life history strategies and migratory patterns over evolutionary time, as well as serving as proximate cues initiating migratory behavior in any given year.

1

Weather-related factors (temperature, wind), length of growing season, and daylength all affect how far north or south particular shorebirds breed or overwinter (see Pitelka *et al.*, 1974; Pienkowski and Evans, this volume). High winds, fog, and cold weather all affect reproductive behavior (see Miller, 1979). Similarly, abiotic factors affect the movements and behavior of migrant shorebirds on their stopover areas. Migration can either involve long, nonstop flights to wintering grounds (Dick *et al.*, 1976) or a series of large jumps interspersed with protracted periods of resting and feeding [presumably to replenish fat stores (Hamilton, 1959)]. The hourly or daily behavior and movements of shorebirds at staging areas are influenced by abiotic factors. Abiotic factors affect shorebirds directly by causing injury or mortality, influencing their location and habitat usage, and determining behavior (feeding or resting, type of feeding); and indirectly by affecting prey availability (see below, and chapters by Puttick, and Pienkowski and Evans, this volume). Other factors such as density (Goss-Custard, 1977) and the presence of territorial conspecifics (see Myers, this volume) also influence the behavior of migratory shorebirds.

In this chapter I review and examine the affects of abiotic factors on migrant shorebirds, and report the results of a year-long study of the abiotic factors affecting the abundance and distribution of shorebirds at a coastal estuary in New York. Abiotic factors examined include temporal (time of day, time of year), tidal (tide time, tide height), weather-related (temperature, wind cloud cover) and habitat variables.

II. ABIOTIC FACTORS AFFECTING MIGRANT SHOREBIRDS

A. Temporal Factors

Temporal factors that could affect shorebirds include time of year, time of day, and daylength. The latter two are distinct factors, as the total amount of light influences available feeding time or types of feeding behavior (where shorebirds feed at night).

1. Seasonality

Seasonal constraints affect shorebirds primarily by setting the broad limits of the timing of reproduction, migration, and overwintering behavior. Most migrant shorebirds breed in the Arctic or north temperate zone.

For north temperate species, spring migration is usually shorter and less protracted than fall migration even in southerly stopover areas, such as Costa Rica (Storer, 1951; Recher, 1966; Smith and Stiles, 1979). Indeed, in many migration areas, fewer species are recorded in the spring compared to the fall (Urner and Storer, 1949; Oring and Davis, 1966), which might also reflect the use of different migration routes. Presumably the constraints of a short breeding season in Arctic areas have made it important for shorebirds to arrive on the breeding grounds as soon as weather and food resources permit (see Pitelka *et al.*, 1974). Such constraints are removed following the breeding season, and many shorebirds cover less distance per time period in the fall (see Myers, 1981).

In many migratory stopover areas, shorebird numbers may build up to several hundred thousands for a few days or weeks (see papers in Pitelka, 1979; Prater, 1981). Although the time of the year influences the number and species composition of shorebirds at any staging area, their distribution and abundance among habitats at that staging area are less influenced by time of year. For a discussion of seasonal effects on shorebirds, see the chapter by Puttick in this volume.

2. Time of Day

Daily patterns of activity have been noted for several species of birds [see Burger (1982) for review]. However, it is difficult to determine if behavior is influenced by time of day for birds that are clearly responding to tide rhythms (see below). Careful analyses of data, often involving computer analysis, can result in separating the effects of tides from those of time of day, although few authors have undertaken such analyses.

In a year-long study on sandy beaches (Cape Town, South Africa), McLachlan *et al.* (1980) reported that White-fronted Sandplover (*Charadrius marginatus*) numbers did not vary by time of day, although Sanderling (*Calidris alba*) numbers did vary. Although Sanderlings fed all day, a maximum number were present in the early morning and late afternoon. Regardless of tide stage, there was a lag at midday, suggesting a temperature effect. Ehlert (1964) reported that the duration of feeding varies as a function of time of day. Indeed, the data from many studies indicate that it is daylength, and not time of day (but see below), that influences shorebird behavior, particularly in the winter when daylengths are short.

3. Daylength

Recently, several authors have reported that daylength is an important contributor to shorebird foraging behavior, as available light is limited in the winter. Initially, authors reported that an increased proportion of daylight hours were spent foraging in winter compared to summer. Puttick (1979, and this volume) reported that Curlew Sandpipers (*Calidris ferruginea*) in South Africa spent 55–65% of the available daytime foraging in the spring and summer, but spent 80% of the day foraging in the autumn and winter (at high tide they fed in the marshes). The increase in percent of daylight hours spent foraging is not merely a matter of fewer available hours, as the mean number of minutes spent foraging did not correlate with the mean number of daylight minutes for the 12 months of her studies (Kendall tau = 0.20, Z = 0.9, not significant). The lowest mean daily duration of foraging occurred from September through November (spring in South Africa). Baker (1981) reported that Ruddy Turnstone (*Arenaria interpres*) and Common Redshank (*Tringa totanus*) fed during almost all the available daylight. Goss-Custard (1977, 1979) noted that shorebirds spend only about 70% of the daytime feeding in the autumn, and 95% of the daylight hours feeding in the winter. At high tides Common Redshank roosted in the spring and fall, but fed in fields in the winter. In a study of several species of shorebirds in England (see Evans, 1981), the average proportion of available daylight used for feeding increased for all species, although it increased disproportionately for smaller species (Common Redshank, Red Knot *Calidris canutus*, and Dunlin *C. alpina*) compared to larger species (Eurasian Oystercatcher *Haematopus ostralegus*, Black-bellied Plover *Pluvialis squatarola*, Eurasian Curlew *Numenius arquata*, Bar-tailed Godwit *Limosa lapponica*, and Ruddy Turnstone).

Not only do shorebirds increase the amount of daylight hours spent foraging in the winter, but many species feed at night (Goss-Custard, 1969; Baker and Baker, 1973; Evans, 1976; Goss-Custard *et al.*, 1977). Goss-Custard (1969) reported that Common Redshank did not feed at night in the summer, but they did from November to February on dark nights, and October through March on bright nights. In midwinter, Common Redshank obtained less than 50% of their daily food requirements (energy) from the estuaries during daylight; the rest came from night feeding and field feeding at high tide.

Night foraging has been reported for several species (see Table I; Evans, 1979; Pienkowski, 1982). However, some species (e.g., Black Oystercatcher *Haematopus bachmani*) apparently feed at night sometimes, but not others (Hartwick and Blaylock, 1979). The number of species of shorebirds that feed at night during migration and on the wintering grounds is undoubtedly greater than our current knowledge would suggest. With the availability of image-intensifying night vision scopes, this aspect of shorebird breeding biology will become better known. During the spring migration (1982), I used a Smith and Wesson Image Intensifier to observe

Table I. Shorebirds That Feed at Night (Usually in Winter)

	Habitat	Source
Eurasian Oystercatcher	Cockle beds, bay area	Davidson (1968), Heppleston (1971, 1972), Dare and Mercer (1973), Hulscher (1976)
American Oystercatcher	Tidal sandflats and mudflats	Burger (this study)
African Black Oystercatcher	Flats	McLachlan and Liversidge (1978)
Eurasian Avocet	Flats	Makkink (1936)
Black-bellied Plover	Tidal flats	Dugan (1981), Pienkowski (1982)
Ringed Plover	Tidal flats	Pienkowski (1982)
Piping Plover	Tidal sand beach	Burger (this study)
Common Redshank	Tidal flats	Baker (1981), Goss-Custard (1969, 1970)
Common Greenshank	Mudflats	Tree (1979)
Willet	Tidal flats	Stenzel et al. (1976), Burger (this study)
Spotted Sandpiper	Mangrove mudflats	Gochfeld (1971)
Eurasian Curlew	Mudflat	Evans (1976)
Ruddy Turnstone	Sandflats and mudflats	Burger (this study)
Bar-tailed Godwit	Mudflats	Smith (1975), Smith and Evans (1973)
Red Knot	Mudflats	Prater (1972)
Sanderling	Tidal sandflats	Burger (this study)
Dunlin	Mudflats	Evans (1976), Mascher (1966)
Swinhoe's Snipe	Woods	Frith (1976)
Eurasian Woodcock	Boggy thickets	Bent (1927)
American Woodcock	Meadows	Bent (1927)

shorebirds on five nights in May on coastal Long Island, New York. During this time I observed six species foraging at night on tidal sandflats and mudflats (Table I).

Presumably shorebirds feed at night because they cannot obtain enough food during the day (e.g., Goss-Custard, 1969; Heppleston, 1971; Pienkowski, 1981). Even visual foragers (such as plovers) are able to feed at night by using the low-intensity light or by detecting their prey by touch or sound (Pienkowski, 1981). Nonetheless, the pecking rate of visual feeders might decrease more than that of tactile feeders at night. Pienkowski (1981, 1982) reported that even at twilight the pecking rate of Ringed Plovers (*Charadrius hiaticula*) decreased significantly. Similarly, the pecking rate of Black-bellied Plovers also decreased at night (Pienkowski,

1982). However, prey type and size (and therefore energy values) taken were not compared between day and night. It has been estimated that Red Knot obtained four times the quantity of prey (*Macoma balthica*) in the day compared to at night (Prater, 1972). Contradictory data exist for foraging rates for Eurasian Oystercatchers: Davidson (1968) and Drinnan (1957) found that they took the same number of cockles (*Cardium edule*) at night as during the day, although Heppleston (1971) reported that the night capture rate was only 58% of the daylight value. This difference may well be due to differences in feeding techniques, prey abundance, temperature severities, or available light (see Hulscher, 1976). The foraging behavior of shorebirds at night requires additional study with particular attention directed to quantification of independent variables such as prey density, prey behavior, and available light.

Recently, Dugan (1981) discussed the importance of nocturnal foraging in shorebirds, noting that some prey organisms (especially intertidal and marine polychaetes) are more active at night, and shorebirds feeding on them can obtain a greater rate of biomass intake at night than by day. Dugan (1981) suggests that some Black-bellied Plovers obtained the major part of their food at night rather than by day by feeding on large *Nereis virens* worms on an estuary in England. This is the type of research necessary to evaluate adequately the importance of nocturnal foraging to shorebirds. But species must be examined separately, with detailed studies of their prey abundance, availability, and behavior.

B. Tidal Factors

Tide is the major factor influencing the distribution, abundance, and behavior of estuarine organisms, including shorebirds (Evans, 1979). For shorebirds, tide affects both the amount of available foraging space, and the availability of prey (Recher, 1966; Evans, 1979; Puttick, 1980). Recher (1966) suggested that the abundance and distribution of shorebirds (both individuals and species) are dependent upon the amount of available feeding space (rather than food density). Prater (1981) has shown that the number of shorebirds wintering on British estuaries increases with estuary size but not proportionately. Hence, overall densities of birds are highest on the smaller estuaries.

Most investigators have reported that shorebirds feed on exposed intertidal areas at low tide and roost on the fields, marshes, and bays at high tide (Hope and Shortt, 1944; Wolff, 1969; Thomas and Dartnall, 1971; Puttick, 1979; papers in Pitelka, 1979; Tree, 1979). The list of species reported to show this pattern is lengthy and includes: Black Oystercatcher

(Morrel *et al.*, 1979; Hartwick and Blaylock, 1979), Double-banded Dotterel [*Charadrius bicinctus* (Robertson and Dennison, 1979)], Shore Plover [*Thinornis novaeseelandiae* (Phillips, 1977)], Willet [*Catoptrophorus semipalmatus* (Stenzel *et al.*, 1976)], Long-billed Curlew [*Numenius americanus* (Stenzel *et al.*, 1976)], Bar-tailed Godwit (Smith and Evans, 1973), Marbled Godwit [*Limosa fedoa* (Kelly and Cogswell, 1979)], Ruddy Turnstone (Robertson and Dennison, 1979), Semipalmated Sandpiper [*Calidris pusilla* (Bent, 1927)], and Curlew Sandpiper (Elliott *et al.*, 1976; Puttick, 1979). Gerstenberg (1979) reported similar tidal patterns for 34 species of shorebirds studied in Humboldt Bay, California. Indeed, for only a few species have authors noted a lack of responsiveness to tide cycles in tidal habitats [White-fronted Sandplover (McLachlan *et al.*, 1980), Least Sandpiper, *Calidris minutilla* (M. A. Howe and M. T. Agricola, personal communication), Collared Plover, *Charadrius collaris* (Strauch and Abele, 1979)].

In the above studies, authors usually mention the tide pattern, but do not provide quantitative data either for abundance in each habitat or for activity (feeding, roosting) while in each habitat. Further, generalizations and oversimplification of the effects of tides on feeding behavior fail to take into account the complexities of differences in tide cycles, habitat use, activity patterns, and foraging success among species. For example, the most significant restriction of foraging space and time for shorebirds may well occur during low tides in which only limited area is exposed [as when winds keep water levels high (Recher, 1966)]. As was shown above, short daylengths restrict available foraging time, often forcing birds to feed all day and at night. Thus, clear seasonal patterns should emerge. Similarly, the reports of shorebirds on fields and marshes often state or imply that they are roosting, when many may well be feeding (see Heppleston, 1971). Indeed, Baker and Baker (1973) noted that the six species they studied during the winter continued to feed throughout the tide cycle.

In a survey of tidal areas in the Philippines (April–May 1983), I noted that Snowy Plover (*Charadrius alexandrinus*), Mongolian Sandplover (*C. mongolus*), and Greater Sandplover (*C. leschenaultii*) fed on tidal mudflats at low tides, and moved into rice paddies, fish ponds, and salt ponds during high tides. Thus, these species were observed in inland areas half of the time, where they continued to feed in man-made habitats. Several other shorebird species also moved back and forth between tidal and nontidal areas, but they were not as abundant in Luzon, Mindanao, and Palawan.

Differences in behavior have been noted as a function of tide cycle, although most authors have shown that shorebirds feed just below or above the water's edge (Recher, 1966; but see Pienkowski, 1981). Senner (1979) noted that shorebirds fed on both rising and falling tides. However, Puttick (1979) found that although Curlew Sandpipers made more feeding attempts on an incoming tide, success rates (number captures/number of attempts) were lower. On the contrary, for Black-bellied Plover, foraging rate increased as the tide fell, and the rate rapidly decreased 2 hr after low tide (Baker, 1974). Goss-Custard (1977), however, found no difference in success rates for Common Redshank in the 3 hr before, compared to the 3 hr after low tide. Again, I suspect success rates are dependent on so many factors that direction of tide alone cannot account for these differences.

The complexity of the relationships of responses of individual species to the effects of tides makes it essential for investigators to gather quantitative data on how individuals respond to tide, and four such studies will be mentioned Burger *et al.* (1977, 1979) examined 12 species of shorebirds feeding on three habitats (inner sandy beach, oceanfront sandy beach, and tidal mudflat) in southern New Jersey. They were particularly interested in tidal effects, and divided the tide cycle into hourly time blocks before and after low tide. In general, birds present were feeding. All species feeding on the mudflats were influenced by tide: for most species, increasing numbers came to feed as the tides dropped, and peak numbers fed an hour after low tide. Similarly, tidal effects were noted on the beaches as shorebirds fed immediately after high tide (both beaches), and after low tide (outer beach only, see Fig. 1). As is clear, shorebirds responded both to tide times and to habitats, moving among the habitats as tide conditions changed.

Connors *et al.* (1981) examined the behavior of Sanderlings in a similar study at Bodega Bay, California. Although they sampled for only 4 days, a clear pattern emerged: Sanderlings fed on tidal flats at low tide and on outer sandy beaches at high tide levels (see Fig. 1), although this pattern does not hold for Sanderlings everywhere (see Evans *et al.*, 1980). Howe and Agricola (unpublished data) examined the use of salt marsh areas by Semipalmated Plover (*Charadrius semipalmatus*), Dunlin, Semipalmated Sandpiper, Least Sandpiper, and Short-billed Dowitcher (*Limnodromus griseus*) over complete tide cycles in May in Virginia. Except for Least Sandpipers, the species showed a generalized pattern that was clearly influenced by tide (species used the marsh at high tide; refer to Fig. 1). And finally in this chapter (see below for data) I report results of a study of 31 species of shorebirds at Jamaica Bay Wildlife Refuge (New York) for a 1-year-period. Figure 1 is a diagrammatic representation of the pattern of shorebird use on three areas (mudflats on a tidal bay and

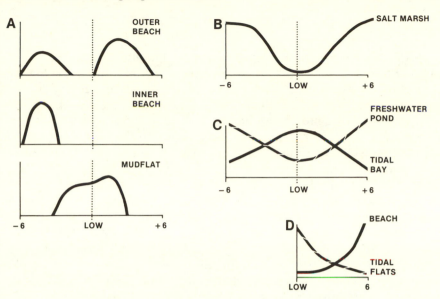

Fig. 1. Migrant shorebird use of habitats influenced by tides. (A) A New Jersey study of migrants in August; all three habitats were tidal [after Burger *et al.* (1979)]. (B) Shorebird use of a salt marsh in Virginia [M. A. Howe and M. T. Agricola (unpublished data)]. (C) Shorebird use of Jamaica Bay Wildlife Refuge over a 1-year period (this study). (D) Sanderling use of Bodega Bay, California [after Connors *et al.* (1981)].

two freshwater ponds). As is clear, tide affected the number of birds on the bay and East Pond, but less so on West Pond (which did not have as extensive mudflats as East Pond).

In examining the four widely separated studies, with different species, it is clear that tide affects shorebird distribution among habitats and that it affects habitats differently (Fig. 1). Habitat considerations will be discussed more fully below. The results from the Jamaica Bay study, where the pattern of change in numbers for East Pond was the inverse of that for the bay, indicate that there is some basis to the usual assumption that roosting numbers are an inverse measure of feeding numbers (Goss-Custard, 1969), except in this case many of the smaller species present in East Pond continued to forage and did not roost (see below, and Goss-Custard, 1969; Heppleston, 1971). Further, all four studies suggest that it would be profitable to simultaneously monitor *all* available habitats (outer beach, mudflat, salt marsh) throughout tide cycles to account for all shorebirds in an area, and to determine when and where they feed and roost.

C. Salinity Factors

Although in general ecologists examine the abundance and distribution of shorebirds by tidal or gross habitat factors, marine biologists frequently use salinity gradients to describe habitat preferences. Although shorebirds may not be directly influenced by salinity, most of the invertebrates that shorebirds prey upon are directly influenced by salinity (Stopford, 1951; Prater, 1981), and thus we might expect that shorebird distribution would be similarly limited. Counter to this hypothesis, however, is the fact that shorebirds forage on a wide diversity of prey items. Thus, I predict that the distribution of shorebirds with narrower ranges of prey species selection might be influenced by salinity. Wolff (1969) examined shorebirds in relationship to salinity and found that Eurasian Oystercatcher, Ruddy Turnstone, and Red Knot fed where salinity exceeded 13%; Eurasian Curlew fed seaward of 1%; and Eurasian Avocet (*Recurvirostra avosetta*) fed from 1 to 10%. Some species, such as Purple Sandpiper (*Calidris maritima*), Sanderling, and Common Redshank, showed no correlation with salinity. Salinity may well be a useful tool to assess habitat widths of shorebirds, in conjunction with prey studies.

D. Weather-Related Variables

Several weather-related variables such as temperature, winds, cloud cover, rain, and storms can be proximate factors affecting the abundance and distribution of migrant shorebirds as well as exerting ultimate effects on the evolution of life history patterns. Many of these variables interact to produce more severe effects than any single variable might inflict. The sum total of weather-related variables is usually more severe in winter. In severe winters, mortality rate in shorebirds may be 10-fold higher than the mortality rate in less severe winters or in spring and fall (Goss-Custard *et al.*, 1977; Prater, 1981). Severe and prolonged periods of gales and cold temperatures can result in mortality as birds are unable to feed (Evans, 1981; Davidson, 1982). Nonetheless, when Evans (1981) computed the yearly survival rate of five species of shorebirds visiting the Tees Estuary, England, he found that even after a severe winter it ranged from a minimum of 66% (Black-bellied Plover) to 88% (Bar-tailed Godwit). Other survival rates were Sanderling = 67%, Eurasian Curlew = 75%, and Ruddy Turnstone = 83% (Evans, 1981). Thus, survival values were relatively high even for species undertaking long migrations to breeding areas (Black-bellied Plover, Bar-tailed Godwit, Sanderling).

1. Temperature

Low temperatures affect migrant shorebirds directly by increasing their energy requirements to maintain body temperatures (Evans, 1976; Puttick, 1979) and indirectly by decreasing prey availability. In the severe winter of 1963 in England, considerable numbers of shorebirds died from starvation, as did many other marine organisms, such as fish, edible crabs, and octopus (Crisp, 1964; Colman, 1964; Pilcher, 1964; Woodhead, 1964).

Several authors have noted that foraging rates decrease at low temperatures (Goss-Custard, 1969; Baker, 1974; Bradstreet *et al.*, 1977). Baker and Baker (1973) reported a positive correlation between percent success and temperatures for Black-bellied Plover, presumably because their prey was less available at lower temperatures. In spring, shorebirds along Lake Erie avoid feeding in pools at low temperatures, and instead they feed on the mud–water interface where food may be more abundant (Bradstreet *et al.*, 1977).

The decrease in feeding rates and foraging success that occurs at low temperatures results from a decrease in prey availability (Goss-Custard, 1969; Smith, 1975; Evans, 1976). As temperature decreases, many prey species become less active (Pienkowski, 1978; Dugan, 1981), and they move deeper into the substrate (Evans, 1981; Townshend, 1981). Invertebrates that continue to rise to the substrate surface at low temperatures do so less often. For example, *Corophium* decrease in frequency of protrusion from their burrows as mud temperatures decrease (Goss-Custard, 1969). Thus, in Common Redshank, feeding rates decrease as temperature decreases, and they feed in different zones as prey availability changes (Goss-Custard, 1969). Pienkowski (1978) presents quantitative data on the effect of temperature decreases on the activity and depth of *Notomastus latericeus* and *Scoloplos armiger* (the prey of Black-bellied Plovers and Bar-tailed Godwits). The decrease in activity of both polychaetes with decreasing temperature is evident even at temperatures above 10°C. Differences in prey availability led to changes in feeding rates of the Black-bellied Plovers and Bar-tailed Godwits feeding on these prey (Pienkowski, 1978).

The reduction in activity of prey at the mud surface exerts pressures on "short-billed" shorebirds. Pienkowski (1981) has developed two models of shorebird foraging (the "sandpiper" strategy and the "plover" strategy). As the types differ with respect to bill length and foraging method (visual vs. tactile foragers), he proposes that the sandpiper strategy (long bill, tactile foragers) is superior at low temperatures, for prey activity is generally more sensitive at low temperatures than is prey depth distribution. Such hypotheses are extremely valuable as they provide paradigms for future study, and indicate mechanisms for observable prey capture rate differences.

2. Wind

Strong winds affect shorebirds directly by increasing their energy requirements (via windchill effects), and indirectly by reducing available foraging areas, decreasing foraging success, and decreasing prey availability. Strong winds can cause windchill (forced ventilation of body surface), leading to increased energy needs (Prater, 1981).

Heavy gales can influence water levels, resulting in less exposed mudflats and lower foraging time (Storer, 1951; Darbyshire and Draper, 1963; Evans, 1981). Evans (1981) reported that westerly gales in the Wattenmeer, northwestern Germany, could assist incoming tide to reach normal high water mark 3 hr before its predicted time. At low winds, shorebirds may simply face into the wind and continue foraging (Bent, 1927). Even at low wind speeds, shorebirds respond behaviorally. Most shorebirds face into the wind while roosting (Robertson and Dennison, 1979). Double-banded Dotterels move to sheltered areas to avoid wind, and stop feeding in high winds (Robertson and Dennison, 1979). White-fronted Sandplovers in Cape Town feed in the dunes when winds are too high to feed on the sand (McLachlan et al., 1980). Dugan et al. (1981) showed that most Black-bellied Plovers on open mudflats stopped feeding at wind speeds of over 21 knots (= 38 kph). However, heavy winds can cause direct problems for shorebirds in that they can be blown about, or have to avoid waves. Oring and Davis (1966) reported that migrants in Oklahoma frequently sought cover in winds of 30 mph. Wishart and Sealy (1980) reported that on windy days, Marbled Godwit feeding rates on exposed sites were higher (23.7/min vs. 18.9/min), but success rate and feeding efficiency were lower than in protected areas. There was no difference between these measures in protected areas in high winds and calm days. Similarly, Pienkowski (1981) reported decreasing prey capture rates by Black-bellied Plovers as wind speeds increased. Dugan et al. (1981) found that Black-bellied Plovers were unable to feed on open mudflats during gales, and Davidson (1981) reported that long-legged waders and those that forage visually have difficulty finding sufficient food during high winds to maintain energy balance. For Purple Sandpipers feeding at sea, high winds can cause them to fly to avoid being hit by whitecaps (Feare, 1966). On a calm day, only 1.7 min/hr was spent avoiding waves, compared to 8.0 min on a rough day. Thus, high winds can cause the birds physical problems in maintaining their balance and avoiding being overwashed by waves.

Lower foraging success of shorebirds in high winds is due not only to the birds having difficulty in foraging, but to reduced prey availability. Winds may dry the substrate, causing prey to come to the surface less frequently (Evans, 1976). At the other extreme, winds sometimes keep more water on the substrate surface (Smith, 1975).

Dugan *et al.* (1981) recently suggested that wind velocity and wind-chill (and not merely low temperatures) were very important factors influencing the adaptive significance of winter fattening of birds. They suggest that wind velocity and windchill are particularly important for plovers and possibly other shorebirds that feed visually. They reported that mean temperatures, when weights of Black-bellied Plover were low, were higher than in equivalent periods when weights were normal. However, wind speed was greater than 45 kph at this low-weight period, and the birds stopped feeding (Dugan *et al.*, 1981). Thus, they suggest that fat reserves might be insurance against periods of gales rather than low temperatures. They cite as evidence that several species of shorebirds maintain highest fat reserves in December and January when wind speeds in England are highest, and not in February when temperatures are lowest.

3. Precipitation

Precipitation alone (not in severe storms) does not seem to affect shorebirds directly, although it does affect prey availability in a number of ways (Evans 1981). Excessive rain creates freshwater habitats, and in estuarine systems it may negatively affect prey abundance (Page *et al.*, 1979). Marine invertebrates may fail to surface from their burrows as frequently. However, heavy rains flood upland fields, forcing worms to the surface, thus providing an abundant food source for Black-bellied Plover, Killdeer (*Charadrius vociferus*), Willet, Marbled Godwit, and Least Sandpiper (Kelly and Cogswell, 1979; Gerstenberg, 1979). Floods may also inundate coastal fields, causing some Eurasian Curlews to forsake nearby mudflats for more profitable feeding on these fields (Townshend, 1981).

4. Storms

Storms may include low temperatures, high winds, and precipitation, often combining to provide severe stress for shorebirds for a number of consecutive days. Storms can result in increased heat loss, which increases energy requirements at a time when it is most difficult to increase food acquisition (Heppleston, 1971; Evans, 1976). Further, storms can restrict feeding activity by making it impossible for shorebirds to forage physically, reducing available foraging space (by keeping tides high), and forcing invertebrate prey to remain inactive and thus less available. Prolonged storms may prevent shorebirds from feeding for several days, and may cause mortality, particularly when high winds and rain are combined with low temperatures (Dobinson and Richards, 1964). Evans (1981) commented that gales preventing shorebirds from feeding for several successive days led to higher mortality than cold weather alone. These factors led Dugan *et al.* (1981) to suggest that the adaptive significance of winter fat storage related to windchill rather than cold alone (see Section II.D.2).

Davidson (1982) suggested that in the absence of lethal effects, in severe weather (low temperatures and high winds), many shorebirds used their internal fat and protein reserves to balance their energy budgets. Species that are visual foragers (e.g., Black-bellied Plover) can use most of their internal reserves during periods of low temperatures and high winds (Davidson, 1982). Further, Pienkowski (1982) reports that high winds apparently depress the extent of nocturnal foraging by Black-bellied and Ringed Plovers. Thus, in severe weather, the inability to add to food reserves by foraging at night would add an additional stress to shorebirds.

5. Cloud Cover

Cloud cover generally appears to have only a minor effect on shorebirds. Some shorebirds [Pectoral Sandpipers, *Calidris melanotus* (Hamilton, 1959)] go to roost earlier in the evening on cloudy days, compared to sunny days. Swinebroad (1964) suggested the cue for roosting groups is light intensity. Presumably, shorebirds would not go to roose early where they were food stressed as they often continue to forage even into the night (see Section II.A.3).

Cloud cover may affect foraging success in that increased cloud cover reduces glare off wet sand. Shorebirds feeding visually thus may have higher foraging success under cloudy skies, assuming light is sufficient for them to see prey.

E. Habitat

Habitat selection in migrating shorebirds can be divided into three broad stages: (1) macrohabitat selection, (2) microhabitat selection, and (3) guild selection. Macrohabitat selection refers to major habitat types such as tidal mudflat, salt marsh, rocky intertidal, or sandy intertidal. The subdivisions are endless, and care must be taken to carefully define and adequately describe habitat types so comparisons can be made across species and geographical areas. The macrohabitat preferences of shore-birds are examined in Burger (Vol. 5 of this series). As is clear, most species have a wide tolerance for macrohabitats. Exceptions may be the "rocky intertidal" species that often remain in these habitats.

Broad preferences of migrating shorebirds species for particular macrohabitats have often been examined (see Baker, 1979; papers in Pitelka, 1979), although such studies usually examine overall preferences without examining the dynamic nature of macrohabitat selection. In any given day, shorebirds may feed or roost on several types of habitats depending upon tidal and weather-related factors. By referring to Fig. 1 it is possible to visualize the dynamic nature of macrohabitat selection. Unfortunately, each of these studies examined only a subset of the available habitats. As mentioned above, simultaneous sampling of all habitat types for several days during the peak of migration and at other times of the year is essential to determining how various habitats are used by shorebirds.

Microhabitat selection refers to selection of specific sites within a habitat, usually as a result of substrate differences. Factors that might be important include particle size, wetness of substrate, or presence of mud or algae. Several investigators have examined microhabitat preferences (Recher, 1966; Ashmole, 1970; Bradstreet et al., 1977; Baker, 1979; Gerstenberg, 1979; Duffy et al., 1981). The technique of examining microhabitat selection is useful in determining differences in the niches of closely related species (i.e., Baker, 1979), and may prove to be correlated with distribution of invertebrate prey items.

In addition to selecting macrohabitats and microhabitats, shorebirds characteristically feed in particular positions relative to tidal waters. Recher (1966) noted that most foraging activity occurs at the leading edge of the tide. Nonetheless, certain species select to feed above or below the water. For example, Dunlins often feed in the water at the water's edge while Least Sandpipers feed above the water's edge (Bengston and Svensson, 1968), Rufous-necked Stints (*Calidris ruficollis*) feed above the water's edge, and Curlew Sandpipers feed below the water's edge (Thomas and Dartnall, 1971). Brooks (1963) noted that there were three categories of shorebirds: large species that feed exclusively along the shore, smaller species that feed on or near the water's edge, and larger species that feed in water up to their belly. I suggest that it might be useful to examine shorebird feeding by guild type, where each guild denotes a particular feeding position relative to the tide. Shorebird guilds could be defined as:

1. On dry areas above high water mark.
2. Above the water on damp substrate.

3. Above-water guild: In the wet area at and above the waves or incoming tidal water.
4. Wave guild: In the edge of the waves or incoming tide.
5. Below-water guild: In the waves or incoming water just below the edge of the wave or incoming tidal water.
6. Deep-water guild: In the water away from the incoming surface, may go up to their belly in water (usually over 12 cm).

This classification will allow comparisons among macro- and micro-habitats, geographical areas, and species. Guild divisions also require careful considerations of prey exploitation methods of members. Further, interactions among guild members, and potential for switching among guilds may lead to testable hypotheses about shorebird foraging competition.

F. Model for the Interaction Effect of Abiotic Factors on Migrant Shorebirds

The distribution of shorebirds geographically, and within or among local areas, is a function of seasonal constraints that have influenced the evolution of breeding, migratory, and wintering ranges. Other factors, such as prey abundance and availability, have also influenced shorebird distribution (see Wolff, 1969; Goss-Custard, 1970; Prater, 1981). At any time of the year, the distribution of shorebirds within a local area, and the behavior of shorebirds, is directly influenced by abiotic factors such as time of day, tide, wind, temperature, and stormy weather. These factors interact to affect habitat selection and activities, and their combined effect may be greater than the summation of the individual effects. For example, high winds and low temperatures combined have a higher probability of being lethal than either factor alone (see above).

Table II indicates the levels of shorebird behavior affected by abiotic factors. Clearly, seasonal constraints affect the distribution of shorebirds geographically. But seasonal constraints also affect habitat selection. For example, Bradstreet et al. (1977) documented differences in habitat use for shorebirds at Long Point on Lake Erie. Storms likewise have exerted an effect evolutionarily in influencing shorebird ranges, but they also affect the distribution of shorebirds proximally. Time of day, tide, and winds all affect habitat selection, as well as whether shorebirds feed or roost. Cloud cover and temperature primarily affect activity patterns. Activity and habitat selection influence each other in that shorebirds must choose appropriate habitats for specific activities, and when in a particular habitat

Table II. Model for Interaction of Abiotic Factors on Migrant Shorebirds

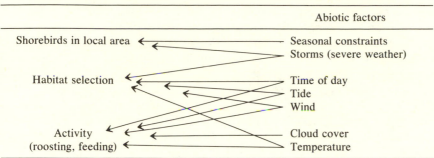

(like a tidal mudflat) shorebirds may continue to feed when they might not if they were on a sand spit with roosting birds.

III. JAMAICA BAY STUDIES

A. Introduction

Biologists interested in shorebird behavior have not usually concentrated on examining how abiotic factors affect their distribution and abundance within local areas. Yet abiotic factors affect shorebirds in numerous ways. Recently, British scientists have been examining how specific factors, such as wind and temperature, affect prey availability, prey abundance, and shorebird feeding behavior (see Evans and Smith, 1975; Evans, 1976; Pienkowski, 1981; Dugan *et al.*, 1981). Nonetheless, no one has examined all the abiotic factors affecting the abundance and distribution of shorebirds for an extended period of time. Without such a study it is difficult to determine the relative importance of particular factors.

In the rest of this chapter I examine the importance of season, time of day, tides, wind, rain, cloud cover, and temperature to the abundance and distribution of shorebirds over a 1-year period at Jamaica Bay Wildlife Refuge on Long Island, New York. Jamaica Bay Refuge contains a tidal bay (with mudflats, sand beaches, and salt marshes) and two freshwater ponds. I was particularly interested in whether abiotic factors were equally important influences on shorebird use of these three areas.

Although several authors have mentioned or examined one or two of these variables, no one has examined all these variables for a 1-year period. Even where detailed analyses of a variable such as tide have been performed, data were gathered for only 4 days (Connors *et al.*, 1981) up to a month (Burger *et al.*, 1977; Myers *et al.*, 1979). Partially, this is because differences in shorebird numbers over the season make analyses difficult (but see methods below). Such difficulties, however, do not obviate the need for such a study.

I had predicted that shorebird numbers would be affected most by tidal and wind variables, and least by temperature and time of day. Second, I predicted that tidal factors would affect shorebird use of both the tidal bay and the ponds although the effect would be opposite: shorebirds should use the tidal bay at low tide (when there is the greatest exposure of the mudflats) and the freshwater ponds at high tide. As one of the ponds contained extensive, exposed, freshwater mudflats (East Pond), I thought that the shorebirds would use this pond more often than the other (West Pond), which had limited beach along the shoreline. Further, I wanted to determine if small-sized shorebirds responded differently to abiotic factors (such as wind) than did larger species. Small-sized shorebirds (= peep) included Semipalmated, Least, Baird's, White-rumped, Western, and Pectoral Sandpipers. If wind and temperature are the severe stresses suggested (see Sections II.D.2–4), then small-sized shorebirds should have greater problems because of their larger surface area/volume ratio and metabolic considerations, compared to larger species.

The results presented below are part of an extensive study on waterbird use of Jamaica Bay Refuge, and data on other groups are presented elsewhere (Burger, 1982, 1983a,b). Further, details on specific shorebirds will also be presented elsewhere (Burger and Gochfeld, in press).

B. Study Area

Jamaica Bay Wildlife Refuge (3600 ha), part of the Gateway National Recreational Area (National Park Service), is located on the south shore of western Long Island, New York. The bay is a tidal lagoon containing many salt marsh islands (Fig. 2). Most of the bay is shallow (less than 3 m deep at low tide) except for dredged channels. The amount of mudflats exposed at low tides varies with the lunar cycle and seasons. However, during late summer and fall the daily tidal fluctuation in Jamaica Bay averages 1.4 m (range 0.9–2.13 m). An average mean low tide exposes 142 ha of tidal mudflats consisting of three major soil types: sand, mud, and peat-muck (Bridges, 1976). There are approximately 374 ha of low salt-marsh (containing primarily *Spartina alterniflora*), submerged at mean high tide, and exposed at mean low tide. High salt-marsh (213 ha,

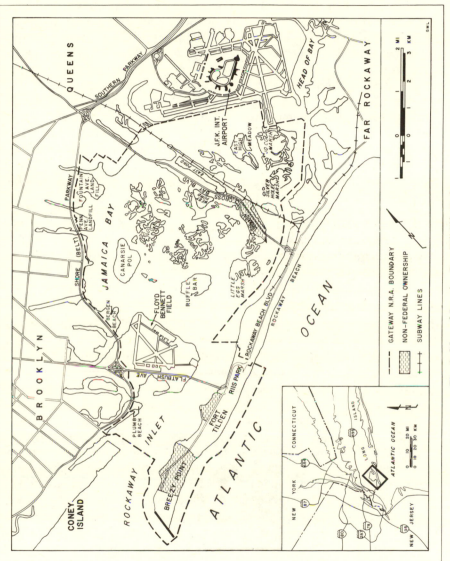

Fig. 2. Map of Jamaica Bay Wildlife Refuge.

containing mostly *S. patens* and *Distichlis spicata*) occurs in well-drained areas above the mean high tide limit.

There are two freshwater impoundments that were created by the deposition of spoil in 1953 (West Pond = 17 ha, East Pond = 39 ha). The National Park Service personnel artificially lower the water levels in West Pond in mid April, and in East Pond after 1 July each year. Large mixed sand–mudflats are present on East Pond. The West Pond water level rises with natural rainwater. Both impoundments contain considerable stands of *Phragmites communis*.

Jamaica Bay Refuge is surrounded by J. F. Kennedy International Airport, residential communities, several active sanitary landfills, and expressways (Fig. 2). Human disturbance is minimal in most areas of the bay (Burger, 1981). A path around West Pond provides easy access to bird watchers and others, but people usually do not disturb the birds. Few people ever visit East Pond because it is surrounded by *Phragmites*, and is generally inaccessible.

C. Methods

I divided Jamaica Bay Refuge into three census areas: the Bay (all tidal areas), East Pond, and West Pond. Shorebirds were censused during daylight hours from 1 May 1978 to 1 May 1979, by a field assistant (Wade Wander) or by us both. One week the ponds were censused for 4 days; the next week the ponds were censused for 2 days and the bay was censused for 2 days. We alternated weeks throughout the year. It usually required 8–10 hr to census the Bay only once, and 7–8 hr to census the ponds twice.

The ponds were censused by walking around the edge and plotting the location of all shorebirds on grid maps of each pond. The Bay was censused from 17 stops around the perimeter (see Fig. 2). The 17 census areas provided coverage of the whole Bay. Locations of shorebirds were plotted on detailed maps of each census stop. The 17 maps contained grids, and together they covered all visible areas of the bay. Any flock of shorebirds was recorded on only one map, although the flock may have been visible from several locations.

Census data reflect counts and not estimates. Shorebirds were identified as to species, except for a few identified as peep in late fall. Peep flocks contained primarily Semipalmated (97%) and Least Sandpipers (2–3%), with a few White-rumped, Western, Baird's, and Pectoral Sandpipers (scientific names of birds in this study are given in Table III).

Table III. Species of Shorebirds Seen at Jamaica Bay from 1 May 1978 to 1 May 1979

Common name	Scientific name	Total No. of individuals	Percent of total
Dunlin	*Calidris alpina*	75,649	32.9
Black-bellied Plover	*Pluvialis squatarola*	40,175	17.5
Semipalmated Sandpiper	*Calidris pusilla*	31,522	13.7
Greater Yellowlegs	*Tringa melanoleuca*	23,834	10.2
Short-billed Dowitcher	*Limnodromus griseus*	19,871	8.6
Sanderling	*Calidris alba*	14,707	6.4
Semipalmated Plover	*Charadrius semipalmatus*	12,666	5.5
Red Knot	*Calidris canutus*	6,579	2.9
Lesser Yellowlegs	*Tringa flavipes*	1,205	0.5
Least Sandpiper	*Calidris minutilla*	1,054	0.4
Ruddy Turnstone	*Arenaria interpres*	909	0.4
Stilt Sandpiper	*Calidris himantopus*	751	0.3
Long-billed Dowitcher	*Limnodromus scolopaceus*	345	0.2
Hudsonian Godwit	*Limosa haemastica*	204	—[a]
Lesser Golden Plover	*Pluvialis dominicus*	135	—
Piping Plover	*Charadrius melodus*	120	—
White-rumped Sandpiper	*Calidris fuscicollis*	102	—
American Oystercatcher	*Haematopus palliatus*	106	—
Pectoral Sandpiper	*Calidris melanotus*	60	—
Western Sandpiper	*Calidris mauri*	59	—
Marbled Godwit	*Limosa fedoa*	22	—
Willet	*Catoptrophorus semipalmatus*	15	—
Northern Phalarope	*Phalaropus lobatus*	7	—
Whimbrel	*Numenius phaeopus*	5	—
American Avocet	*Recurvirostra americana*	5	—
Baird's Sandpiper	*Calidris bairdii*	4	—
Wilson's Phalarope	*Phalaropus tricolor*	4	—
Common Snipe	*Capella gallinago*	3	—
Buff-breasted Sandpiper	*Tryngites subruficollis*	2	—
Ruff	*Philomachus pugnax*	1	—
Spotted Sandpiper	*Actitis macularia*	1	—

[a] Dashes = too few to compute.

By plotting the location of all shorebirds on maps, it was possible to examine the number of flocks of shorebirds as well as the number of individuals. For analyses, I was interested both in the number of species, and in species diversity [H, defined as the sum over i species of P_i (log P_i), where P_i is the proportion of individuals in the ith species]. At each census stop we recorded date, time, tide time, tide height, tide direction, wind direction, wind velocity, cloud cover, precipitation, and temperature as well as the number of individuals of each species in each flock. The independent variables were defined and measured as follows:

Temporal
    1. Date: the day of the year where 1 January = 1, and 31 December
       = 365.
    2. Time: time of the census on a 24-hr clock.
Tidal
    1. Tide cycle: number of hours before ($-$) or after ($+$) low tide.
    2. Tide height: a relative value of the water level of the bay, derived
       from tide tables.
    3. Tide direction: rising ($+$) or falling ($-$) tide.
Weather-related
    1. Wind direction: direction of the wind, including N, NE, E, SE,
       S, SW, W, NW.
    2. Wind velocity: speed of wind recorded at J. F. Kennedy Airport
       on the edge of the refuge.
    3. Cloud cover: estimated at each census location, recorded as a
       percent.
    4. Precipitation: scored from 0 (none) to 8 (heavy rain or snow).
    5. Temperature: measured by thermometer.

To determine the importance of these variables to the dependent measures [number of flocks, number of individuals, number of species, and species diversity (H)], I used stepwise multiple regression models to determine the variables that should be entered in the model, and general linear models multiple regression procedures to determine the variables (and interactions of variables) that contributed to the variations in the dependent variables (Barr *et al.*, 1976). The model selection procedure determines the "best" model, gives R^2 values and levels of significance for each of the contributing variables. The procedure first selects the variable that contributes the most to explaining the variability, and then selects that next variable that contributes the most to the coefficient of determination (R^2). This procedure is continued until all variables that have not been included in the model are not statistically significant. If variables are correlated, only the one that contributed the most to the R^2 would be used.

Most variables could be entered directly (i.e., wind speed, temperature). However, wind direction is not ordinal, and the procedure examined each dependent variable by wind direction X compared to non-X; examining separately each of the eight wind directions. If it found no significant differences, wind was not included in the model. Negative data were not entered in the model: if no shorebirds were present, we did not enter the independent variables. Thus, correlations between variables differed in the three census areas because shorebirds used them differently (see Table IV). Significant correlations for particular habitats indicate preferences, as frequency analyses of sampling indicated no biases in the sampling procedures. That is, all three census areas were sampled equally with respect to variations in the independent measures. The statistical design was formulated before the initiation of the study. All statistical procedures were performed by R. Trout of Rutgers University Statistics Department. All statistical procedures were performed on log-transformed data [log $e(x + 1)$]. On graphs I plot logs for the number of shorebirds and number of flocks.

In this chapter I present the best models for each dependent variable and the levels of significance for the independent variables. The relative value of the probability levels are indicative of the contribution made by each variable (i.e., a variable significant at the 0.0001 level contributes more to the observed variation in the dependent variable than one that is significant at only the 0.01 level). A variable whose contribution to the variance of the model has a probability greater than 0.05 is not included in the model.

I compared census areas with χ^2 tests where the expected value was derived from the mean value for all census areas combined. Other statistical procedures will be discussed where appropriate.

D. Results

1. Species Presence and Abundance

During the year, over 230,000 shorebirds were censused on Jamaica Bay Refuge, including the ponds (Table V). Adjusting for differences in censusing procedures (the Bay was censused a third as often), 66% were on East Pond, 27% were on the Bay, and less than 7% were on West Pond. The relationship between the two ponds is accurate for they were always censused on the same day. Because birds were not individually marked, it is impossible to determine the actual number of different individuals that used the refuge.

Overall abundance for the year is shown in Table V. Dunlin, Black-bellied Plover, Semipalmated Sandpiper, Greater Yellowlegs, and Short-billed Dowitcher accounted for over 80% of all shorebirds observed at Jamaica Bay Refuge. However, not all species used the three census areas (Bay, East Pond, West Pond) equally (Table V). Pectoral Sandpiper primarily used West Pond; Black-bellied Plover, Semipalmated Sandpiper, Semipalmated Plover, Knot, Greater and Lesser Yellowlegs, dowitchers, and Stilt Sandpiper (of the species having more than 500 individuals pres-

Table IV. Correlation Coefficients of Independent Variables for Sightings of Shorebirds on the Bay, East Pond, and West Pond at Jamaica Bay[a]

	Date, time	Wind velocity	Wind direction	Cloud cover	Temperature	Tide height	Tide cycle
				Bay ($N = 106$)			
Date, time	—	0.03	-0.12	0.06	-0.90***	-0.56***	0.47***
Wind velocity		—	0.82***	-0.67***	-0.04	0.02	-0.18
Wind direction			—	0.42***	0.21*	0.23*	-0.18
Cloud cover				—	0.00	0.19*	0.5
Temperature					—	0.42***	-0.67***
Tide height						—	-0.18
Tide cycle							—
		West Pond ($N = 213$), East Pond ($N = 793$)[b]					
Date, time	—	-0.16***	-0.18***	-0.30***	-0.80***	0.06	0.08
Wind velocity	-0.07***	—	-0.02	0.03	0.17***	-0.13	-0.07
Wind direction	-0.44	0.60***	—	0.09**	0.29***	0.34***	0.08***
Cloud cover	0.07	0.20**	-0.23***	—	0.32***	0.15***	0.24***
Temperature	-0.84***	-0.18***	0.59***	-0.37***	—	0.05	0.01
Tide height	0.01	-0.18**	-0.01	0.07	-0.15*	—	0.02
Tide cycle	0.24***	-0.20**	0.14	-0.07	-0.29***	0.11	—

[a] Probability levels: * 0.05, ** 0.01, *** 0.001.
[b] West Pond values below the diagonal; East Pond values above the diagonal.

Table V. Abundance of All Species of Shorebirds Observed at Jamaica Bay Wildlife Refuge

	West Pond		East Pond		Bay[a]		Total	
Dunlin	6,915	(9.1)	31,378	(41.5)	37,356	(49.4)	75,649	(32.9)
Black-bellied Plover	1,499	(3.7)	32,841	(81.8)	5,835	(14.5)	40,175	(17.5)
Semipalmated Sandpiper[b]	2,665	(8.5)	25,656	(81.3)	3,201	(10.2)	31,522	(13.7)
Greater Yellowlegs	1,171	(4.9)	21,628	(90.8)	1,035	(4.3)	23,834	(10.2)
Short-billed Dowitcher	1,152	(5.8)	17,150	(86.3)	1,569	(7.9)	19,871	(8.6)
Sanderling	16	(0.1)	3,990	(27.1)	10,701	(72.8)	14,707	(6.4)
Semipalmated Plover	794	(5.3)	10,543	(82.2)	1,329	(10.5)	12,666	(5.5)
Red Knot	584	(8.9)	5,667	(86.1)	328	(5.0)	6,579	(2.9)
Lesser Yellowlegs	186	(15.4)	968	(80.4)	51	(4.2)	1,205	(0.5)
Least Sandpiper	132	(12.5)	824	(78.2)	98	(9.3)	1,054	(0.4)
Ruddy Turnstone	204	(22.4)	3	(0.3)	702	(77.3)	909	(0.4)
Stilt Sandpiper	15	(2.0)	736	(98.0)	0	(0.0)	751	(0.3)
Long-billed Dowitcher	1	(0.3)	344	(99.7)	0	(0.0)	345	(0.2)
Hudsonian Godwit	1	(0.5)	203	(99.5)	0	(0.0)	204	(—)
Lesser Golden Plover	4	(3.0)	38	(28.2)	93	(68.8)	135	(—)
Piping Plover	0	(0.0)	3	(2.5)	117	(97.5)	120	(—)
White-rumped Sandpiper	11	(10.8)	91	(89.2)	0	(0.0)	102	(<0.1%)
Pectoral Sandpiper	34	(56.7)	26	(43.3)	0	(0.0)	60	(—)
Western Sandpiper	0	(0.0)	44	(74.6)	15	(25.4)	59	(—)
Marbled Godwit	2	(9.1)	20	(90.9)	0	(0.0)	22	(—)
Other species	4	(12.1)	26	(78.8)	3	(9.1)	33	(—)
Total	15,410	(6.7)	152,232	(66.1)	62,481	(27.2)	230,123	

[a] Bay figures corrected (multiplied by 3) for sampling schedule (except for species observed only once or twice).
[b] Unidentified peep were assigned to Semipalmated (97%) or Least (3%) Sandpipers on the basis of such groups identified from other vantage points or by helicopter.

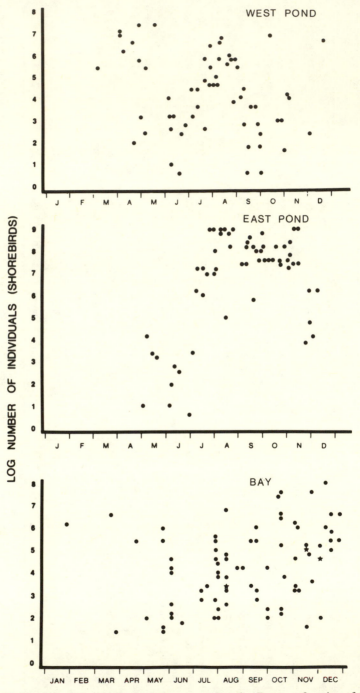

Fig. 3. Number of shorebirds on study areas at Jamaica Bay as a function of date.

ent) concentrated on East Pond, and Dunlin, Ruddy Turnstone, and Sand-erling concentrated on the Bay.

During the course of the study, 31 species of shorebirds used Jamaica Bay (Table III). At any one sample site (site on the Bay, or census around the ponds) the number of species present usually was less than 15. In general the number of species on East Pond was 16 or less, on West Pond 10 or less, and on the Bay 9 or less.

2. Seasonal Distribution of Shorebirds

Shorebirds were usually present on Jamaica Bay Refuge from March through January, with peaks in May (spring migration) and July through November (fall migration; Table VI, Fig. 3). Because the bay was cen-sused four times a month, one data point for January and none for Feb-ruary indicates no shorebirds were observed there for the February census times (see Fig. 3). Table VI lists the number of birds observed during spring and fall migration (defined as after 1 July) for the 10 most abundant species. Most shorebirds migrated north through the bay in May and June, and migrated south through the bay from July through November. A few species, such as Dunlin, Black-bellied Plover, and Sanderling, sometimes overwinter at Jamaica Bay Refuge although heavy ice and cold temper-atures in late January 1979 resulted in most shorebirds leaving the area. Thus, fall migration was prolonged, and generally higher numbers were present than in spring. Variability in the number of shorebirds using the bay reflects predominantly tidal effects (see below) whereas the lack of variability on East Pond reflects that shorebirds used this area for foraging and roosting throughout the tide cycle (see Section III.E). Peep were present at Jamaica Bay Refuge from early May until November, although five Least and Semipalmated Sandpipers were present in early December.

I then examined the fall migration period for peep as a function of census area (Fig. 4). Peak fall migration of peep occurred during July and August, and it ended in early November. This pattern differs from the general fall migration period for all shorebirds (refer to Fig. 3) in starting and ending earlier in the fall. Further, peep primarily used the ponds and did not roost or feed on the Bay.

3. Models for Factors Influencing Shorebird Numbers and Locations

The main objective of the study was to determine the factors that influence shorebird numbers on an east coast bay used extensively by migrating shorebirds in the spring and fall. The dependent measures were number of individuals (per day for the Bay, and per census around each

Table VI. Numbers of Shorebirds at Jamaica Bay Refuge during Spring and Fall (after 1 July) Migration as a Function of Census Area. Migration Periods Given in Parentheses

		Spring migration	Fall migration
Dunlin	East Pond	25 (May)	31,353 (August–December)
	West Pond	5,728 (March–May)	1,187 (October–December)
	Bay	1,023 (May–June)[a]	35,133 (October–January)
Black-bellied	East Pond	26 (May–June)	32,815 (July–December)
Plover	West Pond	1,324 (March–May)	175 (August–October)
	Bay	957 (April–June)	4,878 (August–December)
Semipalmated	East Pond	20 (June)	25,636 (July–November)
Sandpiper	West Pond	278 (May–June)	2,387 (July–September)
	Bay	780 (May–June)	2,421 (July–October)
Greater	East Pond	3 (May)	21,625 (July–November)
Yellowlegs	West Pond	37 (April)	1,134 (July–November)
	Bay	0	1,134 (July–November)
Short-billed	East Pond	0	17,150 (July–November)
Dowitcher	West Pond	35 (May–June)	1,940 (July–October)
	Bay	43 (June)	17,931 (July–August)
Sanderling	East Pond	40 (May)	3,950 (July–November)
	West Pond	0	16 (August)
	Bay	45 (May)	10,656 (July–January)
Semipalmated	East Pond	33 (May)	10,508 (July–November)
Plover	West Pond	72 (May)	722 (July–September)
	Bay	105 (May–June)	1,227 (July–October)
Red Knot	East Pond	0	5,667 (July–November)
	West Pond	200 (May)	384 (August)
	Bay	0	328 (August–December)
Lesser	East Pond	0	968 (July–November)
Yellowlegs	West Pond	18 (April–May)	33 (July–September)
	Bay	0	51 (July)
Least Sandpiper	East Pond	0	824 (July–September)
	West Pond	71 (April–May)	61 (July–August)
	Bay	10 (May–June)	88 (July–October)
Total		10,873	215,056
Percent		4.8	95.2
Percent without Dunlin		2.7	97.3

[a] 400 present in January, none in February, 700 in March.

pond), number of flocks, number of species and species diversity (H); independent measures included temporal (time of day, date), tidal (tide cycle, tide height, tide direction), weather (wind velocity, wind direction, cloud cover, precipitation, temperature), and interaction variables (combinations of variables such as wind velocity × wind direction). In this

Table VII. Factors Contributing to the Variability in the Number of Shorebirds at Jamaica Bay. Given Are Probability Levels for Each Variable

	Bay	East Pond	West Pond
Model			
R^2	0.65	0.67	0.60
F	4.33	37.82	6.69
p	0.0003	0.0001	0.0002
df	14, 33	7, 93	6, 27
Factors contributing to variability			
Temporal variables			
Date	0.01	0.0001	
Date and time		0.05	0.004
Tidal variables			
Tide cycle	0.01	0.01	0.05
Tide height	0.01		
Tide direction		0.001	
Weather variables			
Wind direction	0.05		
Wind velocity and direction	0.001		
Temperature	0.001		0.001
Temperature and wind velocity	0.001		
Cloud cover	0.01		
Cloud and wind velocity	0.001	0.05	
Cloud and temperature	0.01		
Interaction variables			
Date and tide height		0.001	
Temperature and tide height	0.01		
Temperature and date	0.001		0.05
Date and wind direction		0.05	0.01
Tide direction and wind direction	0.05		

section I present models accounting for the variability in the dependent measures, and in succeeding sections I will discuss each set of variables.

The models accounted for over 60% of the variability in the number of shorebirds present in all three census areas (Table VII). On the Bay 65% of the variability was accounted for by tide cycle, tide height, wind direction, wind velocity × wind direction, temperature, cloud cover, and interactions of temperature and cloud cover with other variables (Table VII). Fewer variables explained the variability in numbers of individuals on the two ponds. On West Pond 60% of the variability in shorebird numbers was explained by date × time of day, tide cycle, wind velocity × direction, tide direction × wind direction, temperature × date, and date × wind direction. On East Pond 67% of the variability was explained by date, tide direction, cloud × wind velocity, date and

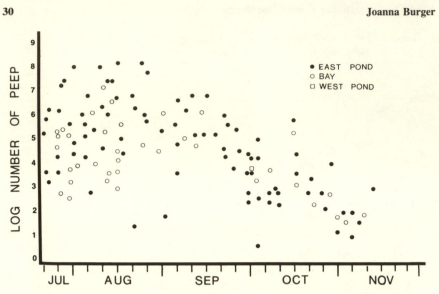

Fig. 4. Number of "peep" on Jamaica Bay Refuge during the fall migration.

Table VIII. Factors Contributing to the Variability in the Number of Peep at Jamaica Bay Refuge. Given Are Probability Levels for Each Factor

	Bay	East Pond	West Pond
Model			
R^2	0.86	0.69	0.90
F	9.31	26.7	20.06
p	0.0001	0.0001	0.0001
df	11, 32	8, 65	6, 22
Factors contributing to variability			
Temporal variables			
Date	0.01	0.01	0.0001
Date and time			0.006
Tidal variables			
Tide cycle	0.009	0.0002	0.05
Tide height	0.002	0.0001	0.001
Tide direction			0.02
Tide height and direction	0.004		
Weather variables			
Wind direction	0.002		
Wind direction and velocity	0.05	0.03	
Cloud cover and wind velocity		0.01	
Temperature		0.05	
Interaction variables			
Temperature and date	0.04	0.01	
Date and tide height		0.0001	

Table IX. Factors Contributing to the Variability in the Number of Shorebird Flocks at Jamaica Bay. Given Are Probability Levels for Each Variable

	Bay	East Pond	West Pond
Model			
R^2	0.65	0.67	0.51
F	4.27	27.0	16.2
p	0.0003	0.0001	0.0001
df	14, 33	7, 93	2, 31
Factors contributing to variability			
Temporal variable			
Date		0.001	0.001
Tidal variables			
Tide cycle	0.01	0.001	
Tide height		0.001	
Weather variables			
Wind velocity		0.01	
Wind direction	0.001		
Wind velocity and direction	0.001		
Temperature and wind velocity	0.001		
Cloud and wind velocity	0.001	0.001	
Interaction variables			
Tide direction and wind direction	0.001		0.001
Temperature and tide height	0.001		
Temperature and date	0.001	0.001	
Date and wind direction		0.05	

wind direction, and date × tide height. Thus, for all three census areas, date, tide, and wind variables significantly entered the models accounting for variability in the number of shorebirds present on Jamaica Bay Refuge.

A similar models procedure for peep alone indicated that more of the variability in the number of peep was explained by the models (Table VIII). Generally, date, tide cycle and height, and wind velocity entered all models.

Between 51 and 67% of the variability in the number of shorebird flocks was explained by the independent variables (Table IX). Tidal and wind variables entered the models for all census areas. For peep, the variables that entered the models (for all census areas) explaining the variability in number of flocks were tide cycle, tide direction, wind direction, wind velocity, and cloud cover (Table X). Again, more of the variability in the number of peep flock was explained by the model compared to the model for all shorebirds.

Between 51 and 70% of the variability in number of shorebird species present was explained by the models (Table XI). Date, wind direction,

Table X. Factors Contributing to the Variability in Number of Peep Flocks at Jamaica Bay Refuge. Given Are Probability Levels for the Factors Contributing to the Best Model

	Bay	East Pond	West Pond
Model			
R^2	0.85	0.77	0.73
F	6.29	11.60	5.30
p	0.0006	0.0001	0.006
df	11, 32	15, 55	6, 22
Factors contributing to variability			
Temporal variables			
Date		0.001	
Date and time	0.0007		
Tidal variables			
Tide cycle	0.0007	0.03	0.0005
Tide height		0.02	
Tide direction		0.05	0.0006
Tide direction and height	0.001		
Weather variables			
Wind direction		0.03	
Wind velocity and direction	0.0007	0.001	
Temperature	0.007	0.04	
Cloud			0.0008
Cloud and wind velocity	0.0007	0.02	0.02
Cloud and temperature	0.01		
Interaction variables			
Tide direction and wind direction	0.0007	0.002	
Temperature and tide height		0.009	
Date and tide height	0.006	0.01	

cloud cover, and temperature entered the models for all census areas explaining the variability in number of species present. No single tide variable entered all models, although either tide direction or tide height entered the models for all three census areas. Generally, less variability in H (species diversity) than in number of species was accounted for by the models (Table XI). The variables that significantly entered the models explaining the variability in H were date, tide height, and tide direction. Similar models for peep indicated that tide cycle and wind velocity or direction entered all models for the number of species and species diversity (Table XII).

From the above models it is clear that abiotic variables influenced the number and distribution of shorebirds at Jamaica Bay Refuge, although the effect varied in the different census areas, and in peep com-

Table XI. Factors Contributing to the Variability in Species Diversity of Shorebirds at Jamaica Bay

	Bay		East Pond		West Pond	
	H	No. of species	H	No. of species	H	No. of species
Model						
R^2	0.52	0.70	0.42	0.54	0.62	0.51
F	5.31	6.14	8.46	11.96	11.90	10.41
p	0.002	0.0001	0.0001	0.0001	0.0001	0.0001
df	8, 39	13, 34	8, 92	9, 91	4, 29	3, 30
Factors contributing to variability						
Temporal variable						
Date		0.001	0.0003	0.001	0.001	
Tidal variables						
Tide cycle		0.001	0.01	0.01		
Tide height				0.001		
Tide direction	0.01					
Tide height and direction	0.05		0.01			
Weather variables						
Wind velocity					0.05	
Wind direction				0.001		
Wind velocity and direction		0.001	0.01	0.05	0.05	
Temperature		0.001				
Temperature and wind velocity						
Cloud cover and wind velocity		0.001	0.05	0.01		
Cloud cover and temperature						0.05
Interaction variables						
Tide direction and wind velocity	0.01	0.001	0.01		0.001	0.01
Temperature and tide height		0.0001				
Temperature and date	0.05	0.001			0.05	
Date and wind direction		0.05	0.001			

Table XII. Factors Contributing to Species Diversity of Peep at Jamaica Bay Refuge

	Bay		East Pond		West Pond	
	H	No. of species	H	No. of species	H	No. of species
Model						
R^2	0.70	0.74	0.36	0.41	0.64	0.91
F	2.03	4.41	5.56	4.25	5.31	16.29
p	0.01	0.003	0.0001	0.0005	0.01	0.001
df	15, 30	11, 34	6, 64	10, 60	5, 23	7, 21
Factors contributing to variability						
Temporal variable						
Date			0.001			
Tidal variables						
Tide cycle	0.05	0.003	0.03	0.04	0.002	0.002
Tide height		0.03	0.02			
Tide direction					0.05	0.02
Tide height and direction	0.04	0.003				
Weather variables						
Wind velocity			0.004	0.03		
Wind direction				0.006	0.05	
Wind velocity and direction						0.0001
Temperature						
Temperature and wind velocity	0.002					
Temperature and tide height		0.05	0.008			0.0001
Cloud cover and temperature		0.05				
Cloud cover and wind velocity						
Interaction variables						
Tide direction and wind direction		0.02	0.005	0.02		

pared to all shorebirds. The effect of the major variables will be discussed below.

4. Temporal Variables

As mentioned above (Section III.D.2), shorebirds were not present on all census days throughout the year. Date did enter many of the models explaining variability in number of shorebirds at Jamaica Bay Refuge.

Table XIII. Species That Were Strongly Tidally Influenced and Some That Were Not

Strongly tidally influenced[a]		Percent in bay
Ruddy Turnstone		77
Sanderling		72
Dunlin		49
American Oystercatcher		45
Black-bellied Plover		15
Semipalmated Plover		11
Semipalmated Sandpiper		10
Nontidally influenced	Primary feeding area	Percent in bay
Pectoral Sandpiper	West Pond	0
Lesser Yellowlegs	West Pond	4
Long-billed Dowitcher	East Pond	0
Stilt Sandpiper	East Pond	0
Least Sandpiper	East Pond	9

[a] Defined as species (or individuals of that species) that left the ponds and flew to the bay as mudflats became exposed by tides.

Time of day usually was a significant variable in the models as an interaction with date, possibly because daylength varies seasonally. For example, the hour before sunset occurs at a slightly different time each day. Furthermore, time of day influenced activity rather than presence. In general, shorebirds were present somewhere on Jamaica Bay all day, although Sanderling may have left the refuge to feed on oceanfront beaches. Date × time interaction entered the model for individuals (for peep and all shorebirds) on West Pond, largely because shorebirds frequently roosted there in the middle of the day (West Pond had no extensive mudflats for foraging).

5. Tidal Variables

For all dependent measures (number of individuals, number of flocks, number of species, and species diversity), tidal factors entered the models significantly for all shorebirds combined (all three census areas) and for peep (all three census areas). Not all species left the ponds when mudflats became exposed, and some species seemed to ignore tidal variables, feeding in a variety of habitats on the ponds and Bay (Table XIII). Tidal variables, however, include tide cycle (hours since low), tide height, and tide direction. All three tidal variables did not enter all models for all areas (refer to Tables VII–XII).

Tide cycles clearly had an effect on shorebird distribution: all shorebirds combined (Fig. 5) and peep (Fig. 6) used the Bay at low tide, and

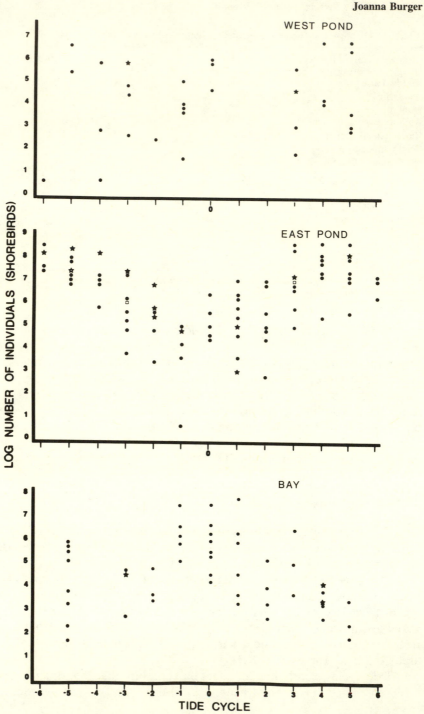

Fig. 5. Number of shorebirds as a function of study site and tide cycle time on Jamaica Bay Refuge.

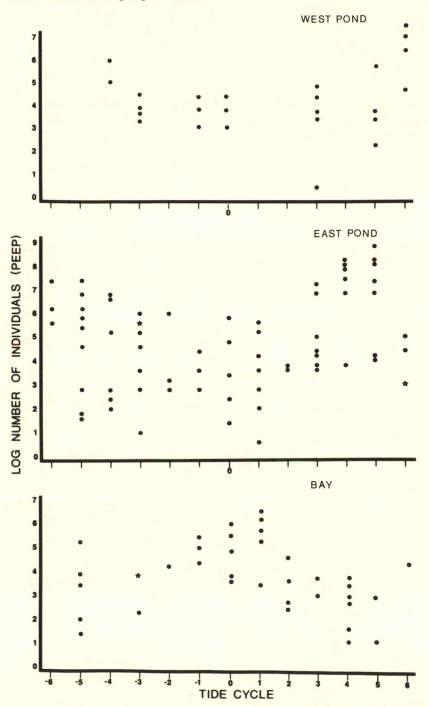

Fig. 6. Number of "peep" as a function of study site and tide cycle time on Jamaica Bay Refuge.

the ponds at high tide. The pattern was clearer for all shorebirds combined than for peep, possibly because peep were more affected by wind variables (see below). The effect of tide cycle on shorebirds is clear in Fig. 5 for the Bay and East Pond, but is less so for West Pond. Although the models procedure indicated tide cycle had an effect, presumably it is masked by combining the data for the entire year. Similarly, tide cycle influenced the number of flocks in all areas (Table IX): more flocks were present on the Bay at low tide, and on East Pond at high tide. On West Pond the greatest number of flocks occurred 2–3 hr before low tide and 4–5 hr after low tide. The number of peep flocks was similarly related to tide cycle time, and the effect was clearer: more flocks were present on the Bay at low tide when the peep could spread out over the mudflats in small, rather than large flocks (Fig. 7). The number of shorebird species present was similarly influenced by tide cycle; more species occurred on the Bay at low tide and on East Pond at high tide with no significant trend on West Pond. On the Bay the number of species at low tide was 9 or less, whereas at high tide on East Pond the number of species present was as high as 16. Species diversity showed a similar pattern as the number of species.

Tide height is an important tidal variable as it affects the amount of mudflat exposed on the Bay although not on the ponds. On the Bay, tide height was not correlated with tide cycle ($r = -0.18$, Table III) for the data. Tide height significantly influenced shorebird numbers on the Bay (tidal) and East Pond (nontidal; Table VII); more birds were present on the Bay at low tide heights and on East Pond at high tide heights (Fig. 8). The pattern was particularly noticeable on East Pond where all birds could be counted. Although all visible birds were counted on the Bay, it required all day to census, shorebird flocks shifting locations could be missed, and some areas of the Bay were not visible. The relationship between tide height and number of peep was more variable (Fig. 9). Although larger shorebirds did use the Bay at tide heights over 5 feet, peep did not. Further, the number of peep using East Pond was variable at all tide heights.

Tide height significantly influenced the number of flocks for all shorebirds combined on the Bay and East Pond (Fig. 10); and for peep flocks on the bay and East Pond (Fig. 11), although the relationship was not as clear for peep. Tide height also significantly affected the number of shorebird species present (Fig. 12), and species diversity (Fig. 13) as a single variable (entered as tide height) or by interactions for all census areas. But clearly the relationship between tide height and both number of species present and species diversity was clearer on East Pond, where shorebirds came as tidal waters rose on the Bay. Thus, there were more species on East Pond at high than low tides, and more species in the Bay at

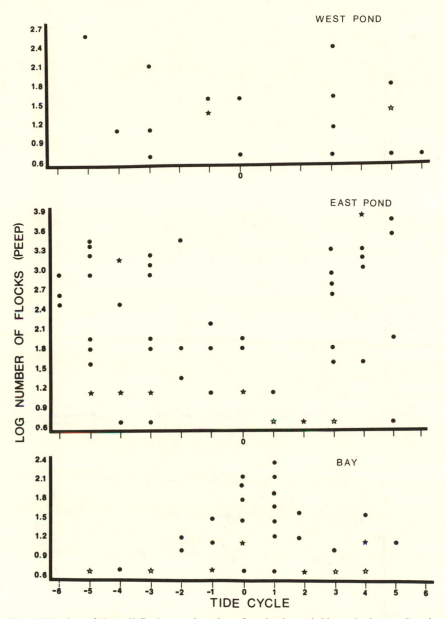

Fig. 7. Number of "peep" flocks as a function of study site and tide cycle time on Jamaica Bay Refuge.

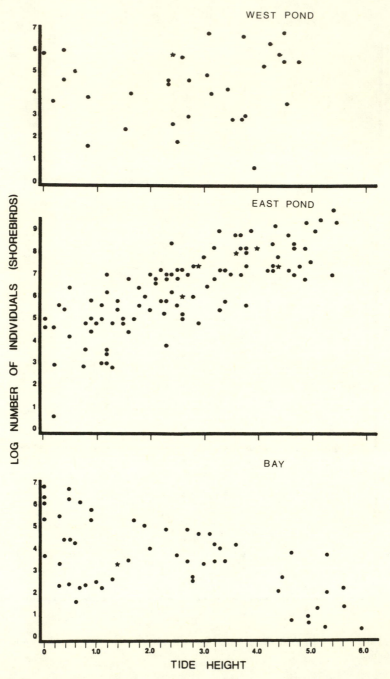

Fig. 8. Number of shorebirds as a function of study site and tide height on Jamaica Bay Refuge.

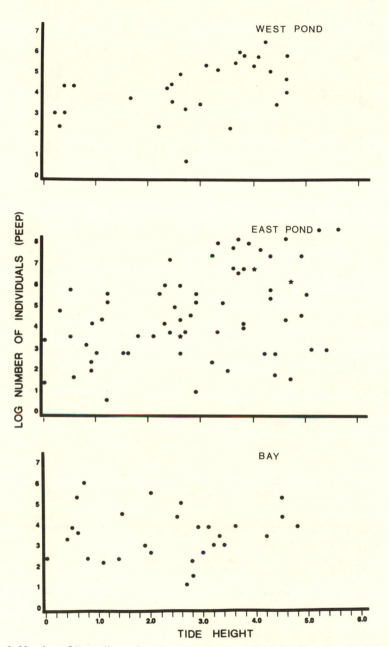

Fig. 9. Number of "peep" as a function of study site and tide height at Jamaica Bay Refuge.

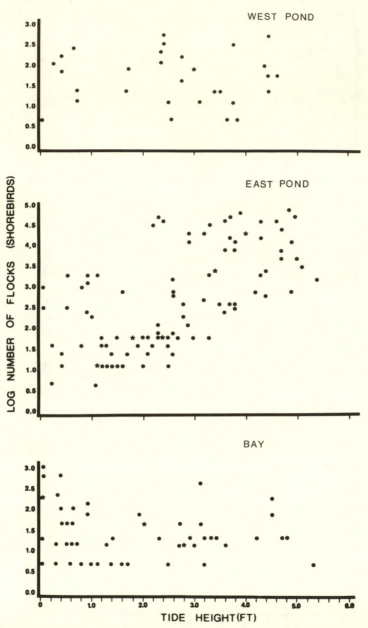

Fig. 10. Number of shorebird flocks as a function of study site and tide height on Jamaica Bay Refuge.

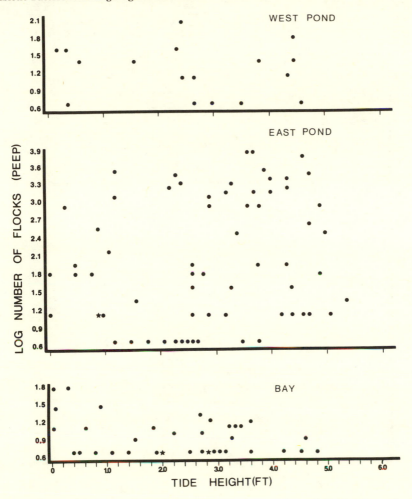

Fig. 11. Number of "peep" flocks as a function of study site and tide height on Jamaica Bay Refuge.

low tide, although the difference was not as great on the Bay. It seemed that although most shorebirds moved into East Pond at high tide, a few individuals of most species remained on the Bay.

Tide direction did not enter as many of the models explaining variability in the number of individuals, number of species, and species diversity for either shorebirds or peep as did the other tidal variables (tide cycle and tide height; see Tables VII, VIII). However, tide direction (and tide direction × tide height) significantly entered all models for number

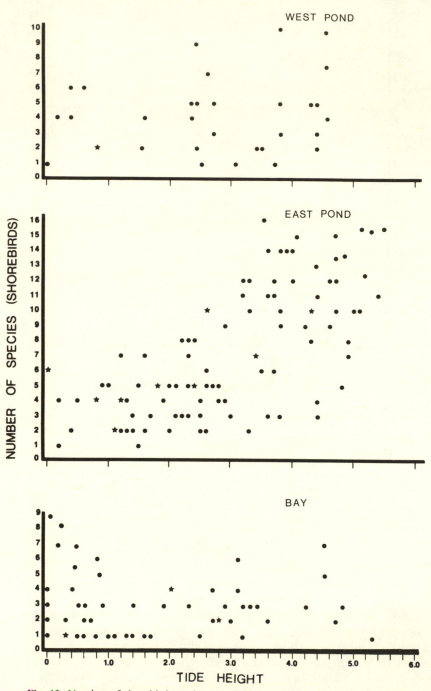

Fig. 12. Number of shorebird species as a function of tide height (given in feet).

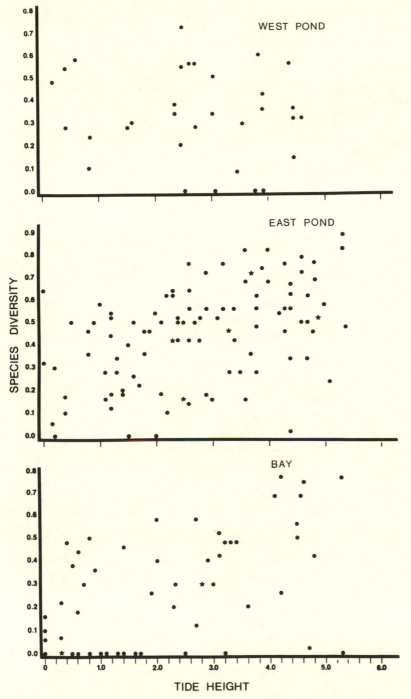

Fig. 13. Species diversity as a function of tide height (given in feet).

Table XIV. Effect of Tide Direction on Flocks of Peep on the Bay Transect at Jamaica Bay Refuge

		Tide direction		
	N	Falling	Low tide	Rising
Fall only				
Number of flocks[a]	40	12	8	20
$\bar{x}$ log number of individuals		3.0	4.9	3.7
Entire year				
Number of flocks[b]	57	13	8	36
$\bar{x}$ log number of individuals		3.7	6.0	4.7

[a] $\chi^2 = 10.55$, $df = 2$, $p < 0.005$.
[b] $\chi^2 = 14.35$, $df = 2$, $p < 0.005$.

of flocks of peep (refer to Table X). Combining the data from all three census areas for peep indicates that both for fall migration and for the data from the entire year there were significant differences in the number of flocks as a function of tide direction. There were significantly more flocks on rising than falling or low tides (Table XIV). Table XIV gives the number of individuals to show that the number of flocks does not relate directly to the number of individuals present. The lower number of individuals at higher tides may indicate that some peep were roosting on salt marsh islands where they were not visible from the census stops. Helicopter surveys indicate that large flocks of peep did use salt marshes at high tides, although such flocks were not always present.

In examining the models and data it is clear that tidal factors strongly influence shorebird numbers and distribution. However, it is difficult to determine which tidal variable shorebirds are responding to. They could cue in to tide cycle time, tide height, or tide direction. To examine the tidal variables in depth, I used a general linear models procedure (Barr *et al.*, 1976) to determine the factors to include in the model, and a stepwise regression procedure to develop models. The computer was given only tide cycle time, tide height, tide direction, tide height × tide direction, and tide direction × wind direction (because wind was important for peep; see below). For all shorebirds combined, tide cycle time entered more models (11 of a possible 12) than any other variable (Table XV). Tide height entered 7 models, and tide direction entered only 1. However, direction × wind direction entered 7 of a possible 12 models (it did not enter any models for East Pond; Table XV). A similar analysis for peep (Table VI) indicated that tide direction × wind direction entered 11 of a possible 11 models, whereas tide cycle entered only 8 of 11. It should be remembered that when the computer is given different variables (only

Table XV. Relative Effects of Tidal Factors on Occurrence of Shorebirds at Jamaica Bay Refuge

	Model		Tide cycle time	Tide height	Tide direction	Tide height × direction	Tide direction × wind direction
	R^2	p					
Log number of birds							
Bay	21	0.01	0.01	0.01		0.01	0.05
East Pond	58	0.0001	0.009	0.0001			
West Pond	49	0.03	0.05				0.05
Log number of flocks							
Bay	18	0.01	0.01				0.05
East Pond	64	0.0001	0.001	0.0001	0.05		
West Pond	51	0.01	0.05				
Number of species							
Bay	28	0.009	0.001	0.01		0.01	0.009
East Pond	53	0.001	0.001	0.0001			
West Pond	51	0.001	0.001				0.005
Species diversity							
Bay	36	0.002			0.05	0.01	0.01
East Pond	37	0.008	0.01	0.01			
West Pond	62	0.003	0.01				0.003

tidal variables in Tables XV and XVI; all variables in Tables VII–XII), the same variable (tide cycle, etc.) may not be significant in both models, as the added variables may contribute more to the variation than tidal variables alone. Thus, in summary, tidal variables alone have a lesser effect on small shorebirds (peep) than wind direction × tide direction (refer to Table XVII).

6. Weather-Related Variables

Wind. Wind can influence shorebird abundance by its velocity, direction, or both. As mentioned above (Section II.D.2), researchers are beginning to document the adverse effects of high winds on shorebird thermodynamics and foraging behavior, and on their prey. Thus, I expected that wind would influence shorebird distribution among areas at Jamaica Bay Refuge. Wind velocity alone did not influence the numbers of shorebirds on the census area (refer to Table VII), but wind velocity × direction entered the model for variability in the number of individuals on the bay. Generally, there were fewer birds (in any given 2-day census period) on the bay at high winds compared to low winds.

In examining Tables VII–XII, wind direction entered more models than other wind variables for both peep and all shorebirds combined as

Table XVI. Relative Effects of Tidal Factors on Occurrence of Small Sandpipers (Peep[a])

	Model		Tide cycle time	Tide height	Tide direction	Tide height × direction	Tide direction × wind direction
	R^2	p					
Log number of flocks							
Bay	40	0.01	0.01	0.05	0.03		0.006
East Pond	68	0.001	—	—	0.01	—	0.008
West Pond	—[b]	—	—	—	—	—	—
Log number of individuals							
Bay	33	0.04	0.03				0.01
East Pond	67	0.0001	0.001	0.0001			0.001
West Pond	78	0.001		0.001	0.0002		0.05
Number of species							
Bay	50	0.001	0.05	0.03	0.03	0.03	0.001
East Pond	31	0.0004					0.002
West Pond	27	0.04	0.04				0.001
Species diversity (H)							
Bay	45	0.0006	0.04	0.03	0.02		0.002
East Pond	25	0.001	0.01				0.006
West Pond	29	0.05	0.05				0.001

[a] Peep = Semipalmated, Least, Baird's, Western, and Pectoral Sandpipers.
[b] Too few were present to analyze.

Table XVII. Number of Models Each Variable Enters Significantly for Peep and All Shorebirds Combined. There Are a Possible 12 Models[a]

Factor	All shorebirds combined	Small shorebirds[c]
Tide cycle	11	8
Tide height	7	5
Tide direction	1	5
Tide height × tide direction	3	1
Tide direction × wind direction	7	11
Variation explained by above variables[b]	21–58%	33–78%

[a] Three census areas × four dependent measures (refer to Tables XV and XVI).
[b] Range for three census areas for number of individuals.
[c] There are only 11 possible models (see Table XVI).

a significant contributor to explaining the variability in number of individuals and number of flocks. Examining the number of shorebird flocks and peep flocks (Fig. 14) indicates that: (1) all shorebirds combined and peep were not present as often when there were W-NW winds as occurred during the study (prevailing winds = winds present for all census days

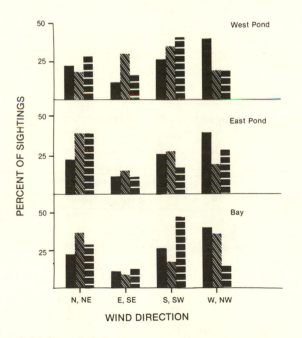

Fig. 14. Percent of sightings of all shorebirds (bars with diagonal lines), peep (interrupted bars), and prevailing conditions (solid bars) as a function of wind direction.

throughout the year), (2) although the winds were similar for all census areas on any given day, all shorebirds combined and peep did not use the areas equally with respect to wind conditions, and (3) all shorebirds combined and peep did not use each area similarly.

The prevailing winds during the year were N-NW, but shorebirds were not present as often during these winds. N-NW winds were concentrated in late fall and winter when almost all shorebirds had migrated from the area. Comparing the distribution of all shorebird species among the three census areas (exposed to the same winds), shorebirds were not equally abundant on each area in all winds (3 × 4 contingency table, χ^2 = 46.4, df = 6, $p < 0.001$). The Bay had an excess of shorebirds on N-NE winds, and a deficit during S-SW winds; West Pond had an excess of shorebirds on E-SE and S-SW winds and a deficit on W-NW winds; and East Pond had an excess during N-NE winds, and a deficit on W-NW winds.

Comparing groups, peep used the Bay differently from all shorebirds combined (χ^2 = 74.94, df = 3, $p < 0.001$), but there were no differences on the ponds (χ^2 tests). Clearly, small-sized shorebirds avoided the Bay in W-NW winds (which were the strongest; see below) whereas other shorebirds often concentrated in the Bay on northerly winds. Few sheltered mudflats existed in the Bay, whereas in the ponds tall *Phragmites* shielded the birds (see below).

The effect of wind direction was enhanced by wind velocity, and this interaction variable entered the models for variability in number of individuals for peep as well as all shorebirds combined (refer to Tables VII and VIII). Generally, when high winds came from the NW, shorebirds avoided the Bay. Wind direction × wind velocity also significantly entered the models accounting for the variation in number of flocks: the number of flocks in the Bay decreased on W-NW winds even with no change in the numbers of individual shorebirds. On East Pond the number of flocks decreased as individuals increased during S-SW winds, but winds from other directions did not influence the number of flocks. The effect in S-SW winds may be due to a piling up of water (decrease in foraging space on the north end of East Pond) with strong S-SW winds, for the winds sweep the length of the pond. Reduced space would result in one large flock rather than several small ones.

Cloud Cover. Cloud cover was a significant variable in the models for the number of shorebirds on the Bay, and entered as an interaction variable (with wind velocity) on East Pond. As cloud cover increased, there were more shorebirds on the Bay and East Pond. Both areas were used extensively for foraging (whereas West Pond was not used as extensively for foraging, and cloud cover did not enter the model for this pond), and cloud cover may decrease glare, improving foraging success. Cloud cover was not as important for peep (it did not enter models alone only as an interaction variable), possibly because they are sometimes tactile foragers.

Temperature. Temperature alone, and as an interaction variable with wind velocity entered the models significantly for the number of shorebirds on the Bay, but not for peep on the Bay (Tables VII and VIII). However, temperature did enter for peep on East Pond (Table VIII) where higher numbers were present at high temperatures (Fig. 15). Temperature usually entered as a squared function, for during spring and fall migration the temperature continued to increase (or decrease in the fall) while the shorebirds increased and decreased each spring and fall. In the late fall as temperature decreased, shorebird numbers decreased in the Bay and increased on the ponds (peep migrated out of Jamaica Bay Refuge). Temperature did enter the model for the number of peep flocks on the Bay and East Pond (few peep ever used West Pond): there were fewer flocks at lower temperatures.

Precipitation. Although data were taken on days with rain or light snow, precipitation did not enter any of the models as a variable contributing to the variability in any of the dependent measures (number of species, individuals, or flocks, and species diversity) for either peep or all shorebirds combined. Although heavy rains often resulted in a cessation of feeding, shorebirds remained in the area.

7. Interaction Variables

Several interaction variables entered the models. In general, interactions within a broad category (wind direction × wind speed, tide direction × tide cycle) contributed more to accounting for variability than did interactions among categories. In these cases, the effect of the variables may be additive. For example, high winds from particular directions would be more severe if vegetation did not shelter foraging areas; and stage in the tide cycle interacts with the direction the tide is moving (water levels 2 hr before low are not the same as 2 hr after low tide).

Further, date interacted with a number of variables (date × wind direction, date × temperature) where the response of the birds was to an interaction of the two, and not to the independent variables alone. For example, low temperatures occur in the early spring and late fall, yet the shorebirds respond differently depending upon date: they used the ponds at low temperatures and the bay at high temperatures in the fall; and they reversed this pattern in the spring.

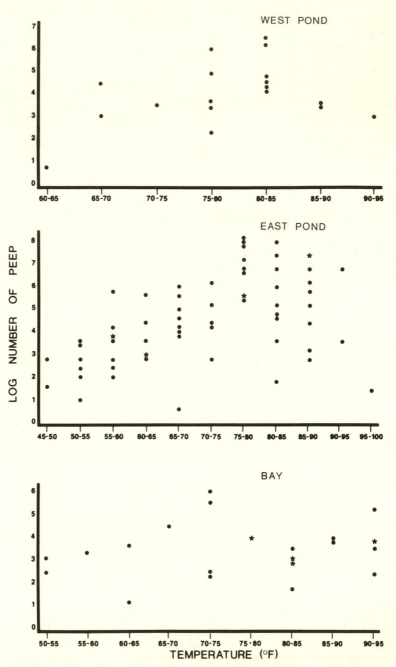

Fig. 15. Number of ''peep'' using each census area as a function of temperature.

E. Discussion

During the year of this study, 31 species of shorebirds were recorded on Jamaica Bay Refuge, similar to the number found along the New Jersey coast by Urner (35 species; Urner and Storer, 1949). Urner's censuses, however, covered a 10-year period and involved eight study areas from Newark to Brigantine. The high number of species concentrating on Jamaica Bay suggests that the area is an important migrating and staging area, particularly critical for shorebirds (see Pitelka, 1979). A similar study on the West Coast yielded 23 species over the course of a year's sampling in several habitats (Storer, 1951).

In the following discussion, I will deal with the effects of temporal, tidal, and weather-related variables, and habitat effects.

1. Effects of Temporal Variables

Jamaica Bay Refuge is used by shorebirds throughout the year, although large concentrations are found only during spring and fall migration. As has been found in other studies (see Storer, 1951; Bradstreet *et al.*, 1977; Isleib, 1979; Harrington and Morrison, 1979; Pitelka, 1979), fall migration is more protracted and involves more and larger flocks of shorebirds. Presumably, the necessity of arrival on the Arctic breeding grounds in spring synchronizes their spring migration, while the fall migration is not contracted by any requirement to time their arrival on the wintering grounds at any precise time. Recent research has indicated temporal differences by age during the fall migration with young migrating later (Jehl, 1963; Bradstreet *et al.*, 1977; Harrington and Morrison, 1979; Page *et al.*, 1979). Such a pattern reduces foraging competition on the breeding grounds (Pitelka, 1959; Holmes and Pitelka, 1968) as well as during migration (see also Myers, 1981).

In this study, date entered almost all the models explaining variability in number of shorebirds, flocks, and species, and in species diversity for the Bay and East Pond. The lower concentrations of birds on West Pond resulted in date being a less significant variable in the models explaining the variability in peep numbers (refer to Table VIII). In general, shorebirds were present from early March until early January when a prolonged period of cold, windy weather caused shorebirds to move out. This result is similar to that reported for England (see Pienkowski, 1981), where severe winds and storms caused the greatest problems for overwintering shorebirds. The small-sized shorebirds, however, did not arrive until May and had almost completely migrated out of Jamaica Bay Refuge by early November. Such a migratory pattern is consistent with Pienkowski's

(1981) hypothesis of great stress from low temperatures and winds (present at Jamaica Bay from December through February).

Time of day entered the models explaining variation in numbers of shorebirds as a significant variable only on the ponds where concentrations were often high at midday or early afternoon. Such flocks often contained predominantly roosting birds while only a few individuals were feeding. Time of day did not significantly affect the variability in number of shorebirds on the Bay, perhaps because the Bay was so tidally influenced (see below). Helicopter surveys indicated that when flocks of shorebirds did roost on the Bay at high tide, they sometimes were on salt marshes not visible from the regular census stops. In contrast, resting shorebird flocks on the ponds were always visible, and those present at midday were always censused. Time entered the models as an interaction variable with date because daylight shifts dramatically over an entire year, shifting the precise time of peak and low numbers of birds. Date × time entered the model for numbers of peep only for West Pond (refer to Table VIII), where number of resting peep built up around midday, particularly when high tide coincided. Generally, however, peep continued to feed throughout the day regardless of tide cycle, and the vast majority fed on East Pond freshwater mudflats.

Diurnal rhythms in shorebird activity have been reported (see Section II.A.2, and Goss-Custard, 1969; Heppleston, 1971; Evans, 1976). Daily activity patterns have also been noted in other marine birds such as gulls [*Larus* (Drent, 1967; Delius, 1970; Spaans, 1971; Galusha and Amlaner, 1978)].

2. Effects of Tidal Variables

Tidal factors are known to affect the behavior and activity patterns in eiders [*Somateria mollissima* (Campbell, 1978)], guillemots [*Uria aalge* (Slater, 1976)], gulls (Drent, 1967; Bianki, 1967; Vernon, 1970; Delius, 1970; Burger, 1976, 1980; Galusha and Amlaner, 1978), terns (*Sterna*) and skimmers [*Rynchops* (Hulsman, 1976; Erwin, 1977)], as well as shorebirds (see Section II.B). In most studies the birds primarily fed at low tide, and rested at high tides. However, as shown by Burger *et al.* (1977), even within any given habitat, the behavior of each species varies, and their use of the habitat peaks at slightly different tide times. Among tidal habitats, peak usage also varied in a coastal New Jersey area (Burger *et al.*, 1977). Usually tidal effects have been noted only from mudflats, bays, or estuaries, although shorebirds frequently move into salt marshes or nontidal areas at high tide. Thus, when all shorebirds in a given area are considered, the curves for abundance of shorebirds on roosting areas should be the inverse of those found on tidal areas (refer to Fig. 1).

In this study tidal variables (tide time, tide height, tide direction, or interactions of these variables) affected the number of individuals, flocks, species, and species diversity for peep and all shorebirds combined at all locations (see Tables VII–XII). Given the importance of date in influencing the numbers of shorebirds at Jamaica Bay Refuge, it is noteworthy that the effect of tides can easily be observed on graphs (Figs. 5–13) where data are lumped regardless of date. The variation at any one point in the tide cycle or at any tide height is partly accounted for by differences in date. Peak numbers occurred on the Bay at low tide in early August and in mid-October, but the numbers available differed markedly (refer to Fig. 3).

Shorebirds at Jamaica Bay Refuge moved from one area to another, and it was possible to stand beside East Pond and observe shorebirds moving into the pond at high tide, and out into the Bay at low tide. The exodus from the pond often required only 15–20 min, although the movement into East Pond as high tide approached was usually more protracted and less dramatic. Many species moved into East Pond to roost, preen, or feed (Dunlin, Sanderling, Greater Yellowlegs, Black-bellied Plover) while others almost always continued to feed on the exposed mudflats in East Pond (Long-billed Dowitcher, Stilt Sandpiper). East Pond and the Bay acted as a unit, with shorebirds frequently moving back and forth between them.

The pattern of tidal effects was much more striking on East Pond than the Bay, largely due to scale and sampling problems. During every census, every bird on East Pond could easily be counted in a few hours. On the contrary, although an attempt was made to count all birds on the Bay, not all areas of the Bay were visible and the censusing usually required 8 hr or more. Because East Pond was entirely visible at one time, the movement of a large group of birds could be noted, whereas the Bay was not entirely visible from one point because it was too large. Nonetheless, the pattern of intensive use of the Bay at low tide and East Pond at high tide was evident throughout the year. Because East Pond was not tidal, shorebird use on it reflected the tidal cycle on the adjacent areas of the Bay. These data indicate that East Pond is an important component of the system, and suggest that coastal bays and estuaries are more usable to shorebirds if freshwater impoundments with exposed mudflats are available for use during high tides.

The number of shorebirds was less affected by tidal variables on West Pond than on East Pond (refer to Figs. 5 and 8). There are several reasons for this difference between the two ponds: (1) significantly fewer shorebirds were present on West Pond compared to East Pond; (2) West Pond did not contain extensive areas of exposed mudflats suitable for feeding as did East Pond in the fall; (3) a path around West Pond usually contained several people, whereas East Pond had no path and usually no people; (4) West Pond did not contain extensive areas for roosting flocks of shorebirds as did East Pond; and (5) West Pond was smaller than East Pond. I believe these factors contributed to lower numbers of shorebirds using West Pond generally, and to their avoidance of West Pond when they left the Bay areas at high tide.

The pattern of census area use by peep was more variable with respect to tide cycle and tide height than that of other shorebirds (refer to Figs. 6 and 9), largely because some flocks of Semipalmated Sandpipers remained in East Pond to feed, while others moved out into the Bay at low tide. Over 80% of the Semipalmated Sandpipers were recorded on East Pond (refer to Table V). Larger-sized shorebirds (such as Dunlin, Sanderling, Turnstone, Lesser Yellowlegs) usually left East Pond at low tide.

The relationship between tidal variables, number of species present, and census area bears comment (refer to Fig. 12). The maximum number of species using West Pond (10) and the Bay (9) was less than on East Pond (16). One might well ask where the additional species come from on East Pond at high tide. There are several reasons for this difference: (1) many of the common species remained on East Pond at all tides (Dowitchers, Greater Yellowlegs) or were not always observed on the Bay or West Pond in any given day, (2) many of the rare species (seen fewer than 50 times during the year) were never observed on West Pond or the bay and, (3) rare or difficult to identify species might be missed if they were on salt marsh islands in the middle of the bay, but were not missed if they were on the ponds. Thus, at low tide certain species left East Pond to feed in the Bay and were no longer present at East Pond. However, at high tides, some individuals of all species present on the bay came into East Pond (not West Pond), resulting in a high number of species. Further, heaviest shorebird use on West Pond occurred during spring migration (when East Pond water levels were high) when the fewest numbers of species were present on Jamaica Bay Refuge. West Pond had fewer species using it at any one time, and fewer used it during the year. That is, during the year, 30 of the 31 species used East Pond but only 21 used West Pond and 18 species were observed on the Bay (refer to Table V).

Most investigators examining the effect of tide on shorebirds select either tide cycle time or tide height because both factors are readily available from local tide charts (see Section II.B). Because most authors were examining whether tide influenced shorebird behavior, their choice of only one tidal measure is appropriate. Yet there are several aspects of tides that could influence the foraging behavior and habitat selection of shorebirds, including the tide cycle, tide height, tide direction, or interactions of these variables. Shorebirds could respond to any of these characteristics, or to different characteristics at different seasons. Tide height could influence shorebird abundance because the water level determines the extent of mudflat exposure. Indeed, Recher (1966) suggested that amount of available habitat was the most important variable determining shorebird numbers (rather than prey availability), although Connors *et al.* (1981) recently showed that Sanderlings switch feeding sites on a tidal schedule to maximize foraging efficiency. Thus, shorebird numbers might be expected to vary directly with tide height: the lower the tide and the more mudflat exposed, the more shorebirds. However, tide height varies monthly and seasonally, as well as daily. The highest high tides and the lowest low tides occur during the new moon, and there is less difference between high and low tide during the full moon (see Wood, 1976). Thus, the same tide height could occur at high tide during a full moon, and 3 hr before high tide during a new moon. Similar differences occur seasonally with higher high tides (and lower low tides) in the spring and fall, compared to summer, although spring low tides are lower than fall low tides (Recher, 1966). If shorebirds waited for a given low tide level before feeding, it might not occur, or mudflats may be exposed for a very short period of time.

Tide cycle time would be an important predictor of available foraging time, as maximum mudflat exposure should occur around low tide. Thus, for example, if shorebirds required 6 hr to obtain sufficient food resources, they could maximize their feeding efficiency by feeding in the 6 hr around low tide when the maximum mudflat area was exposed. If, on the other hand, shorebirds required 12 hr of foraging, then they could either feed on the mudflats for the entire time they were exposed (sometimes 6–8 hr), and move into freshwater ponds or salt marshes to feed the other hours; they could feed on both low tide cycles (even though one might be at night); or a combination of both.

Tide direction might be expected to influence shorebird behavior because invertebrate numbers in the upper sand surface (where they are available as prey) and invertebrate activities vary depending on whether the tide is rising or falling (Stopford, 1951). Thus, tide direction may be important, particularly for species that feed at the wave edge (Recher and Recher, 1969; Burger *et al.*, 1977). For species that feed in the water (yellowlegs, dowitchers, Dunlin), tide direction may be less critical because the water has covered the substrate for some time (but this requires testing).

Interactions between tidal variables could be important where maximum tidal exposure and direction of tidal movement both contribute to increased foraging efficiency. Interactions between tidal variables should be important not only because wind speed and direction can have an effect on tide height (and thus amount of mudflat exposed), but also because wind can adversely affect foraging behavior, thermodynamics, and prey availability (see Section II.D.2). Tidal and wind interactions might be more important for smaller species, which would have more difficulty maneuvering in the wind.

One of the primary objectives of this study was to determine which tidal factors seemed to be a better predictor of shorebird numbers throughout the year, and whether small-sized shorebirds responded differently to tidal variables than larger species. When the computer was given only the tidal variables, and not the other abiotic variables, the models accounted for a greater percentage of the variability in the number of individuals and number of flocks for the small-sized shorebirds (peep) than for all shorebirds combined (refer to Tables XV and XVI). From this I conclude that tidal variables (including tide direction × wind direction) are more important in determining the abundance and local distribution of peep than they are to other shorebirds. Second, less of the variation was explained by the models for the Bay compared to the other census areas for peep and for all shorebirds combined. This may be because shorebirds are concentrated in the ponds at high tide, and disperse over a wider area in the Bay where some are not visible. As many investigators have suggested, roosting numbers are a good inverse measure of feeding numbers in an area (see Goss-Custard, 1969); thus, one might surmise that the large numbers present in East Pond at high tide disperse to tidal areas at low tide. And third, different tidal variables entered the models for peep compared to all shorebirds combined.

For all shorebirds combined, tide cycle entered 11 of 12 possible models (three census areas, four dependent measures), whereas it entered only 8 (of 11) for peep (Table XVII). However, tide direction entered 5 (of 11) models for peep, but only 1 (of 12) model for all shorebirds combined. Because clearly the tidal variables are related (but not linearly), the computer models procedure would select the variable that contributed the most to explaining the variability. These data suggest that both tide direction and tide cycle are important for peep, whereas tide cycle is clearly more important for all shorebirds combined. Further, tide direction × wind direction entered 11 of a possible 11 models for peep, but only 7 of the 12 models for all shorebirds combined. This also suggests that tide direction × wind direction is a more important independent variable contributing to variability in peep dependent measures than it is for all shorebirds, and that it is a more important variable for peep than the other tidal variables are. These conclusions are corroborated by the more highly significant probabiltity levels for the contribution of tide direction × wind direction to the models for peep compared to all shorebirds. The models procedure selects the variable that gives the highest R^2 value and adds that to the model first (and it has the lowest probability levels, i.e., it is most significant). Tide direction × wind direction was the first variable to enter 10 of 11 of the models for peep, but it entered the models first only once for all shorebirds combined (species diversity model on West Pond). Thus, tide direction × wind direction was the single variable that gave the highest R^2 for 91% of the peep models whereas tide cycle gave the highest R^2 for 60% of the models for all shorebirds combined. Thus, no other variable entered first more often. One might argue with specific statistical procedures, but overall it is clear that peep respond more to tide direction × wind direction and that other shorebirds respond more to tide cycle (see below for further discussion of wind variables).

For all shorebirds combined, tide cycle was the best predictor of abundance. I believe tide cycle reflects tide direction and relates to tide height, so it encompasses the other variables. Feeding at low tide maximizes foraging space (and time) regardless of the height of the tide, whereas foraging at a particular tide height might not. The scatter in the figures (Fig. 5) is due to date. For any given day, many of the shorebirds (excluding peep) fed mainly on the Bay at low tide and East Pond at high tide. However, because the number of shorebirds present varied seasonally (refer to Fig. 3), the abundance varied for any point on the tidal scale. The peak number present on the Bay on 1 September and 1 November may both have occurred at tide time 0, but the total present on the two days differed. The computer models procedure handled this by first removing the effect due to date (R. Trout, personal communication).

In this study the largest numbers of peep individuals fed on the Bay at low tide, but the largest number of flocks occurred on a rising tide (refer to Table XIV). This is a result of the formation and behavior of individual flocks. As the tide falls, some peep flocks move out to the Bay, and more flocks arrive as time passes, but none depart, as foraging may be more optimal at low tide. As the tide reaches dead low, a maximum number of birds are on the mudflats, and flocks coalesce (because of available space and numbers of birds). As the tide rises, some individuals leave, decreasing the number of peep foraging on the Bay. The departed individuals then are left in a number of smaller, scattered flocks (see Recher, 1966). In New Jersey the greatest concentraion of shorebirds occurred on tidal mudflats an hour after low tide (Burger *et al.*, 1977).

3. Effects of Weather Variables

Despite the frequent references to the effects of weather in influencing migratory behavior and timing (Brooks, 1963; Holmes, 1966; Recher, 1966) and breeding chronologies (Holmes, 1966; Pitelka *et al.*, 1974), until recently little emphasis was placed on examining the effects of behavior on abundance, local distribution, and foraging strategies of migrant shorebirds (see Section II.D). Recently, Evans (1976), Pienkowski (1981), Dugan (1981), and their colleagues have been studying how temperature and wind velocity affect prey abundance and availability, and how these factors affect shorebird feeding behavior and success. They have clearly shown that low temperatures and high winds decrease prey abundance and availability (see Section II.D for a discussion of mechanisms). Their excellent studies provide testable paradigms for analysis of different species, different foraging behaviors, and different habitats.

The results of the Jamaica Bay study clearly indicate that wind variables (velocity and direction, or as interaction variables) were important contributors to explaining the variability in the number of all shorebirds and of peep on all three census areas. However, when the computer was given only tidal variables and tide direction × wind direction, the latter variable entered more of the models for peep explaining more of the variability in the dependent measures than it did for all shorebirds combined (see above).

In general, shorebirds were affected more by wind direction than by wind velocity (that is, wind direction significantly entered more of the models). I suggest that their response to wind direction (rather than velocity) reflects that within Jamaica Bay Refuge they could not change the ambient wind velocity, but they could mitigate its effects by moving to sheltered areas, or to areas where wind did not combine with the tides to reduce available foraging space. Shorebirds generally avoided the Bay in S-SW winds; winds from these directions are in the direction of rising tides, and when they coincide with particularly high tides (new moon), the tides can be higher than predicted, and can sometimes result in little or no mudflat exposure. Further, one of the mudflat areas used for foraging by shorebirds is on the northwest side of the Bay, where S-SW winds would have the full extent of the Bay to build up increased wave action, and make foraging difficult (particularly for small-sized shorebirds). Shorebirds concentrated (in excess of that predicted by the prevailing winds) on East Pond on N-NE winds, and on West Pond in E-SE and S-SW winds. In general, winds from these directions would be lessened because dense stands of *Phragmites* occupy the edges of each pond in these directions.

Peep were more affected by wind direction variables than were all shorebirds combined (that is, wind direction variables significantly entered more models). The negative effects of wind on foraging ability (maneuverability, disturbed water surface, lower prey abundance and activity; see Bengtson and Svensson, 1968; Dunn, 1973) and thermoregulation (excessive heat loss because of increased windchill leading to higher energy needs) would be greater on smaller-sized shorebirds compared to larger shorebirds. Thus, peep might respond at lower wind velocities by moving into the protected areas of the ponds or the lee side of salt marsh islands in the Bay. Shorebirds that moved into East Pond (almost completely surrounded by dense *Phragmites*) could continue to forage on the extensive freshwater mudflats.

In this study temperature entered some of the models accounting for the variability in the dependent measures but not as many as wind variables. Where it did enter, it also seemed to affect where shorebirds were at low temperatures; use of West Pond and the Bay decreased at lower temperatures even for peep. At freezing temperatures, areas of the Bay and ponds were frozen, and most shorebirds migrated from the area.

4. Overall Habitat Utilization

All three census areas of Jamaica Bay Refuge were used extensively by migrant shorebirds throughout the year (refer to Fig. 3). However, overall, over half of the shorebirds were recorded on East Pond and only about 7% on West Pond. Two factors seemed to be responsible for this relationship: (1) tidally responsive species moved between East Pond and the Bay depending on tide stage, and (2) some species primarily fed in either East Pond or West Pond, but only a few species concentrated on West Pond (notably Lesser Yellowlegs, Pectoral Sandpiper; refer to Table V).

Tidally responsive species moved back and forth in large flocks. They fed in the Bay, and generally rested or fed in East Pond. For other species (primarily peep), some individuals moved into the Bay at low tide, but many remained in East Pond and fed throughout the day. Indeed, most peep fed all day, although larger species frequently roosted in large flocks at high tide. Peep used West Pond infrequently in the fall, mostly because there were no mudflats (thus no feeding areas), and peep usually fed all day. Species that fed solitarily, usually in shallow water, frequently fed in West Pond. Substrate differences between the two ponds (West Pond has a sandier bottom) may also account for differences in habitat use. Further, West Pond is less encircled by dense *Phragmites*, and is more exposed to winds than East Pond.

Table XVIII. Relative Use of Census Areas during Spring and Fall Migration

	Spring migration[a]		Fall migration[b]	
East Pond	147	(1.4%)	150,496	(69.9%)
West Pond	7763	(71.3%)	7,216	(3.4%)
Bay	2963	(27.3%)	57,344	(26.7%)

[a] Spring migration = March–June.
[b] Fall migration = July–January.

In general, West Pond was used more by shorebirds during the spring migration than East Pond; and the reverse pattern existed during the extended fall migration (refer to Fig. 3). Table XVIII, derived from Table VI, indicates the relative use of the census areas for the 10 most abundant species (which account for 98% of the shorebirds on Jamaica Bay). This use pattern is due to management practices. West Pond always contains a small sandy beach and feeding area. The water is drawn down on 1 April, leaving mudflats (never as extensive as those in East Pond) that are used by shorebirds until early summer when rainwater has again filled the pond so that there are no extensive mudflats. However, East Pond contained no mudflats or beaches until July when the National Park Service personnel drew down the water to expose extensive mudflats. The relative preference of shorebirds for East Pond in the fall suggests that it might be used more extensively in the spring if water levels were manipulated earlier.

Another consideration is the relative amount of human disturbance. East Pond is undisturbed because people would have to cut their way through dense *Phragmites* to reach the pond. However, West Pond is easily accessible by a well-cared-for path, and people are almost always present on this path. Although people walking on the path frequently do no flush shorebirds, they may contribute to shorebirds avoiding this pond (see Burger, 1981). When no other freshwater mudflats exist (as in the spring), the shorebirds use West Pond, but when other areas become available (with the summer drawdown of East Pond), they switch to them.

The results of this study indicate that given the opportunity, shorebird species will use different habitats under different environmental conditions. Indeed, in times of severe high wind and low temperature stress, protected foraging areas may be critical for survival. The presence of both freshwater and tidal mudflats provides adequate foraging areas throughout the tide cycle. As Myers and Myers (1979) showed for shorebirds wintering in Argentina, a diversity of tidal and nontidal habitats provides ideal wintering conditions. The importance of freshwater mudflats cannot be underestimated, particularly in view of the extensive use those at Jamaica Bay received from all shorebirds. Some individuals seemed to use the freshwater mudflats exclusively for foraging. Roosting areas are an extremely important requirement for migrant shorebirds (Prater, 1981), and must be present at staging areas or shorebirds will not remain long. Further, the presence of foraging and roosting areas buffered from the wind is important, particularly for small-sized shorebirds, which would be greatly affected by high winds and windchill.

5. Conclusions

In this study over 60% of the variability in the number of shorebirds on the ponds and Bay was explained by temporal, tidal, and weather-related variables. The most important and consistent variables influencing the number of shorebirds on Jamaica Bay Refuge were date, tide cycle (or tide height), and wind (velocity and direction). In general, small-sized shorebirds (peep) were more affected by wind variables than were larger species. Wind direction had the greatest effect as shorebirds could mitigate high winds by selecting particular areas. Tide cycle (rather than tide height or tide direction) was the factor that entered the most models explaining the variability in the dependent measures (number of individuals, species, and flocks, and species diversity) for all shorebirds combined, although tide height and direction entered more models for the small-sized shorebirds. These variables also entered most models explaining the variability in the number of species, number of flocks, and species diversity. In general, the number of shorebirds on East Pond was less affected by wind and temperature variables (it was more protected), and the number of shorebirds on West Pond was less affected by tidal variables. More tidal, temperature, and interaction variables entered the models explaining the variability in shorebird numbers on the Bay than on the two ponds. The data strongly support the hypothesis that diversity of physical habitat as well as the presence of freshwater ponds with exposed mudflats are essential to providing sufficient staging areas for migrating shorebirds. Temporal, tidal, and weather-related variables interacted differently in the three census areas. During the spring migration shorebirds primarily used West Pond and the Bay, whereas in the fall migration they used East Pond and the Bay, a difference that reflects a lack of mudflats in East Pond during spring (see above). Unlike some previous studies (see Evans, 1976; Gerstenberg, 1979), this study indicates that where exposed freshwater mudflats are available (as in East Pond), the number of shorebirds on these areas are tidally dependent as the birds use these areas at high tides when they are forced off the exposed mudflats in the Bay. Further, many shorebirds remained on East Pond even during low tide where they continued to feed. The results indicate that temporal, tidal, and weather-related variables affect the local distribution and abundance of shorebirds, and that small-sized shorebirds respond differently to these variables than all shorebirds combined.

IV. SUMMARY

I studied the effect of abiotic effects on the abundance and local distribution of shorebirds on Jamaica Bay Wildlife Refuge for a 1-year period. I was particularly interested in determining how temporal, tidal, and weather-related factors affected shorebirds, particularly the small-sized shorebirds (peep). I had predicted that date and tidal variable would have the greatest effect on local shorebird distribution, and that peep would be more affected by wind than would larger shorebirds. Further, I wanted to determine what aspect of tide (tide cycle, tide height, tide direction) had the greatest effect on local shorebird movements.

Jamaica Bay is a 3600-ha tidal bay containing numerous salt marshes, mudflats, and two man-made freshwater ponds. The coastal refuge is situated next to J. F. Kennedy International airport, residential communities, and sanitary landfills. The bay areas and East Pond are relatively undisturbed by humans, but an access path around West Pond was almost always used by bird watchers or other people.

Shorebirds were censused 4 days a week for 8–10 hr a day from 1 May 1978 to 1 May 1979. Censusing of the Bay was accomplished by recording all shorebirds on scaled maps of 17 census areas designed to cover all visible areas of the Bay. The Bay transect required a full day to census. The two ponds were censused on the same day by walking around their perimeters, recording birds on gridded maps.

Dependent measures considered were the number of individuals, number of flocks, number of species, and species diversity (H). Independent variables considered included temporal (date, time of day), tidal (tide height, tide cycle, tide direction), weather-related variables (wind velocity, wind direction, temperature, cloud cover, precipitation), and interactions of these variables. I used stepwise regression procedures to develop models to account for the variability in the dependent measures on three census areas (tidal Bay, East Pond, West Pond).

During the study over 230,000 shorebirds were censused: 66% were recorded on East Pond, 27% on the Bay, and only 7% on West Pond. Dunlin, Black-bellied Plover, Semipalmated Sandpiper, Greater Yellowlegs, and Short-billed Dowitcher accounted for over 80% of the observed shorebirds. Overall, 31 species occurred; 30 were reported on East Pond, 18 used the Bay, and 21 used West Pond.

Shorebirds were present from March to January when a prolonged period of freezing weather, high winds, and ice-covered feeding areas forced the wintering species to move out temporarily. Spring migration involved fewer species and fewer individuals, and was largely restricted to April, May, and June whereas fall migration extended from mid-July to December. Only about 5% of the migrants occurred during the spring migration compared to the fall migration.

The regression models accounted for over 60% of the variability in the number of shorebirds on all three census areas and the significant variables in all three models were tidal (tide cycle or tide direction) or wind-related (velocity or direction). Peep were affected differently by the tidal factors.

Although date significantly influenced the overall presence of migrant shorebirds in the area, it also affected their use of habitats. Shorebirds used West Pond during spring and East Pond during fall migration, largely because extensive freshwater mudflats were exposed in East Pond only after 1 July when water levels were drawn down artificially.

For all measures on all census areas, tidal factors significantly entered the models for all three census areas. One of the objectives of the study was to determine which tidal factors shorebirds responded to the most. A models procedure was used in which the computer was given only tidal factors to account for the variability in the measures. Variables given included tide cycle, tide height, tide direction, tide height × tide direction, and wind direction × tide direction. These variables had previously been selected by the computer as the variables that give the highest R^2. When given only the above variables, tide cycle entered 11 of the 12 models (four dependent measures × three census areas) first for all shorebirds combined, indicating that it contributed the most to accounting for the variability in the dependent measures. However, tide direction × wind direction was the first variable to enter 10 of 11 models for peep. These data suggest that tide cycle time has the greatest tidal effect on most shorebirds, but tide direction × wind direction has the greatest tidal effect on peep.

As was expected, shorebirds concentrated on the Bay at low tides, and moved into East Pond at high tide. Although tide significantly affected shorebird abundance on West Pond, the effect was not as dramatic. At low tide up to 9 species were present at one census time on the Bay, and up to 6 were present on the ponds; whereas at high tide a maximum of 16 species were recorded at one time on East Pond, but only 10 on West

Pond and 9 on the Bay. This difference occurs because some species fed almost exclusively in one or the other pond, remaining there at low tide (thus decreasing the number of species on the Bay).

Wind variables entered almost all of the models for peep and for all shorebirds combined. However, wind direction rather than wind velocity entered more models significantly. Peep responded more strongly to wind, avoiding the bay areas in W-NW winds (often the strongest winds). The relative importance of wind direction (rather than velocity) relates to their ability to mitigate the effects of high wind velocities by selecting habitats that are buffered from the winds.

Cloud cover entered some models for shorebirds: shorebirds seemed to concentrate in the Bay in dense cloud cover. This may relate to foraging success: clouds decrease glare, making it easier for visual foragers to locate prey. Cloud cover did not contribute significantly to many of the models for peep, perhaps because visual cues are less important to peep.

Temperatures entered some models significantly, particularly as an interaction with wind variables, suggesting that low temperatures and high winds combine to provide stressful conditions for shorebirds. Precipitation did not enter any models significantly, and observations indicated that, in the absence of high winds, shorebirds ignored rain.

Overall, shorebirds used primarily East Pond and the Bay, moving back and forth depending on tide state. However, the majority of shorebirds remained in East Pond and continued to forage on the freshwater mudflats throughout the day at all tide stages. Some species, however, used West Pond extensively (Pectoral Sandpipers, Lesser Yellowlegs), suggesting that it was filling a particular habitat requirement. The presence of West Pond was particularly crucial during spring migration when over 70% of the shorebirds used this area (mudflats were not exposed on East Pond during spring migration). During the fall migration 70% of the shorebirds used East Pond, and 27% used the Bay.

The results of the Jamaica Bay study clearly indicate that tidal and wind-related variables have the greatest effect in accounting for variations in numbers of individuals, flocks, and species in different habitats. The wide variety of habitats ensures that shorebirds can partially mitigate the adverse effects of tides, high winds, and low temperatures.

ACKNOWLEDGMENTS

I would like to thank Wade Wander for extensive field assistance in the collection of these data, and R. Trout for help in the statistical design of this study prior to its inception, and statistical and computer analysis throughout this study. I thank P. A. Buckley, P. R. Evans, M. Gochfeld, M. A. Howe, and B. G. Murray for discussions and comments on the manuscript or ideas contained herein. The personnel at Jamaica Bay Wildlife Refuge made the work pleasant. This research was supported by Contract CX 1600-8-0007 from the North Atlantic Region of the National Park Service. I especially thank Paul A. Buckley (Chief Scientist, Northeastern Region) for his continued interest, support, and advice concerning the project.

REFERENCES

Ashmole, M. J., 1970, Feeding of Western and Semipalmated Sandpipers in Peruvian winter quarters, *Auk* **87**:131–135.

Baker, J. M., 1981, Winter feeding rates of Redshank *Tringa totanus* and Turnstone *Arenaria interpres* on a rocky shore, *Ibis* **123**:85–87.

Baker, M. C., 1974, Foraging behavior of Black-bellied Plovers (*Pluvialis squatarola*), *Ecology* **55**:162–167.

Baker, M. C., 1979, Morphological correlates of habitat selection in a community of shorebirds (*Charadriiformes*), *Oikos* **33**:121–126.

Baker, M. C., and Baker, A. E. M., 1973, Niche relationships among six species of shorebirds on their wintering and breeding ranges, *Ecol. Monogr.* **43**:193–212.

Barr, A. J., Goodnight, J. H., Sall, J. P., and Helwig, J. T., 1976, *A User's Guide to SAS*, Sparks Press, Raleigh, N.C.

Bengtson, S. A., and Svennsson, B., 1968, Feeding habits of *Calidris alpina* L. and *C. minuta* Leisl. (Aves) in relation to the distribution of marine shore invertebrates, *Oikos* **19**:152–157.

Bent, A. C., 1927, Life histories of North American shorebirds, Part 1, *U.S. Natl. Mus. Bull.* **142**.

Bianki, V. V., 1967, Gulls, shorebirds and alcids of Kandalaksha Bay, *Proc. Kandalaksha State Res* **6**:1–247.

Bradstreet, M. S. W., Page, G. W., and Johnston, W. G., 1977, Shorebirds at Long Point, Lake Erie, 1966–1971: Seasonal occurrence, habitat preference, and variation in abundance, *Can. Field Nat.* **91**:225–236.

Bridges, J. T., 1976, Vegetation survey, soils map and estimate of plant health conditions at Jamaica Bay Wildlife Refuge, Report of National Park Service and New York Botanical Gardens.

Brooks, W. S., 1963, Effect of weather on autumn shorebird migration in East-central Illinois, *Wilson Bull.* **77**:45–54.

Burger, J., 1976, Daily and seasonal activity patterns in breeding Laughing Gulls, *Larus atricilla*, *Auk* **93**:308–323.

Burger, J., 1980, Age differences in foraging Black-necked Stilts in Texas, *Auk* **97**:633–636.

Burger, J., 1981, The effect of human activity on birds at a coastal bay, *Biol. Conserv.* **21**:231–241.

Burger, J., 1982, Jamaica Bay studies. I. Environmental determinants of abundance and distribution of Common Terns (*Sterna hirundo*) and Black Skimmers (*Rynchops niger*) at an East Coast estuary, *Colon. Waterbirds.* **4:**148–160.

Burger, J., 1983a. Jamaica Bay studies. II. Effect of tidal, temporal and weather variables on distribution of ibises, egrets and herons on a coastal estuary, *Acta Oecol.* **4:**181–189.

Burger, J., 1983b. Jamaica Bay studies. III. Abiotic determinants of distribution and abundance of gulls (*Larus*) at an East Coast estuary, *Estuarine Coastal Shelf Sci.* **16:**191–216.

Burger, J., and Gochfeld, M., in press. Jamaica Bay studies. IV. Flocking associations of shorebirds at an Atlantic coastal estuary, Biology of Behavior.

Burger, J., Howe, M. A., Hahn, D. C., and Chase, J., 1977, Effects of tide cycles on habitat selection and habitat partitioning by migrating shorebirds, *Auk* **94:**743–758.

Burger, J., Hahn, D. C., and Chase, J., 1979, Aggression in mixed species flocks of shorebirds, *Anim. Behav.* **27:**459–469.

Campbell, J. H., 1978, Diurnal and tidal behaviour patterns of eiders wintering at Leith, *Wildfowl* **29:**47–52.

Colman, J. S., 1964, Fish mortality off the Isle of Man and in the Port Erin Aquarium, *J. Anim. Ecol.* **33:**210.

Connors, P. G., Myers, J. P., Connors, C. S. W., and Pitelka, F. A., 1981, Interhabitat movements in Sanderlings in relation to foraging profitability and the tidal cycle, *Auk* **98:**49–64.

Crisp, D. J., 1964, The effect of the severe winter of 1962–63 on marine life in Britain, *J. Anim. Ecol.* **33:**165–210.

Darbyshire, M., and Draper, L., 1963, Forecasting wind-generated sea waves, *Engineering* **195:**482–484.

Dare, P. J., and Mercer, A. J., 1973, Foods of the oystercatcher in Morecambe Bay, Lancashire, *Bird Study* **20:**173–184.

Davidson, N. C., 1981, Survival of shorebirds (Charadrii) during severe weather: The role of nutritional reserves, in: *Feeding and Survival Strategies of Estuarine Organisms* (N. V. Jones and W. J. Wolff, eds.), pp. 231–249, Plenum Press, New York.

Davidson, N. C., 1982, The effects of severe weather in 1978/79 and 1981/82 on shorebirds at Teesmouth: A preliminary view, *Wader Study Group Bull.* **34:**8–10.

Davidson, P. E., 1968, The oystercatcher—A pest of shellfisheries, in: *The Problems of Birds as Pests* (R. K. Murton and E. N. Wright, eds.), pp. 141–155, Academic Press, New York.

Delius, J. D., 1970, The effect of daytime, tides and other factors on some activities of Lesser Black-backed Gulls, *Larus fuscus, Rev. Comp. Anim.* **4:**3–11.

Dick, W. J. A., Pienkowski, M. W., Waltner, M., and Minton, C. D. T., 1976, Distribution and geographical origins of Knot *Calidris canutus* wintering in Europe and Africa, *Ardea* **64:**22–47.

Dobinson, H., and Richards, A. J., 1964, The effects of the severe winter of 1962/63 on birds in Britain, *Br. Birds* **57:**373–434.

Drent, R. H., 1967, Functional aspects of incubation in the Herring Gull (*Larus argentatus* Pont.), *Behaviour Suppl.* **17.**

Drinnan, R. E., 1957, The winter feeding of the oystercatcher (*Haematopus ostralegus*) on the edible cockle (*Cardium edule*), *J. Anim. Ecol.* **26:**441–469.

Duffy, D. C., Atkins, N., and Schneider, D. C., 1981, Do shorebirds compete on their wintering grounds?, *Auk* **98:**215–229.

Dugan, P. J., 1981, The importance of nocturnal foraging in shorebirds: A consequence of increased invertebrate prey activity, in: *Feeding and Survival Strategies of Estuarine Organisms* (N. V. Jones and W. J. Wolff, eds.), pp. 251–260, Plenum Press, New York.

Dugan, P. J., Evans, P. R., Goodyer, L. R., and Davidson, N. C., 1981, Winter fat reserves in shorebirds: Disturbance of regulated levels by severe weather conditions, *Ibis* **123**:359–363.

Dunn, E. K., 1973, Changes in fishing ability of terns associated with windspeed and sea surface conditions, *Nature (London)* **244**:520–521.

Ehlert, W., 1964, Zur Ökologie und Biologie der Ernährung einiger Limikolen-Arten, *J. Ornithol.* **105**:1–53.

Elliott, C. C. H., Waltner, M., Underhill, L. G., Pringle, J. S., and Dick, W. J. A., 1976, The migration system of the Curlew Sandpiper *Calidris ferruginea* in Africa, *Ostrich* **47**:191–213.

Erwin, R. M., 1977, Foraging and breeding adaptations to different food regimes in three seabirds: The Common Tern, *Sterna hirundo*, Royal Tern, *S. maxima*, and Black Skimmer, *Rynchops niger*, *Ecology* **58**:389–397.

Evans, P. R., 1976, Energy balance and optimal foraging strategies in shorebirds: Some implications for their distribution and movement in the non-breeding season, *Ardea* **64**:117–139.

Evans, P. R., 1979, Adaptations shown by foraging shorebirds to cyclical variations in the activity and availability of their intertidal invertebrate prey, in: *Cyclic Phenomena in Marine Plants and Animals* E. Naylor and R. G. Hartnoll, eds.), Pergamon Press, Elmsford, N.Y.

Evans, P. R., 1981, Migration and dispersal of shorebirds as a survival strategy, in: *Feeding and Survival Strategies of Estuarine Organisms* (N. V. Jones and W. J. Wolff, eds.), pp. 275–290, Plenum Press, New York.

Evans, P. R., and Smith, P. C., 1975, Studies of shorebirds at Lindisfarne, Northumberland. 2. Fat and pectoral muscle as indicators of body condition in the Bar-tailed Godwit, *Wildfowl* **26**:64–76.

Evans, P. R., Brearey, D. M., and Goodyer, L. R., 1980, Studies on Sanderlings at Teesmouth, N. E. England, *Wader Study Group Bull.* **30**:18–22.

Feare, C. J., 1966, The winter feeding of the Purple Sandpiper, *Br. Birds* **59**:165–179.

Frith, H. J., 1976, *Reader's Digest Complete Book of Australian Birds*, Reader's Digest, Sydney.

Galusha, J. G., Jr., and Amlaner, C. J., Jr., 1978, The effects of diurnal and tidal periodicities in the numbers and activities of Herring Gulls, *Larus argentatus* in a colony, *Ibis* **120**:322–328.

Gerstenberg, R. H., 1979, Habitat utilization by wintering and migrating shorebirds on Humboldt Bay, California, in: *Studies in Avian Biology No. 2* (F. A. Pitelka, ed.), pp. 33–40, Cooper Ornithological Society, Allen Press, Lawrence, Kans.

Gochfeld, M., 1971, Notes on a nocturnal roost of Spotted Sandpipers in Trinidad, West Indies, *Auk* **88**:167–168.

Goss-Custard, J. D., 1969, The winter feeding ecology of the Redshank *Tringa totanus*, *Ibis* **111**:338–356.

Goss-Custard, J. D., 1970, Feeding dispersion in some overwintering wading birds, in: *Social Behaviour in Birds and Mammals* (J. H. Crook, ed.), pp. 3–35, Academic Press, New York.

Goss-Custard, J. D., 1977, The ecology of the Wash. III. Density-related behaviour and the possible effects of a loss of feeding grounds on wading birds (Charadrii), *J. Appl. Ecol.* **14**:721–739.

Goss-Custard, J. D., 1979, The energetics of foraging Redshank, *Tringa totanus*, in: *Studies in Avian Biology No. 2* (F. A. Pitelka, ed.), pp. 247–257, Cooper Ornithological Society, Allen Press, Lawrence, Kans.

Goss-Custard, J. D., Jenyon, R. A., Jones, R. E., Newbery, P. E., and Williams, R. L., 1977, The ecology of the Wash. II. Seasonal variation in the feeding conditions of wading birds (Charadrii), *J. Appl. Ecol.* **14**:701–719.

Hamilton, W. J., III, 1959, Aggressive behavior in migrant Pectoral Sandpipers, *Condor* **61**:161–179.

Harrington, B. A., and Morrison, R. I. G., 1979, Semipalmated Sandpiper migration in North America, in: *Studies in Avian Biology No. 2* (F. A. Pitelka, ed.), pp. 83–99, Cooper Ornithological Society, Allen Press, Lawrence, Kans.

Hartwick, E. B., and Blaylock, W., 1979, Winter ecology of a Black Oystercatcher population, in: *Studies in Avian Biology No. 2* (F. A. Pitelka, ed.), pp. 207–215, Cooper Ornithological Society, Allen Press, Lawrence, Kans.

Heppleston, P. B., 1971, The feeding ecology of oystercatchers *Haematopus ostralegus* L. in winter in northern Scotland, *J. Anim. Ecol.* **41**:651–672.

Heppleston, P. B., 1972, The comparative breeding ecology of oystercatchers (*Haematopus ostralegus* L.) in inland and coastal habitats, *J. Anim. Ecol.* **41**:23–51.

Holmes, R. T., 1966, Breeding ecology and animal cycle adaptations of the Red-backed Sandpiper (*Calidris alpina*) in northern Alaska, *Condor* **68**:3–46.

Holmes, R. T., and Pitelka, F. A., 1968, Food overlap among co-existing sandpipers on northern Alaskan tundra, *Syst. Zool.* **17**:305–318.

Hope, C. E., and Shortt, T. M., 1944, Southward migration of adult shorebirds in the west coast of James Bay, Ontario, *Auk* **61**:572–576.

Hulscher, J. B., 1976, Localisation of cockles (*Cardium edule* L.) by the oystercatcher (*Haematopus ostralegus* L.) in darkness and daylight, *Ardea* **64**:292–310.

Hulsman, R., 1976, The robbing behavior of terns and gulls, *Emu* **76**:143–149.

Isleib, M. E., 1979, Migratory shorebird populations on the Copper River Delta and Easter Prince William Sound, Alaska, in: *Studies in Avian Biology No. 2* (F. A. Pitelka, ed.), pp. 125–130, Cooper Ornithological Society, Allen Press, Lawrence Kans.

Jehl, J. R., Jr., 1963, An investigation of fall-migrating dowitchers in New Jersey, *Wilson Bull.* **75**:250–261.

Kelly, P. R., and Cogswell, H. L., 1979, Movements and habitat use by wintering populations of Willets and Marbled Godwits, in: *Studies in Avian Biology No. 2* (F. A. Pitelka, ed.), pp. 69–82, Cooper Ornithological Society, Allen Press, Lawrence, Kans.

McLachlan, G. R., and Liversidge, R., 1978, *Roberts Birds of South Africa*, John Voelcker Bird Book Fund, Cape Town.

McLachlan, G. R., Wooldridge, T., Schramm, M., and Kuhn, M., 1980, Seasonal abundance, biomass and feeding of shorebirds on sandy beaches in the Eastern Cape, South Africa, *Ostrich* **51**:44–52.

Makkink, G. F., 1936, An attempt at an ethogram of the European Avocet, *Ardea* **25**:1–74.

Mascher, J. W., 1966, Weight variations in resting Dunlin (*Calidris a. alpina*) on autumn migration in Sweden, *Bird-banding* **37**:1–34.

Miller, E. H., 1979, Function of display flights by males of the Least Sandpiper, *Calidris minutilla* (Vieill.), on Sable Island, Nova Scotia, *Can. J. Zool.* **57**:876–893.

Morrell, S. H., Huber, H. R., Lewis, T. J., and Ainley, D. G., 1979, Feeding ecology of Black Oystercatchers on South Farallon Island, California, in: *Studies in Avian Biology No. 2* (F. A. Pitelka, ed.), pp. 185–186, Cooper Ornithological Society, Allen Press, Lawrence, Kans.

Myers, J. P., 1981, A test of three hypotheses for latitudinal segregation of the sexes in wintering birds, *Can. J. Zool.* **59**:1527–1534.

Myers, J. P., and Myers, L. P., 1979, Shorebirds of coastal Buenos Aires Province, Argentina, *Ibis* **121**:186–200.

Myers, J. P., Connors, P. G., and Pitelka, F. A., 1979, Territory size in wintering Sanderlings: The effects of prey abundance and intruder density, *Auk* **96**:551–561.

Oring, L. W., and Davis, W. M., 1966, Shorebird migration at Norman, Oklahoma: 1961–1963, *Wilson Bull.* **78**:166–174.

Page, G. W.; Stenzel, L. E., and Wolfe, C. M., 1979, Aspects of the occurrence of shorebirds on a central California estuary, in: *Studies in Avian Biology No. 2* (F. A. Pitelka, ed.), pp. 15–32, Cooper Ornithological Society, Allen Press, Lawrence Kans.

Phillips, R. E., 1977, Notes on the behavior of the New Zealand Shore Plover, *Emu* **77**:23–27.

Pienkowski, M. W., 1978, Differences in habitat requirements and distribution patterns of plovers and sandpipers as investigated by studies of feeding behaviour, *Verh. Ornithol. Ges. Bayern* **23**:105–124.

Pienkowski, M. W., 1981, How foraging plovers cope with environmental effects on invertebrate behaviour and availability, in: *Feeding and Survival Strategies of Estuarine Organisms* (N. V. Jones and W. J. Wolff, eds.), pp. 179–192, Plenum Press, New York.

Pienkowski, M. W., 1982, Diet and energy intake of Grey and Ringed Plovers, *Pluvialis squatarola* and *Charadrius hiaticula*, in the non-breeding season, *J. Zool.* **197**:511–549.

Pilcher, R. E. M., 1964, Effects of the cold weather of 1962–63 on birds of the north coast of the Wash, *Wildfowl Trust Annu. Rep.* **15**:23–26.

Pitelka, F. A., 1959, Numbers, breeding schedule, and territoriality in Pectoral Sandpipers of northern Alaska, *Condor* **61**:233–264.

Pitelka, F. A. (ed.), 1979, *Studies in Avian Biology No. 2*, Cooper Ornithological Society, Allen Press, Lawrence, Kans.

Pitelka, F. A., Holmes, R. T., and MacLean, S. E., Jr., 1974, Ecology and evolution of social organization in Arctic sandpipers, *Am. Zool.* **14**:185–204.

Prater, A. J., 1972, The ecology of Morecambe Bay. III. The food and feeding habits of Knot (*Calidris canutus* L.) in Morecambe Bay, *J. Appl. Ecol.* **9**:179–194.

Prater, A. J., 1979, Shorebird census studies in Britain, in: *Studies in Avian Biology No. 2* (F. A. Pitelka, ed.), pp. 157–166, Cooper Ornithological Society, Allen Press, Lawrence, Kans.

Prater, A. J., 1981, *Estuary Birds of Britain and Ireland*, Poyser, Calton, England.

Puttick, G. M., 1979, Foraging behaviour and activity budgets of Curlew Sandpipers, *Ardea* **67**:111–122.

Puttick, G. M., 1980, Energy budgets of Curlew Sandpipers at Langebaan Lagoon, South Africa, *Estuarine Coastal Mar. Sci.* **11**:207–215.

Recher, H. F., 1966, Some aspects of the ecology of migrant shorebirds, *Ecology* **47**:393–407.

Recher, H. F., and Recher, J. A., 1969, Some aspects of the ecology of migrant shorebirds. II. Aggression, *Wilson Bull.* **81**:140–154.

Robertson, H. A., and Dennison, M. D., 1979, Feeding and roosting behaviour of some waders at Farewell Spit, *Notornis* **26**:73–88.

Senner, S. E., 1979, An evaluation of the Copper River Delta as a critical habitat for migrating shorebirds, in: *Studies in Avian Biology No. 2* (F. A. Pitelka, ed.), pp. 131–146, Cooper Ornithological Society, Allen Press, Lawrence, Kans.

Slater, P. J. B., 1976, Tidal rhythm in a seabird, *Nature (London)* **264**:636–638.

Smith, P. C., 1975, A study of the winter feeding ecology and behaviour of the Bar-tailed Godwit (*Limosa lapponica*), Unpublished Ph.D. thesis, University of Durham, U.K.

Smith, P. C., and Evans, P. R., 1973, Studies of shorebirds at Lindisfarne, Northumberland. I. Feeding ecology and behavior of the Bar-tailed Godwit, *Wildfowl* **24**:135–139.

Smith, S. M., and Stiles, F. G., 1979, Banding studies of migrant shorebirds in northwestern Costa Rica, in: *Studies in Avian Biology No. 2* (F. A. Pitelka, ed.), pp. 41–47, Cooper Ornithological Society, Allen Press, Lawrence, Kans.

Spaans, A. L., 1971, On the feeding ecology of the Herring Gull *Larus argentatus* Pont. in the northern part of The Netherlands, *Ardea* **59**:73–188.

Stenzel, L. E., Huber, H. R., and Page, G. W., 1976, Feeding behaviour and diet of the Long-billed Curlew and Willet, *Wilson Bull.* **88**:314–332.

Stopford, S. C., 1951, An ecological survey of the Cheshire foreshore of the Dee Estuary, *J. Anim. Ecol.* **20**:103–122.

Storer, R. W., 1951, The seasonal occurrence of shorebirds on Bay Farm Island, Alameda County, California, *Condor* **53**:186–193.

Strauch, J. G., and Abele, L. G., 1979, Feeding ecology of three species of plovers wintering on the Bay of Panama, Central America, in: *Studies in Avian Biology No. 2* (F. A. Pitelka, ed.), pp. 217–230, Cooper Ornithological Society, Allen Press, Lawrence, Kans.

Swinebroad, J., 1964, Nocturnal roosts of migrating shorebirds, *Auk* **76**:155–159.

Thomas, D. G., and Dartnall, A. J., 1971, Ecological aspects of the feeding behaviour of two *Calidridine* sandpipers wintering in southeastern Tasmania, *Emu* **71**:20–26.

Townshend, D. J., 1981, The importance of field feeding to the survival of wintering male and female curlews *Numenius arquata* on the Tees Estuary, in: *Feeding and Survival Strategies of Estuarine Organisms* (N. V. Jones and W. J. Wolff, eds.), Plenum Press, New York.

Tree, A. J., 1979, Biology of the Greenshank in southern Africa, *Ostrich* **50**:240–257.

Urner, C. A., and Storer, R. W., 1949, The distribution and abundance of shorebirds on the north and central New Jersey coast, 1928–1938, *Auk* **66**:177–194.

Vernon, J. D. R., 1970, Food of the Common Gull on grassland in autumn and winter, *Bird Study* **17**:36–38.

Wishart, R. A., and Sealy, S. G., 1980, Late summertime budget and feeding behaviour of Marbled Godwits (*Limosa fedoa*) in southern Manitoba, *Can. J. Zool.* **58**:1277–1282.

Wolff, W. J., 1969, Distribution of non-breeding waders in an estuarine area in relation to the distribution of their food organisms, *Ardea* **57**:1–25.

Wood, F. J., 1976, The strategic role of Perigean spring tides, U. S. Department of Commerce, National Oceanic and Atmospheric Administration, Washington, D.C.

Woodhead, P. M. J., 1964, The death of fish and sub-littoral fauna in the North Sea and the English Channel during the winter of 1962–63, *J. Anim. Ecol.* **33**:169–173.

Chapter 2

MIGRATORY BEHAVIOR OF SHOREBIRDS IN THE WESTERN PALEARCTIC

Michael W. Pienkowski and Peter R. Evans

Department of Zoology
University of Durham
Durham DH1 3LE, England

Migration is a behavioral adaptation that permits animals to breed in areas where they could not survive throughout the year, or that allows individuals to produce more offspring in the breeding areas than they would have been able to rear if they had stayed to breed in areas used at other seasons. Migrations during the nonbreeding season are presumed to enhance the chances of survival of the adults so that they may have the opportunity to breed again. In this chapter we shall be concerned mainly with *coastal* waders (shorebirds), both because these are the species of Charadrii on which most work has been done and because this series of books is concerned with marine animals. This concentration of work on coastal shorebirds has happened partly for practical reasons and partly because conservation studies have been needed more particularly in coastal habitats (see chapters by Senner and Howe, and Evans and Pienkowski, Vol. 5 of this series). We will, however, refer to noncoastal species where studies on these have helped to answer certain problems raised by the phenomenon of migration. We shall concentrate on the western Palearctic migration system, because more information is available from this system than from elsewhere, relating to the general questions raised by shorebird migration. The migration patterns in the Americas are treated separately by Morrison (this volume).

I. GENERAL MIGRATION PATTERNS OF SHOREBIRDS IN THE WESTERN PALEARCTIC

The most strongly migratory species are those that breed in the Arctic. Waders from breeding grounds stretching between Ellesmere Island, Canada, in the west and eastern Siberia in the east utilize coastal areas in western Europe and Africa in the nonbreeding seasons. Their wintering areas cover a very wide range of latitudes, from South Africa (35°S) to the British Isles (50–60°N). A few species survive the winter even in Norway (>60°N), Iceland (65°N), and southwestern Greenland (>60°N). The importance of the coasts of western Europe to wintering shorebirds results from the great extent of wetland (particularly intertidal) areas arising from the geomorphologically complex European coastlines. These lie in an area affected by the warm waters of the North Atlantic Drift (Gulf Stream).

The availability of large habitable areas so far north in winter leads to high west–east and east–west components in the migrations of many species to this area. At least two geographical races (from Canada/Greenland/Iceland and from Eurasia, respectively) can be separated in several shorebird species. In some, the distributions of the races may overlap throughout much of their winter range; this may be true of Sanderling *Calidris alba*, although the situation for this species is not yet entirely clear. In others, one race winters mainly in Africa, while the other stays mainly in Europe. The degree of overlap of the races differs between species; it is considerable in Dunlin *C. alpina* (Pienkowski and Dick, 1975) and probably nil in Red Knot *C. canutus* (Dick *et al.*, 1976). However, there is no general rule as to which breeding population moves further (Table I). Red Knots and Ruddy Turnstones *Arenaria interpres* from the Eurasian breeding areas tend to winter further south than those from the northwest, but the reverse is true for Dunlins (Dick *et al.*, 1976; Branson *et al.*, 1978; Pienkowski and Dick, 1975).

In total, more shorebirds winter in northwestern Africa than in the whole of western Europe. The Banc d'Arguin (Mauritania) is a particularly important wintering area (NOME, 1982). The variety of species and numbers reaching southern Africa are, by comparison, very small. Nevertheless, these individuals regularly perform some of the longest migrations known among land-birds—each journey covering some 13,000 km.

In central Asia and eastern Siberia, shorebird migration studies have been restricted mainly to work in the Soviet Union (although most of these are not readily available). Those species and populations wintering in Africa have been studied en route in Iran (Argyle, 1975), in east Africa

Table I. Summary of the Main Migration Patterns of Some of the Common Coastal Shorebirds in the Western Palearctic[a]

Species	Breeding areas	Timing of main autumn migration	Autumn molting areas	Wintering areas	Timing of main spring migration
Eurasian Oystercatcher (*Haematopus ostralegus*)	Iceland, Faeroes, N Britain	Late July to September (differs in different areas)	Mainly at winter site; some near breeding areas	Coasts near breeding areas, both coasts of Scotland, Irish Sea coasts	Late January to April
	Britain	"	"	Britain; France, Iberia; a few NW Africa	"
	Netherlands	"	"	Netherlands, France, Iberia, NW Africa; a few to E Britain	"
	Norway	"	Mainly at winter areas	Coasts of North Sea; a few to W Britain and further south	"
	Baltic & White Sea coasts	"	"	Probably Waddenzee southwards on mainland European coast; a few to Britain	To early May
	E Europe, central Asia	"	"	E Mediterranean, Red Sea, Arabian Gulf, India; sparsely E Africa	March, April to May, June
Ringed Plover (*Charadrius hiaticula*)	Greenland, & probably N Canada	August to September	Probably W Africa, some possibly N of wintering areas	Probably W Africa	May to early June
	Iceland	Probably July onwards	As above, but some start molt on breeding grounds and arrest molt for migration	Probably W Africa	April to May

(Continued)

Table I. (*Continued*)

Species	Breeding areas	Timing of main autumn migration	Autumn molting areas	Wintering areas	Timing of main spring migration
	Britain	(September)	Some start while breeding; near breeding site or winter quarters	Some near breeding areas; elsewhere in Britain (mainly W & S), Ireland, France, N Spain	(February to May)
	S Scandinavia, Baltic coasts, W mainland Europe	June to September	Some on breeding grounds; some arrest; some on winter grounds or intermediate molting areas	W Britain, Ireland to N Africa	February to May, tending to be latest in E breeding areas
Black-bellied (or Grey) Plover (*Pluvialis squatarola*)	N Scandinavia, USSR	August to October	"	Africa	May to June
	Arctic USSR	Late July to November	At winter areas or molting sites within and extending N & E of winter range (e.g., Baltic coasts). Some may start molt at or near breeding grounds and arrest	Europe & Africa from Britain & Denmark southwards to South Africa	February to March (South Africa) April to early June (Europe)
Red Knot (*Calidris canutus*)	Greenland & Arctic Canada	July to September	Within winter range (tending to E of winter sites)	W Europe	May to early June (often preceded by movement to premigratory sites in March to April)

Arctic USSR to unknown extent eastwards (but probably including at least some birds from New Siberian islands) [Farther E in USSR [S Canadian Arctic	"	Mainly in winter range	W & S Africa	April (South Africa) May to early June (Europe)
	?	?	SW Pacific, especially Australasia South America]	?]
Sanderling (*Calidris alba*) Greenland & possibly parts of Arctic Canada	Late July to October	Mainly near winter range but some onward movement after molting (e.g., in NW Africa)	W Africa but possibly further N & S also	April to early June (earlier in Africa)
USSR	"	"	W Europe & S Africa & possibly intermediate areas	"
The extents of the wintering grounds and degree of overlap of these populations are not yet well determined				
Little Stint (*Calidris minuta*) N Scandinavia & USSR	July to early November	Near winter range but some movement after molt (e.g., in Mediterranean & NW Africa)	Mediterranean coasts southwards in Africa	May to early June (earlier in Africa)
Curlew Sandpiper (*Calidris ferruginea*) Arctic USSR	July to September	NW Africa, southwards, with onward movement from northern areas, e.g., Morocco, after rapid molt	Mauritania, inland W Africa, Guinea-Bissau	April to early June

(Continued)

Table I. (*Continued*)

Species	Breeding areas	Timing of main autumn migration	Autumn molting areas	Wintering areas	Timing of main spring migration
Dunlin (*Calidris alpina*)	NE Greenland	August to September	Probably on or near winter quarters	Probably NW Africa	May to early June
	Iceland & SE Greenland	July to September	NW Africa	NW Africa	April to early May
	British Isles & mainland Europe	June to October	Near winter quarters, plus few in W Europe, some during migration	Most in NW Africa, a few in W Europe	March to April
	Scandinavian mountains, Svalbard, USSR	July to October	Mainly in winter range, but with some tendency to N & E of winter sites. A few start on or near breeding grounds and arrest	W Europe from Denmark & Britain southwards, Mediterranean, Morocco	March to May (after some movement to premigratory sites in March)
Bar-tailed Godwit (*Limosa lapponica*)	N Scandinavia & USSR	July to October	Mainly in winter range but some movement after molting	N Europe to W & S Africa	March to early June, with some previous movement to premigratory areas
Eurasian Curlew (*Numenius arquata*)	Ireland	June to November	Mainly on winter grounds	Probably resident, some moving to coast	February to March to W & S breeding areas, April to May to NE

N Britain	(differing in different areas, E populations tending to be latest)	"	British W coast & Ireland	"
S England	"	"	SW England, France, Iberia	"
Norway	"	"	N Britain, Ireland, W France	"
Sweden	"	"	Britain, Ireland, Denmark to N France	"
Poland, Baltic states, Finland, NW USSR	"	"	European mainland coasts; some Britain & Ireland	"
Germany, Netherlands	"	"	W France; some Britain, Ireland to Morocco	"
SE Europe, Crimea, S Russia	"	"	Mediterranean, S to N Africa	"
Asian USSR	"	"	Africa, S Asia	"
Common Redshank (*Tringa totanus*) Iceland	June to August (differing in different areas, NE populations to October)	In winter range, but some onward movement after molting	Iceland, SW Norway, Britain, W France; some to NW Africa	February to March, NE populations to May
Britain & Ireland	"	Near breeding areas or near winter quarters	British Isles, France; some S to NW Africa	"

(Continued)

Michael W. Pienkowski and Peter R. Evans

Table I. (*Continued*)

Species	Breeding areas	Timing of main autumn migration	Autumn molting areas	Wintering areas	Timing of main spring migration
	Netherlands	"	Most populations on or near wintering areas, with some movement after molting. Some of the Baltic and North Sea coast populations molt during migration along Atlantic coasts	France, Iberia, Mediterranean	"
	Denmark, S Sweden, Germany, Belgium	"		Denmark to Iberia, Mediterranean, Morocco; a few Britain & W Africa	"
	N Sweden, Norway	"	"	Overlap with previous group but most in NW Africa	"
	Finland	"	"	British Isles southwards; most in NW Africa	"
	E Europe	"	"	Mediterranean & N Africa	"
	Asia	"		S Asia	"
Ruddy Turnstone (*Arenaria interpres*)	N Canada & Greenland	July to September	Within winter range, but some movement after molting	British Isles & Norway to Iberia; some NW Africa	April to early June, probably after some movement to premigratory sites
	N Europe, Scandinavia, & USSR	"	Near winter areas	A few North Sea coasts, but mainly S of 20°N in Africa	April to early June

[a] Based on articles by Bainbridge (for Curlew) and Pienkowski (other species) in Cramp and Simmons (1983), which list the numerous original sources. More recent information from Ferns (1980–81) and Fournier and Dick (1981) is also incorporated. Movements within winter (which may be extensive—see Section IV) are not noted in this table.

(by, e.g., Pearson *et al.*, 1970), and in South Africa (by, e.g., Elliott *et al.*, 1976; Summers and Waltner, 1979).

Our knowledge of shorebird migration in the eastern Palearctic and the Pacific islands remains very incomplete, despite studies in some areas (e.g., McClure, 1974; Melville, 1981; Johnston and McFarlane, 1967; Johnson, 1979; Thomas and Dartnall, 1970, 1971; Ali and Hussain, 1981; Veitch, 1978).

II. WHY SHOREBIRDS MIGRATE—FACTORS AFFECTING DISTRIBUTION IN THE BREEDING SEASON

Of the two major groups within the shorebirds, the sandpipers (Scolopacidae) contain many more extensively migratory species than the plovers (Charadriidae). The former have evolved chiefly in the Arctic and north temperate areas and the latter chiefly in the tropics (e.g., Larson, 1957). The discussion that follows concerns chiefly the Scolopacidae but also the Arctic breeding species of other groups. Taken to its extreme, the question may be posed, what are the advantages for these species of breeding in the Arctic over breeding in temperate or tropical areas?

The extreme abundance of certain terrestrial invertebrates, principally insects, during the Arctic summer is often cited as a reason for the greater variety of shorebirds that breed there (e.g., Welty, 1962; Thomson, 1964; Lack, 1968; Dorst, 1974). Densities of breeding shorebirds are related to prey densities in some areas [e.g., Dunlin in Alaska (Holmes, 1970)]. Production of young is also timed to coincide with peak insect abundance in some areas [e.g., Alaska (Holmes, 1966a,b; Holmes and Pitelka, 1968), Arctic Canada (Nettleship, 1973, 1974)], but not in all [e.g., eastern Greenland (Green *et al.*, 1977)]. It has often been assumed that the continuous Arctic daylight provides favorable feeding conditions, leading to fast growth of young. However, in Arctic regions, shorebird chicks show a diurnal pattern of foraging activity, probably related to the lower temperatures at "night" when the sun is low (Safriel, 1975; Pienkowski, 1980, 1984b). Growth rates and fledging times of Ringed Plover *Charadrius hiaticula* chicks did not differ between High Arctic (Greenlandic) and temperate (British) breeding areas (Pienkowski, 1984a).

The summer is also the main period of growth and reproduction of the littoral invertebrates that form most of the prey of shorebirds in north temperate areas in the nonbreeding season, yet relatively few shorebird species stay to breed in these areas. Indeed, most of those that do stay to breed utilize tundra-like or marshy habitats (e.g., Dunlin, Greater Golden Plover *Pluvialis apricaria*, Common Redshank *Tringa totanus*), rather than coastal areas. Of those that nest at the coast, the young feed mainly on Diptera (e.g., Ringed Plover) or, atypically for shorebirds, are fed by the adults [e.g., Eurasian Oystercatcher *Haematopus ostralegus* (Norton-Griffiths, 1969)]. Foraging on concealed intertidal prey, such as are present in mudflats, is probably not profitable for shorebird chicks in the first few days after hatching, when they require abundant, obvious prey (Pienkowski, 1984b). Extensive use of intertidal areas for foraging by chicks would probably also require colonial nesting on nearby shores. Sites safe from predators (essentially islands) may be limited, especially within short distances of feeding areas. Usage of more distant feeding areas would probably require a more seabird-like life-style, i.e., feeding of young by the parents and taking of much larger prey than is usual, because of the energetics of food carrying.

Throughout their range, most shorebird species depend primarily on camouflage (and sometimes concealment) for protection of their nests and sitting birds from predators (but see Gochfeld, Vol. 5 of this series). The intensity of nest-predation alone may limit the southern extent of nesting ranges, as argued by Larson (1960) and Pienkowski (1980, 1984c). Larson argued that the southern edge of the breeding range of many Arctic breeding shorebirds is limited by the intensity of predation by Arctic Fox *Alopex lagopus*, against which protective adaptations and their success vary between species, so that limits of distribution vary also. Similar arguments apply to other predators, and generally the numbers and variety of predators increase toward lower latitudes (see Pienkowski, 1980, 1984c). The limits of distribution may be extended by naturally protected sites, such as islands. Pienkowski (1984c) argued in a similar way concerning the southern breeding limit of Ringed Plovers. Additionally (and particularly relevant to this species), the extent of suitable bare ground for nesting decreases to the south of the Arctic, so that nests are more easily found by predators.

The breeding range of Ringed Plovers extends from the High Arctic to the temperate, allowing comparisons of breeding performance at different latitudes (as summarized below from Pienkowski, 1984c). The shorter season in the High Arctic may prevent opportunities for any replacement layings, compared with up to five (mean 2.6) nesting attempts per year in Britain. Despite this, an average of 0.7 young/pair per year were fledged at the Arctic site (and this is thought to be an underestimate—probably by 50%—for the area in general, because of factors associated with nearby human habitation). This figure was exceeded at some naturally or artificially protected sites in Britain, but at other sites the estimate was 0.05 fledged young/pair per year. This difference was due to the difference in the rate of predation on eggs. The probability of survival to hatching of clutches in Arctic areas varied between 38% and 85%. In Britain, at protected sites, it reached 46%, but in other areas was as low as 1.4% (Pienkowski, 1984c). Järvinen and Väisänen (1978) argued that predation did not appear to be a factor involved in the increasing variety of breeding shorebird species found as one travels northward in Fennoscandia. However, they were apparently looking mainly for a northward increase in predation, which might have allowed coexistence by preventing competitive exclusion. For this they assumed that competition was important in limiting the variety of species present in any one area.

Against the advantages of nesting in the Arctic must be set the disadvantages faced by most individuals that undertake long migrations at least twice each year, namely that they must cope with major changes in foraging habitats, and increasing risk of predation; as well as the risks of migration. There has been considerable debate as to whether natural selection has acted on particular morphological features primarily in the breeding or the nonbreeding areas (e.g., Salomonsen, 1954; Hale, 1973, 1980). Studies are needed of the problems faced by birds in making these switches of habitat, particularly by young birds on their first migration, which they generally undertake in the absence of their parents and, indeed, of most adults of the same species.

III. WHY SHOREBIRDS MIGRATE—FACTORS AFFECTING DISTRIBUTION IN THE NONBREEDING SEASON

The evolved "aim" of shorebird behavior in the winter is to enable an individual to survive to the next breeding season, and then be in a fit condition to breed successfully. In some other Arctic breeding migrants, notably geese, feeding conditions in the winter quarters largely determine body condition, and possibly breeding success, the following summer (e.g., Ankney, 1977; Drent et al., 1981). At present, there is little evidence of any such long-term cross-seasonal link in shorebirds. Most migrants rapidly build up fat reserves shortly before, or at staging posts during, migration. This is particularly so in spring (see Section VI). In at least some cases, flight muscles also increase in mass at these times, not earlier. Therefore, we consider here mainly the requirements for over-winter survival, and how easily these can be met in different wintering areas.

Many species of shorebirds in winter feed principally on intertidal benthic invertebrates living in soft sediments (see, e.g., Pienkowski, 1982). In winter, many polychaete worms and some bivalve molluscs burrow deeper in the substrate [e.g., *Arenicola marina* (Smith, 1975), *Nereis diversicolor* (Muus, 1967), *Macoma balthica* (Reading and Mc-Grorty, 1978)], or move to lower tidal levels [e.g., some *Arenicola marina* (Brady, 1943)]. Several crustaceans even migrate to sublittoral areas [e.g., *Carcinus maenas*, *Crangon vulgaris* (Naylor, 1963; Swennen, 1971)]. These phenomena were reviewed by Evans (1979a), who also summarized information on the widespread tendency for the activity of such animals at the substrate surface to decline markedly as temperatures fall (Goss-Custard, 1969; Smith, 1975; Pienkowski, 1981b, 1983b; see also Goss-Custard, this volume). These changes in prey behavior in winter lead to increased difficulties for shorebirds feeding upon them (see, e.g., Evans, 1976; Pienkowski, 1981a, 1982, 1983c). In northwestern Europe, the incidence of gales increases in winter. These may also cause difficulties for feeding shorebirds; by holding water over the feeding areas (see Section IV); creating waves, making prey localization through shallow water difficult (Evans, 1976); drying the substrate above the tide edge and so reducing surface activity of prey (Smith, 1975; Pienkowski, 1980); and impeding foraging movements of the birds (Pienkowski, 1983a). In such conditions, birds may abandon their preferred feeding areas and/or stop feeding (Evans, 1976; Davidson, 1981a; Dugan *et al.*, 1981). Rainfall may also reduce surface activity of prey [see Goss-Custard (1969) for the amphipod *Corophium*] and reduce prey capture rates [see Pienkowski (1983c) for Ringed and Black-bellied Plovers *Pluvialis squatarola*].

All these changes tend to reduce the rate of food intake by shorebirds in winter, at a time when energy demands for thermoregulation are greatest (Evans, 1976; Pienkowski *et al.*, 1984). The situation is exacerbated in some situations in the Northern Hemisphere because the daylight period also shortens in winter. Many shorebirds search for prey visually or use vision to direct their subsequent tactile foraging. Several authors have reported that, in many (but not all) situations, shorebirds fulfill their food requirements by daytime if they can [Goss-Custard (1969) for Common Redshank, Heppleston (1971) and Goss-Custard *et al.* (1977b) for Eurasian Oystercatcher, Smith (1975) for Bar-tailed Godwit *Limosa lapponica*, Pienkowski (1982) for Ringed and Black-bellied Plovers]. Pienkowski (1983a) showed that the average distance to which Black-bellied Plovers ran to take prey was reduced at night, and that prey intake rate was reduced in some circumstances. The latter does not always happen, however, as the activity of some prey animals increases at night (presumably to avoid many of the diurnal predators) (see Pienkowski, 1980, 1983b;

Dugan, 1981b). Also, the prey that are available at night may be larger than by day. This may more than offset the reduced foraging abilities of the birds as shown, for example, for Black-bellied Plover at the Tees Estuary, northeastern England, by Dugan (1981b), and for some field-feeding species by McLennon (1979).

Overall, winter conditions become increasingly adverse with increasing latitude, and with an increasingly continental type of climate (e.g., from western Ireland to Denmark; see section IV). The limits that these impose on wintering distributions of different shorebird species are discussed in relation to prey densities and availabilities by Pienkowski (1981a, 1982). Birds relying totally or mainly on visual foraging (notably the plovers) depend on the activity of prey animals at the sediment surface in order to detect these. Prey activity (such as the frequency of brief visits to the sediment surface) is more sensitive to temperature than to the depth in the substrate at which such animals spend most time. This depth distribution is the feature of principal importance to tactile foragers (including many sandpipers). Thus, the minimum acceptable temperatures for feeding sufficiently fast (so as to meet energetic requirements) tend to be at higher temperatures for birds using the "plover" method than for those foraging as "sandpipers." Conversely, when temperatures are high, "plovers" can tolerate a lower absolute prey density than can "sandpipers," because of the larger area of detection of prey by visual foraging than by tactile methods. This leads to more southerly wintering distributions of plovers than of many sandpipers, and the presence of the former on beaches of lower productivity. The relative importance of average winter conditions and of the occasional very severe winter in determining the northern limits of distribution has yet to be assessed critically.

Shorebirds buffer some of the effects of short-term shortage of food in winter by depositing, in autumn, fat reserves of up to about 30% of lean body-weight (Evans and Smith, 1975; Pienkowski, 1981a; Davidson, 1981a). Protein reserves may also be held in the muscles, but these are used as an energy source only when fat reserves are exhausted (Evans and Smith, 1975; Davidson, 1981a; Davidson and Evans, 1982). Fat reserves, and therefore total body mass, of most species reach a peak in midwinter (Minton, 1975; Pienkowski et al., 1979; Davidson, 1981a). The subsequent decline appears generally to be due to a matching of the declining need for such reserves against the costs of carrying them, rather than to an inability to maintain them in late winter (Evans and Smith, 1975; Pienkowski et al., 1979; Dick and Pienkowski, 1979; Dugan et al., 1981). The need for such an "insurance," in the form of fat, changes seasonally as the energetic demands and the risk of not being able to meet them change. These vary both geographically and between species. Most species spending the nonbreeding season in northwestern, eastern, and southern Africa, South America, and Australia do not appear to need such reserves (Dick and Pienkowski, 1979; Pearson *et al.*, 1970; Middlemiss, 1961; Elliott *et al.*, 1976; Summers and Waltner, 1979; McNeil, 1970; Thomas and Dartnall, 1970, 1971). This is presumably because the availability of prey during the austral summer, or in equatorial regions, is always adequate to allow birds to forage effectively. For Dunlin, the need for winter reserves is already apparent in southwestern Britain and increases northeastwards. Within Britain, the size of the midwinter fat reserve increases with decreasing mean midwinter temperature, which is probably also a measure of the risk of exceptionally severe conditions (Pienkowski *et al.*, 1979). A similar pattern is found in Common Redshanks wintering on the west coast of Britain; but on the east coast, where weather is more severe, the relationship breaks down, presumably because the birds are unable to maintain fat reserves appropriate to the risk of needing them (Davidson, 1982). Indeed, Common Redshank is the shorebird species that suffers highest mortality in severe winters (Dobinson and Richards, 1964; Pilcher, 1964; Pilcher *et al.*, 1974; Goss-Custard *et al.*, 1977b; Baillie, 1980; O'Connor and Cawthorne, 1982).

Clearly then, there are severe risks in wintering too far north everywhere, and too far east in Europe. The difficulties encountered in meeting energy needs (as modified by reserves) may be aggravated by competition between and within species (see below). Predation levels may also affect the balance of costs and benefits differently in different areas (see Page and Whitacre, 1975). We shall discuss competition between species here and within species in Section IV.

Pienkowski (1981a) argued that the space-demanding, visual foraging strategy of plovers may be impaired (and plovers thereby excluded from the best feeding areas near the tide edge) by tactilely foraging shorebirds (which are less affected by crowding). This apparently resulted in Black-bellied Plovers exploiting better feeding areas after Bar-tailed Godwits had left Lindisfarne, northeastern England, in spring. Zwarts (1974) gives several further examples of possible interference and avoidance between species. Indirect competition for food supplies may occur by the predation of different age classes of prey. At Teesmouth, Dunlins and Black-bellied Plovers fed mainly on small (0- to 1-year class) *Nereis diversicolor* (but the two species generally used different feeding situations) whereas Eurasian Curlews *Numenius arquata* and Bar-tailed Godwits took mainly 1 + -year-class animals. The intensity of predation on both was high, possibly because reclamation schemes restricted large shorebird populations to a small area of feeding habitat (Evans *et al.*, 1979; Evans, 1981a). At Schiermonnikoog in The Netherlands, Eurasian Oystercatchers fed primarily on 3-year-old mussels *Mytilus edulis* but Herring Gulls *Larus argentatus* fed on first-year spat. There, local changes in feeding distribution of the two species (but not in total numbers) have been described by Zwarts and Drent (1981). Food-robbing within and between species may also affect the profitability of feeding in an area.

The metabolic consequences of size differences between bird species may also be important. Generally, energy requirements per unit mass increase with decreasing body size, because of the increasing surface area : volume ratio and the decreasing thickness of plumage (Lasiewski and Dawson, 1967; Kendeigh, 1970; Ebbinge *et al.*, 1975). Also, the sizes of prey items taken by small shorebirds are usually smaller than those taken by larger shorebirds. Thus, the time taken each day to meet food requirements by small shorebirds is often longer than by large ones (Fournier, 1969; Pienkowski, 1973, 1981a; Goss-Custard *et al.*, 1977b). Davidson and Evans (1982) cite evidence suggesting that small individuals of a species may be at greater risk of dying in very severe weather. This may be because they are unable to catabolize protein reserves (after fat reserves are exhausted) fast enough to meet their high metabolic rate per unit mass. Thus metabolic considerations suggest that smaller shorebirds should winter further south, or in warmer areas, than larger shorebirds. Evans (1981b) reviews some evidence that this may, indeed, be the situation.

There is evidence for several species that feeding rates of juveniles are, in some circumstances, lower than those of adults [e.g., Black-necked Stilt *Himantopus mexicanus* (Burger, 1980); other examples below]. This may occur when the young encounter new habitats on migration or in winter quarters [Ruddy Turnstone (Groves, 1978)], or as the onset of winter conditions makes feeding more difficult [Ringed Plover (Pienkowski, 1984b)]. After migration, young birds may need to learn what are suitable foraging areas within new types of habitats (Dick and Pienkowski, 1979). Young Eurasian Oystercatchers may take 3 years to become as efficient as adults at feeding on mussels *Mytilus edulis* (Norton-Griffiths, 1967, 1969).

Direct evidence that young birds are excluded from some areas by competition from adults is difficult to obtain. However, this is suggested by some established cases of spatial differences in distribution of the two age groups [W. J. A. Dick (personal communication) for the Wash; van der Have and Nieboer (1984) and Swennen (1984) for the Waddenzee; Pienkowski (1975) and Dick (1976) for northwestern Africa]. Further information on Eurasian Oystercatchers (Goss-Custard *et al.*, 1982) is noted in Section IV.

The reasons for racial differences in wintering areas (Table I) remain a subject for speculation. In several species, populations perform leapfrog migrations, the southernmost breeding populations migrating least, if at all, and the northernmost furthest. The first examples of this behavior in shorebirds to be discussed were Ringed Plover and Common Redshank (Salomonsen, 1955). Hale (1980) pointed out that the southernmost breeding populations are probably hybrids, and suggested that they may be capable of surviving harsher winter conditions. Alerstam and Högstedt (1980) pointed out several further examples of leapfrog migration in related species-pairs of shorebirds (e.g., Eurasian Curlew and Whimbrel *Numenius phaeopus*; Common Snipe *Gallinago gallinago* and Great Snipe *G. media*; Dunlin and Curlew Sandpiper *Calidris ferruginea*) and argued that leapfrog migration is widely found in migratory birds. At least one well-studied shorebird, Dunlin, does not fit the classical pattern in the western Palearctic (Pienkowski and Dick, 1975). Alerstam and Högstedt reanalyzed data on this species and suggested that leapfrog migration does occur; however, this is still disputed by Pienkowski *et al.* (in press).

Alerstam and Högstedt (1980) suggested that leapfrog migration arises from a great selective advantage for birds breeding in temperate areas if they remain during winter within the same climatic regime as, and close to, their breeding sites. Such individuals can respond directly to the development in spring of favorable breeding conditions, which occur at rather different dates from year to year. They argue further that the advancement of the Arctic spring is more regular (allowing Arctic birds to migrate far from their breeding sites and still return at an optimal breeding time by use of their circannual clock). This is disputed by Slagsvold (1982) and Pienkowski *et al.* (in press), who point out that the date of onset of spring is not more predictable in the Arctic (see also Green *et al.*, 1977); that too early an arrival there may carry much higher risks of death through starvation (see Morrison, 1975; Marcström and Mascher, 1979); that breeding as early as possible does not necessarily lead to highest production of young [Byrkjedal (1980) and Pienkowski (1980, 1984c) both show higher nest predation early in the season than later]; and that Alerstam and Högstedt may underestimate the cost of migration (see Dick and Pienkowski, 1979). Pienkowski *et al.* (in press) dispute also that birds wintering in the same climatic region as the breeding area detect suitable conditions before migrating to the breeding grounds; they present evidence that, contrary to Alerstam and Högstedt's prediction, the departure dates in spring of birds heading for an arctic area are not less variable than those of birds moving within the temperate zone.

Pienkowski *et al.* (in press) suggest other possible reasons for the occurrence or not of leap-frog migration. Assuming that it is advantageous to winter as close to the breeding grounds as possible, competition for, e.g., feeding sites could lead to expulsion of less dominant individuals. Many of the latter would probably be smaller individuals. In a species in which individuals of arctic breeding populations are generally smaller than those of temperate breeding populations, arctic birds might be forced to move further south in winter. In contrast, in species in which arctic-breeding birds did not tend to be smaller than their more southerly breeding conspecifics, leap-frog migration would not result. Migration patterns in the eastern Atlantic region (where the geography and the climate allow scope for a wide range of wintering latitudes) support this alternative model, for both leap-frog and non-leap-frog species.

Moreau (1967) noted the large gaps in knowledge concerning both the physiology of shorebird migration and the differences in extents of migration between individual shorebirds from the same breeding population. Although, since then, work on physiological aspects of migration (e.g., increase in mass and fat levels before movement) in shorebirds has made considerable progress (see Section VI), little has been done to elucidate why some move further than others. Pienkowski (1984a,b) found that some Ringed Plovers breeding in northeastern England stayed to winter in the same area while others left to winter elsewhere in northern England and southern Scotland. Individuals tended to behave in the same way from year to year, but not exclusively so. The differences did not appear to be age-related, and differences existed even between members of the same brood. There were some indications that birds staying in the area had better feeding performances than those departing, but this was not proven conclusively (Pienkowski, 1984a,b). There could be similarities with Kestrel *Falco tinnunculus*, another partial migrant as far as British breeding birds are concerned (Snow, 1968). Snow found some indications for Kestrels that food shortage leads to an increase in long-distance migrations; also, young birds tend to be more migratory than adults. Burger (1981) argued that a behavioral dichotomy between short- and long-distance migrants in Herring Gulls *Larus argentatus* could be accounted for by differences in food availability and climatic conditions, but this comparison concerned birds from different breeding areas, rather than individuals from the same area.

In the choice of wintering areas, there is likely to be a "trade-off" between the costs of migration (presumably increasing with distance traveled) and the chance of over-winter survival (presumably higher further south for the reasons discussed above). There are, as yet, very few data with which to investigate this trade-off, as estimates are required of the timing of mortality in the annual cycle. Those estimates of overall annual survival that are available indicate survival to be high (Evans and Pienkowski, Vol. 5 of this series). In view of the poor condition of some young birds after their first migration (Section V), it is possible that the main advantage in residency or short-distance migration over long-distance migration lies in the avoidance of a long flight and/or a major change in habitat by newly fledged birds, rather than its avoidance each year by adults. If this were the case, it might be reflected in a difference between short- and long-distance migrant populations in the survival rates from fledging to 1 year of age. Although such data are available for temperate-breeding, short-distance migrant populations of Eurasian Oystercatcher (Harris, 1967, 1975) and Ringed Plover (Pienkowski, 1984a), there appear to be no such directly measured data on any Arctic breeding shorebird population for comparison. If the distance to be moved by young birds were a problem, one might expect young birds to move less far than adults. This could, however, place them in a less favorable winter environment. Competition between individuals, age groups, breeding populations, and other species might also present problems, as discussed elsewhere in this chapter.

IV. ITINERANCY AND MOVEMENTS IN THE NONBREEDING SEASON

The use of staging posts during migration has long been known: some studies have used these sites to document the timing of migration (e.g., Mason, 1969; Netterstrøm, 1970; Kaukola and Lilja, 1972; Meltofte *et al.*, 1972; Edelstam, 1972; Harengerd *et al.*, 1973; Johnson, 1974; Folkestad, 1975; Plath, 1976). The positioning of staging posts in relation to flight ranges is considered further in Section VI.

There is increasing evidence, from systematic counts, recoveries of banded birds, and observations of birds with marks visible in the field, that several further migrations occur within the general "wintering area" during the nonbreeding season (Evans, 1976, 1981b; Pienkowski and Knight, 1977; Dugan, 1981a; Townshend, 1982a, Pienkowski and Pienkowski, 1983; see also below). The first site of residence in late summer and autumn is used in many cases for the annual complete molt of adults, although some populations may start molt on the breeding grounds or elsewhere and arrest it during migration before resuming it later (Hoffmann, 1957; Pienkowski *et al.*, 1976; Boere, 1976). Populations of a few species undergo the whole molt near the breeding areas before migration, e.g., North American races of Dunlins (Holmes, 1971). A few populations, e.g., some Common Redshank and Dunlin in western Europe and north-western Africa, usually those using a series of closely adjacent suitable staging posts en route, migrate in active wing-molt (Pienkowski and Dick, 1975; Pienkowski *et al.*, 1976). Some populations wintering in the South-ern Hemisphere (or near the equator) delay molt until their arrival there, to make use of the favorable summer conditions in their "winter" quarters (Pienkowski *et al.*, 1976).

The phenomenon of movements within the nonbreeding areas in western Europe is the subject of intensive investigation at present (see, e.g., Pienkowski and Pienkowski, 1981, 1983). The project arose in re-sponse to increasing development of coastal lands in Europe, partly as-sociated with petroleum exploitation. Clearly, if some coastal sites are to be safeguarded, a knowledge is required of the networks of sites on which particular species or races of birds depend. The preliminary findings given below result from sightings of color-marked shorebirds by a network of about 300 observers at about 200 sites around the European coastline, this study forming one part of the project. The data have not yet been corrected for numbers of birds checked or numbers of birds present in each area. However, they establish several patterns of movements, some previously unsuspected. They also emphasize the considerable extent of within-season movements, information previously unavailable except from recaptures of banded birds. Such recaptures, within the same non-breeding season, are infrequent, even with the high intensity of shorebird-banding studies in western Europe. Also, because of regional variation in bird-catching activity, which affects not only the provision of marked birds but also the chances of recapturing them, quantitative information on within-season movements from banding studies is subject to large biases.

In the 2 years of study completed by the time of writing, many thou-sands of birds have been dye-marked (chiefly in autumn), particularly in parts of the Vadehavet (western Denmark), the Wattenmeer (northwest-ern Germany), the Waddenzee (The Netherlands), and the Wash (eastern England). Movements have been recorded from molting areas in the Wat-tenmeer and Waddenzee, used in late summer and early autumn, to win-tering areas in Britain, Ireland, the southern Netherlands, and France for Ringed Plover, Black-bellied Plover, Dunlin, Bar-tailed Godwit, and Com-mon Redshank. In spring, the Wattenmeer areas have been used as staging posts by Black-bellied Plovers, Sanderlings, and Dunlins marked in winter in Britain. Similarly, movements from the Wash, used in autumn as a staging post and molting area by several species, have been documented to wintering areas on the east, south, and west coasts of Britain, the

Channel Islands, France, Spain, and even Morocco. Two species are discussed in more detail below, to illustrate both the complexities of movements within a single species and the differences in behavior of two species present in the same geographical area at the same times of year.

Dunlin *Calidris alpina*

Most birds marked by this study in Western Europe belonged to the nominate race *C. a. alpina*, which breeds in the Scandinavian mountains and eastwards across the USSR. The sightings in winter of birds marked in autumn 1980 or 1981 at molting sites or staging areas in the Wattenmeer/Waddenzee, the Zeeland Delta, and the Wash (Fig. 1) showed a large overlap in wintering distributions of birds from these areas. Furthermore, the regular coverage by observers indicated that the first birds reaching estuaries in western Britain from marking areas in eastern England, The Netherlands, and Germany arrived at about the same date, shortly after completion of molt in autumn (Pienkowski and Pienkowski, 1980, and unpublished).

The birds from the Wash were marked distinctively according to age. The resulting sightings showed no difference in the midwinter ranges of adults and juveniles. The changes in numbers of marked birds seen at many sites indicated that some movements continued throughout the winter. This was particularly apparent at small sites; as most birds present at these could be checked on each visit, arrivals and subsequent departures of single marked birds were clear. In 1980–81 at least, there were some indications that the movements of adults led to a northeastwards shift toward spring molting areas as early as January and earlier than those of juveniles. Most of the movements of this race of Dunlin to spring molting and staging areas take place in March. The general nature of such movements is shown in Fig. 2 by sightings in spring of birds marked in winter. (Note that Figs. 1, 2, and 3 do not show birds remaining near the marking site: some birds do behave in this way, and later analyses will attempt to estimate proportions following the various behavioral patterns.)

We do not know if birds molt into spring plumage at the same places as they undergo autumn molt, but evidence from this species and Blackbellied Plover (see below) suggests that some individuals do. This probably does not apply to all, however, as some sites used in autumn are probably unsuitable in spring (see below).

Black-bellied (or Grey) Plover *Pluvialis squatarola*

Branson and Minton (1976) summarized available banding recoveries for Black-bellied Plover. These indicated that many birds on their mi-

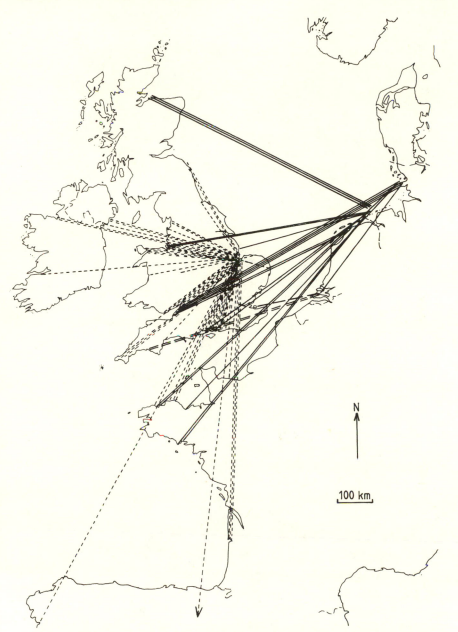

Fig. 1. Some of the sightings away from the marking area in the following winter of Dunlins *Calidris alpina* marked in autumn 1980 or 1981 at the Wattenmeer/Waddenzee (——), the Zeeland Delta (– – –), or the Wash (-----). Lines join the site of each sighting to the marking area and do not necessarily indicate routes taken. The exact site of marking of some birds from the Wattenmeer/Waddenzee could not be determined, and the lines for these are drawn from the center of the Wattenmeer/Waddenzee complex. The arrow indicates a sighting near Tangier, Morocco.

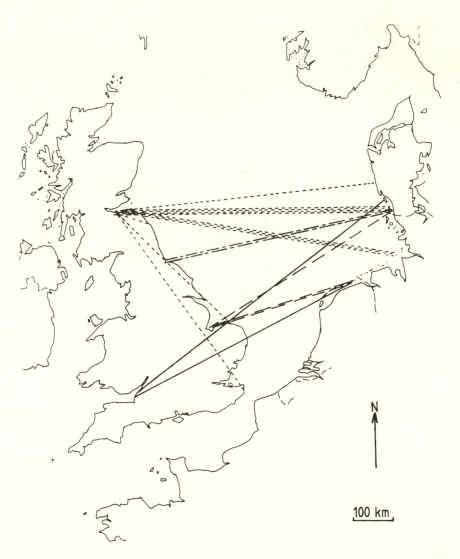

Fig. 2. Some of the sightings away from the marking area in the following March and early April of Dunlins *Calidris alpina* marked in winter (1978/9 to 1981/2) at the Severn Estuary (———), the Wash (– – –), the Tees Estuary (---), or the Firth of Forth (-----). Lines join the location of each sighting to the marking area and do not necessarily indicate routes taken.

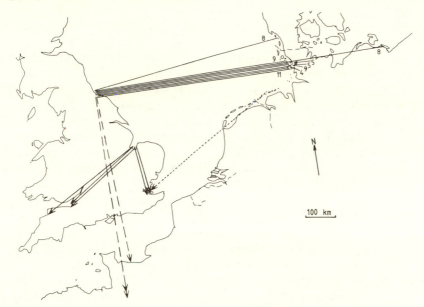

Fig. 3. Some of the sightings away from the marking areas of Black-bellied Plovers *Pluvialis squatarola*, in 1978 to 1982. Arrowed lines indicate sightings later in the same winter of birds marked in autumn. ———, Adults from the Wash; – – –, juveniles from the Tees Estuary (double arrow indicates sighting in Vendée, western France); -----, bird of unknown age from Wattenmeer or Waddenzee. The sightings in Denmark and Germany all relate to individually color-banded birds that spend the winter (January, or earlier, to March) at Teesmouth. For these, the numerals indicate the months of sighting at the other sites.

gration from breeding grounds in Siberia stopped to refuel or molt in northwestern Europe, before moving on to southwestern Europe and western Africa for the winter. Relatively small numbers stay in northwestern Europe.

At Teesmouth this species has been the subject of detailed behavioral studies since 1975, based on color-marked individuals. Some results of that study, in combination with information from the observer network of the present project, are summarized in Fig. 3. These conform with the general outline from banding recoveries, but add considerable detail. Birds dye-marked at the Wash and the Wattenmeer during autumn molt moved later to the Thames Estuary, and from the Wash also to southwestern England. Some young Black-bellied Plovers, marked at Teesmouth in early autumn, subsequently moved further south in the same autumn to France. A notable difference between the movement patterns recorded for this species and those of the Dunlin is the lack of any element of northward movement of the plovers from the autumn molting sites to the wintering areas.

Several adult Black-bellied Plovers that had spent the autumn, and in some cases the early winter, on the Baltic coast of the German Democratic Republic, or on the tidal areas of the west coasts of Denmark and the Federal Republic of Germany moved to Teesmouth during the winter. Some marked birds from Teesmouth were seen again in the same area of the Wattenmeer in spring. Observations during the severe weather of January to March 1979 and December 1981 have shown that some Black-bellied Plovers visit Teesmouth only in such severe winters. Whether these birds also spend the rest of the nonbreeding seasons and all mild winters in the Wadden Sea is unknown but seems likely. This situation is reported in more detail by Townshend (1982b).

As can be seen from the Black-bellied Plover example, further movements occur later in the winter, not only just after the completion of molt in late autumn. Most of these movements are not made in direct response to weather, although they may be adaptive responses to normal climatic patterns (see below). Townshend (1982a,b) described regular arrivals of Black-bellied Plovers at Teesmouth in late July–August (adults that molt at Teesmouth), late September–October (juveniles), November (adults that had molted elsewhere), late December–January (adults, probably from less clement areas), and February (adults and juveniles, probably returning northwards). Dugan (1981a) reported northward and westward movements of Red Knots in Britain at various times throughout the winter.

The bases for choosing particular staging and molting areas remain largely unknown. While some species such as Red Knot generally perform long-distance migrations with few staging sites (Dick *et al.*, 1976; Dick, 1979; Håland and Kålås, 1980), other species appear to show more variable behavior, depending upon local geography. Several species (including some Common Redshanks and Dunlins), whose migration route lies along the coastline of western Europe and northwestern Africa, appear to use the many wetlands along this coast for a more leisurely migration by a series of short flights, in some cases even molting concurrently (see above). The same species in other areas, and probably the same individuals in other parts of their migration, are able to deposit reserves for longer flights (e.g., Minton, 1975; Pienkowski *et al.*, 1979).

Clearly, suitable feeding habitat is a basic requirement whatever sites are being used. This has been considered by, for example, Wolff (1969), Zwarts (1974), Evans (1976), and Pienkowski and Knight (1977).

Major molting sites appear to be rather fewer in number than sites used in midwinter. Large intertidal sites [such as the Vadehavet–Wat-

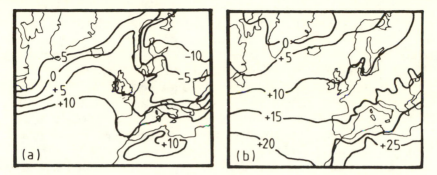

Fig. 4. Mean surface air temperatures (°C) in (a) January and (b) June, 1931–1960. [From Meteorological Office and Dugan (1981a)].

tenmeer–Waddenzee complex (herein termed "Wadden Sea") along the coast of western Denmark, northern Germany, and The Netherlands; and the Wash, eastern England] appear to be favored. This may be because, in the warmer weather of early autumn in western Europe, birds encounter fewer problems in obtaining food. This allows them to be more choosy about the sites they use. At this time most species of birds, not only of shorebirds, perform their annual complete molt of body and flight feathers. During molt, besides a reliable source of food, reasonable safety from predators (while the shorebirds' flight may be impaired) is of obvious importance. As most predators are terrestrially based, shorebirds are probably safest feeding on large intertidal areas rather than small estuaries. However, by the time molt is ending (generally in late October), weather conditions are rapidly worsening and hours of daylight shortening markedly. The decrease in monthly mean temperature is more marked further away from (i.e., to the east of) the Atlantic Ocean, so that by winter the isotherms run north–south, with an average temperature difference of 5°C between western Ireland and the Wattenmeer (Fig. 4a). Because of the reduced activity of prey at colder temperatures as well as the risk of the mud and sand surface freezing (and making prey totally unavailable), there may well be advantages for birds to move west by the winter.

At the same time that colder temperatures increase the energy requirements of birds and, together with longer nights, make the catching of prey more difficult, the incidence of gales, particularly of westerly gales, increases. These may markedly raise the water level in the Wattenmeer areas, preventing exposure of the birds' feeding areas for long periods. For example, in the Wattenmeer in August 1980, westerly gales caused the tide to pass the predicted high-water mark 3 hr before the time of high water. The mudflats remained submerged for much of the following low-water period, forcing Dunlins to feed on the salt marsh. Gales are, of course, at least as common around the British Isles and in western France, but in these countries the tidal range is much greater (Fig. 5). Thus, although water levels in these areas are also affected by gales and atmospheric pressure, the effect is much smaller in proportion to the tidal range, and so the exposure times of feeding areas change much less.

By spring, gales have subsided and the continental coasts tend to warm faster than the more oceanic coasts (Fig. 4b; Dugan, 1981a). Many birds return to the Wadden Sea areas (e.g., Figs. 2 and 3; Smit and Wolff, 1981) for spring molt of their body feathers and fattening for migration before returning to the Arctic breeding areas (Boere, 1976). The spring fattening sites could be areas of early growth and/or reproduction of prey or of high densities of prey. The latter seems more likely. High densities of prey might survive in sites to the north and east of the European mid-winter feeding areas, because birds had to leave them in autumn (when prey became unavailable as temperatures fell) before they had reduced appreciably the absolute densities of prey there (Evans, 1979b; Pien-kowski, 1981a). A notable difference between sites chosen in autumn and in spring is the use of the Baltic coasts by many molting birds in the former but not the latter season. This may be due to over-winter reduction in prey density on the Baltic coasts where prolonged and severe icing results from the low salinity (see Evans, 1981b). Similarly, low densities of *Corophium volutator* in spring in the Bay of Fundy, and consequent lower usage by shorebirds than in autumn, have also been attributed to ice-scour (see Hicklin and Smith, 1979).

We have just discussed the ultimate factors leading to movements within the nonbreeding season. Until recently there had been little evidence, in coastal shorebirds, for movement in direct response to cold weather or gales as proximate factors. Hard-weather movements have long been recognized of several inland-wintering shorebirds, such as Northern Lapwing *Vanellus vanellus* (Evans, 1976). However, Towns-hend (1982b) found that, in addition to the many Black-bellied Plovers that use the Tees Estuary regularly in each winter, a further group of color-banded individuals visited the area only in particularly severe winter weather in 1976–77, 1978–79, and 1981–82. The birds had probably spent the milder winters, and the early part of the severe winters, in The Neth-erlands Waddenzee, where winter conditions tend to be more severe than in Britain (see above). Some evidence has also been gathered of hard-weather movements by Common Redshanks and some other shorebirds species in the severe winters of 1978–79 and 1981–82 (Davidson, 1981a; Clark, 1982).

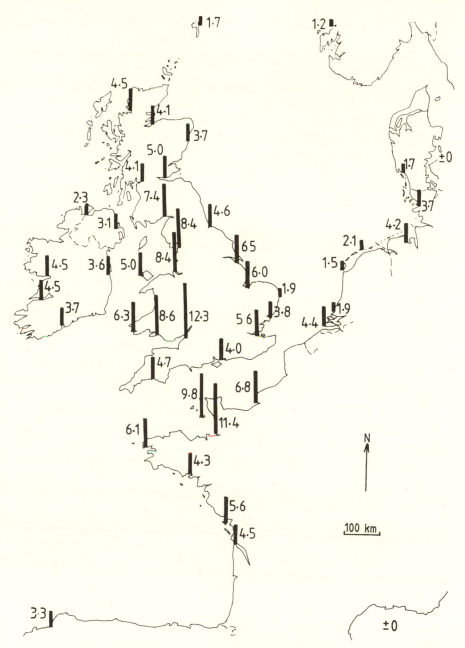

Fig. 5. Mean range of spring tides (m) at various sites in western Europe. [Source: *Admiralty Tide Tables*, Vol. 1, 1982.]

A notable feature of the hard-weather movements of Black-bellied Plovers described above is that, in these movements, at least some birds moved to sites of which they had prior experience. Such reconnoitering of alternative areas may occur in other species. There is evidence from two areas, California (Myers, 1980) and northeastern England (Evans, 1981b), that wintering Sanderlings may adopt two contrasting movement strategies, which may be extremes of a continuum. Birds may remain for prolonged periods (possibly the whole winter) at one site. Presumably such a behavior may provide advantages in terms of local knowledge, particularly when food is hard to come by. Individual birds using the second strategy are seen for various periods during the nonbreeding season at a particular site but are absent on intervening dates, when they may be at other sites, up to 60 km away. Possibly, these birds are sampling several areas, so that if some become unsuitable, the bird can return to others. Sanderlings are birds of sandy beaches where the food supply may depend on weather conditions: storms may deposit wrack (containing bivalve spat) on the shore, enhancing feeding conditions: or scour the sand surface, removing most of the potential prey; or open or close openings to lagoons. More recent information from the Californian study is reviewed by Myers (Chapter 6).

The most mobile species in winter in Britain is the Red Knot. Individuals move rapidly from one feeding area to another within large estuaries (Minton, 1975; Pienkowski and Clark, 1979; Symonds *et al.*, 1984); and also from one estuary to another (Dugan, 1981a). The reasons for these movements are far from clear, but could involve the sampling of many areas, for there is limited evidence that birds may retrace their flight path. As the prey of this species includes a very high proportion of bivalve molluscs (Ehlert, 1964; Hofmann and Hoerschelmann, 1969; Davidson, 1971; Prater, 1972; Evans *et al.*, 1979; Goss-Custard *et al.*, 1977a; Pienkowski, 1982), it is possible that sampling of spatfall could be involved. For many bivalves, spatfall is irregular both in time and in space (see, e.g., Green, 1968).

There has been a decline in the population of Red Knot wintering in western Europe [i.e., those from the High Arctic Canadian–Greenlandic breeding population (Dick *et al.*, 1976)] in recent years (Prater, 1982). This has been paralleled by a reduction in usage of some areas, e.g., the estuaries of western France (R. Mahéo, in Prater, 1982). Possibly, such areas are used only when population levels are high, because some birds are then forced from, or choose not to stay in, the more traditional wintering areas. Because of the very variable breeding success of High Arctic populations (see Evans and Pienkowski, Vol. 5 of this series), considerable variations in population size, and possibly therefore in the extent of movements, are to be expected.

In all situations in which some individuals stay in an area and others leave, the question arises whether those leaving do so from choice or after eviction by others. Zwarts (1974, 1976) and Goss-Custard (1977) have demonstrated that, as the total number of shorebirds on an estuary increase, a decreasing proportion tend to settle on those feeding sites occupied by the first arrivals. This could arise if individuals arriving later chose to feed in less crowded sites, which might be more profitable (Fretwell and Lucas, 1970; Zwarts and Drent, 1981). Alternatively, they might be excluded from the sites occupied first, by means of depletion of food resources, interference with feeding, some form of dominance system, or territoriality.

Competition through depletion of food resources seems possible, in some situations. Zwarts and Drent (1981) argued that Eurasian Oystercatchers and Herring Gulls severely "overfished" mussel stocks at Schiermonnikoog, The Netherlands. Major changes in prey abundance may lead to changes in bird distribution between estuaries, as proposed by Sutherland (1982), working on Eurasion Oystercatchers on the east shore of the Irish Sea (although the changes in prey abundance there were not necessarily caused by predation). Generally, competition resulting from depletion seems more likely to act through a reduction in the feeding rate achieved by the bird rather than by the total removal of cohorts of prey (see Goss-Custard, 1980). Interference in foraging behavior between species has been discussed above (Section III), and may influence distributions, at least within an estuary if not on a wider scale. Within a single species, an inhibitory effect on feeding performance of a high density of conspecifics has been demonstrated for Common Redshanks (Goss-Custard, 1976), Eurasian Curlews (Zwarts, 1980), and Eurasian Oystercatchers (Vines, 1980; Zwarts and Drent, 1981; Goss-Custard *et al.*, 1981).

Smith and Evans (1973) described a situation that may lead to competition between the sexes in Bar-tailed Godwits, in which females are larger than males. Females fed equally well in shallow or deep water, whereas males performed better in shallow than in deep water. All birds fed better in flocks than solitarily. As flocks become compressed in downshore distribution in cold weather (due to the reduction in prey availability and concentration of this in the water, rather than on the exposed sand) (Smith, 1975), there may be less space for males. Sexual differences may act through the physical limitations related to size rather than via intraspecific competition. Townshend (1981) found that short-billed male Eurasian Curlews were less able than the long-billed females to reach the polychaete *Nereis diversicolor* at low temperatures, probably when these prey burrowed deeper in the mud. In such conditions, males tended to resort to fields, where they fed on earthworms.

Territoriality on the part of shorebirds in the nonbreeding season has been noticed in many species in recent years (e.g., Panov, 1963; Recher and Recher, 1969; Goss-Custard, 1970; Myers *et al.*, 1979a; Ens, 1979; Townshend *et al.*, 1984). Although territoriality could theoretically impose a limit on numbers wintering on a particular estuary, it has not been shown to do so. In the cases of Black-bellied Plover at Teesmouth (Townshend *et al.*, 1984) and Sanderlings at Bodega Bay, California (Myers *et al.*, 1979b), the failure of territoriality to limit numbers could arise because not all potential territory sites are occupied; because nonterritoriality provides an acceptable alternative use of space in most weather conditions; or because, at high density, flocks can displace territorial birds; or some combination of these. Despite these features, some Black-bellied Plovers, especially young birds, do leave Teesmouth after displacement from territories by adults, and some of these individuals have moved on at least as far as France (Townshend, submitted).

Dugan (1981a) postulated some form of dominance among the nonterritorial Black-bellied Plovers at the same Teesmouth site but his evidence was slight. If dominance does play a part in the social system of wintering Black-bellied Plovers, its form must differ greatly in different places: in some areas, individuals show frequent aggressive interactions (e.g., Burger *et al.*, 1979), but elsewhere aggression is rare (e.g., Stinson, 1977). The variation in social structure in this species and its underlying reasons are discussed by Townshend *et al.* (1984). In the estuarine Shelduck *Tadorna tadorna* (which is ecologically a shorebird), regulation of numbers in a winter feeding flock, arising from exclusion of subordinate individuals from feeding areas, was proposed by Jenkins *et al.* (1975). However, in further studies on the same flock, Evans and Pienkowski (1982) were unable to find evidence for food shortage, for consistent dominance relationships, or for regulation by processes within the flock. Timings of arrival and departure of marked individuals tended to be similar from year to year, and departure did not result from exclusion through intraspecific competition. There is, therefore, little proof at present to support any limitation of numbers in shorebird flocks, and hence little evidence of forced eviction leading to movements within a winter, as a result of dominance mechanisms within the flocks. A current study of Eurasian Oystercatchers (Goss-Custard *et al.*, 1982) is, however, producing evidence that immature birds tend to be subordinate to adults on the feeding areas and abandoned these areas as adults returned in autumn from the breeding areas. The situation is, however, complex and is discussed further by Goss-Custard (this volume).

Although several of the studies cited above, and others such as that by Baker and Baker (1973), suggest that competition between shorebirds is widespread in winter, not all studies reach this conclusion and some argue that competition may be negligible, at least in the Southern Hemisphere "wintering" grounds (Duffy et al., 1981).

We can conclude that many shorebird populations are comprised of individuals that depend on more than one area in the nonbreeding season. This dependence may involve a sequence of sites required each year (for migration staging, molt, and for various periods of the winter) or a set of alternative sites to be used according to winter conditions, or some combination of these two groups. Much further work remains to be done to determine which factors are critical in determining the sites chosen.

V. ROUTES AND NAVIGATION

Other factors being equal (but see below), one would expect shorebirds to travel between staging posts by the shortest possible routes. The shortest route between any two points on the earth's surface lies along a great circle (i.e., a line that if extended, divides the earth into two equal hemispheres). Lines of longitude are all great circles, but of lines of latitude, only the equator is. Therefore, for birds with a mainly north-south component in their migration, there is little difference between a compass-line and a great-circle route. For many shorebirds wintering in western Europe and northwestern Africa, however, there is a strong east–west component in their migrations (see Moreau, 1972; and Section I). Grimes (1974) pointed out that the departure directions from Ghana in spring of many shorebirds fitted well to great-circle routes to their breeding grounds. Dick et al. (1976) also noted that records of Red Knot migrating between Siberia, northern Europe, and western Africa fitted a great-circle route better than a single compass direction. Indeed, Lack (1962) had had to propose a "dog-leg" migration route to account for the autumn arrival of Red Knot in eastern England from the NNE, as seen by radar. But this is the direction in which they would have been expected to arrive if they had flown on a great circle from their Siberian breeding areas (Dick et al., 1976). Banding recoveries of several species of shorebirds in between Siberian breeding grounds and South African winter quarters also tend to fit great-circle routes (Summers and Waltner, 1979). To test critically the idea that great-circle routes are followed would require long-range radar studies in northern Scandinavia or northern Russia. At present, such information is not available.

Shorter-range radar studies have been made of migration by shorebirds between estuaries in northern Britain (Evans, 1968). Here, birds generally moved overland on direct routes between sites rather than following topographical features, except for the coastline, and then only when this approximated closely to the direct route. In autumn, radar studies have shown that shorebirds tend to migrate with following winds (Lack, 1962–63; Evans, 1968; Richardson, 1979). [Numerous studies on *visible* migration of a range of bird families have recorded more migrants when winds were contrary, but this may be because birds stop or fly low in such conditions (see Richardson, 1978).] In western Europe in some autumns, unusually large numbers of Siberian breeding migrants occur, especially juveniles of species such as Curlew Sandpiper, Little Stint *C. minuta*, and Red Knot. Their arrival often follows the movement of complex atmospheric depressions over eastern Europe, causing strong easterly winds across the north of the continent (Stanley and Minton, 1972; Dick *et al.*, 1976; Wilson *et al.*, 1980). It is not clear to what extent the birds are deflected westwards from a chosen, more southerly, route and to what extent the birds are adapted to make use of these wind systems to help them toward their destinations, albeit by a longer route than usual in terms of ground distance.

Because so many of these arrivals are juveniles, it may be that they have departed on migration in response to inappropriate weather cues and subsequently been blown off-course; alternatively, the adults may have departed earlier (for their molting grounds) at a time of summer when the track of the depressions passes further north, so that winds are favorable for more southerly movements.

In most species of shorebirds, juveniles are not accompanied on migration by their parents, or even by other adults of the same species, as most of these leave the breeding areas before the young are ready to migrate (see Cramp and Simmons, 1983). (In some North American populations, adults remain to molt near the breeding grounds while juveniles precede them on migration; see earlier.) It is not known what criteria determine the sites used by young birds on their first journeys in autumn. Juveniles tend to be seen at far more sites than are adults (e.g., Wilson *et al.*, 1980). Both coastal and inland sites are used, including some with "unsuitable" habitats. (In Europe, such inland sites are rarely used by adults of coastal-wintering species. This difference from the behavior of some North American populations, of which both adults and juveniles regularly use prairie sloughs on migration from Alaska to the east coast, probably results from the availability of coastal sites at reasonable intervals on the most direct route along the Atlantic coast of Europe.) It is not known whether the wider occurrence of young birds implies that individual juveniles investigate many sites on a preprogrammed migration route or whether different individuals investigate different sites, each bird stopping at only a few. Possibly, some juveniles have to stop at sites short of their "goal" because of an inability to complete the flight, particularly if they have set out under inappropriate weather conditions. Certainly, many birds—particularly young of the year—arrive at certain destinations with severely depleted energy reserves [e.g., in Mauritania (Dick and Pienkowski, 1979)]. Some fail to recover and, in any event, such birds are probably particularly susceptible to predation (Dick, 1976; Dick and Pienkowski, 1979).

We do not know whether young shorebirds inherit information that enables them to migrate to particular localities. In view of the rapidity with which estuarine and coastal habitats change, over a time scale of thousands if not hundreds of years, this seems unlikely. More probably, as in many passerines, they inherit an approximate direction or sequence of directions of migration, any changes being internally time-programmed [cf. Garden Warbler *Sylvia borin* (Gwinner and Wiltschko, 1978)]. Adults apparently fly more directly to sites of which they have previous experience (see above). Studies on Dunlin banded on autumn migration near Gdańsk, Poland, tend to support the idea that adults take more direct routes (Gromadzka, 1983). More juveniles than adults were recovered at shorter distances and at inland sites, and their migration tended to be slower. The migration of adults appeared to be concentrated in particular directions: 40% of all recoveries in the same autumn as that of banding occurred in a westerly direction (261–269°) and 13% in a southwesterly direction, but no such strong concentrations occurred for juveniles. However, the distributions of winter recoveries did not differ between the ages (Gromadzka, 1983).

How young birds initially select sites at which to stop during their first migration is unknown. A number of features of an estuary (length of tideline, substrate types, even usage by flocks of birds) may be evident to them while they are flying by day (Evans and Dugan, 1984), particularly from heights of up to 10,000 feet (3000 m), at which they normally fly over the North Sea (Lack, 1962–63). In other areas, particularly toward the end of the long overwater flight from eastern North America to South and Central America, when migrants may reach 20,000 ft (6000 m) (Richardson, 1976; Williams *et al.*, 1977), birds may find greater difficulty in evaluating habitats while in flight. Alternatively, individuals may sample several sites, as suggested by the more widespread occurrence on coasts, estuaries, and inland wetlands of young birds than adults in autumn. The juveniles may have more time than adults for such sampling before the onset of winter conditions as, in north temperate areas, most adults undergo their annual complete molt in a single site during this period (e.g., Pienkowski *et al.*, 1976).

However sites are selected, once they are chosen, there seems to be a high rate of return to them in later years. Fidelity to breeding areas is well established in many species (see, e.g., review in Soikkeli (1970) and Oring and Lank (Vol. 5 of this series)]. The annual rate of return to specific wintering areas is also high; e.g., at Teesmouth, nearly all the birds surviving return (Evans and Pienkowski, Vol. 5 of this series). See Pitelka (1979) for other examples. Townshend (manuscript submitted) found that the patterns of movement between estuaries and the use of space within an estuary by Black-bellied Plovers were determined for future years largely by the environmental and social conditions experienced by a bird in its first year of life.

At staging sites, such information on return rates is more difficult to obtain, as the briefer period of presence makes it difficult to be certain of seeing all surviving color-banded birds during their period of usage of the site. However, numerous recaptures of banded birds have been recorded in subsequent years at sites on the migration routes in Scandinavia (Nørrevang, 1959), Poland (Gromadzka, 1983), southern France (Hoffmann, 1957), Tunisia (Moreau, 1972), and Morocco (Pienkowski, 1976), as well as at wintering sites in Western Europe also used for migration staging. Pienkowski (1976) showed that the chance of return by Dunlins passing along the coast of Morocco to the migration staging site used in a previous year was higher than the chance of using an alternative site on the same migration route.

The annual return to sites used on the migration route away from the breeding grounds does not, however, necessarily imply that the same route will be used on the return migration. Different routes in autumn and spring have been demonstrated in several species and geographical areas. For example, some Curlew Sandpipers wintering in western Africa take a more easterly route in spring than in autumn (Wilson *et al.*, 1980). Several species migrating to Greenland are seen in larger numbers on the west coast of the British Isles in spring than in autumn (see, e.g., Pienkowski, in Cramp and Simmons, 1983). Semipalmated Sandpipers *Calidris pusilla* in North America tend to use a more westerly route in spring than in autumn (Harrington and Morrison, 1979).

The underlying reasons for the usage of different routes in spring and autumn have not been investigated in detail. Seasonal variation in potential feeding opportunities along the different routes may be involved. For the above examples of Curlew and Semipalmated Sandpipers, the autumn routes of some of the populations are more coastal. This may be an adaptation to exploit the estuarine food resources after summer production. The western route of Greenlandic birds through Britain in spring may help to speed spring migration by using a shorter route (see Section VII). The same could apply to the more easterly route of some Curlew Sandpipers in spring, especially as there is some evidence that, by late winter, these birds may have moved further east and south in western Africa than would be suggested by their first arrival sites in autumn (Wilson *et al.*, 1980; Fournier and Dick, 1981).

Little is known of the methods by which shorebirds navigate, though it must be supposed that these are broadly similar to the methods used by other migrant bird species. Radar observations have shown that, under moderate crosswinds, shorebirds can still keep to a chosen course (Evans, 1968). Hence, they must be able to compensate for wind-drift. They migrate by night and day, often departing shortly before dark (Lack, 1962–63; Evans, 1968), and presumably can use sun, stars, and the earth's magnetic field to provide cues for compass orientation. However, if, as seems probable, some species migrate along great-circle routes, shorebirds may have more sophisticated navigational systems than those so far investigated in passerine migrants (Able, 1980).

VI. PHYSIOLOGICAL PREPARATION FOR MIGRATION

Birds store almost anhydrous fat as a fuel for migration. Metabolism of dry fat provides about 9.2 kcal amount per gram, more than twice the energy provided by the same mass of protein or carbohydrate (see, e.g., Le Maho *et al.*, 1981). The water produced by fat metabolism may be of use in water balance during the journey. Several studies have investigated the rate of fat deposition prior to migration. It is preferable to measure this in marked individuals rather than by taking samples from a population, as heavier birds tend to leave and light ones to arrive between successive sampling dates, so depressing the estimated rate of weight gain. However, as fat deposition may be interrupted by capture (Page and Middleton, 1972; Fuchs, 1973; Davidson, 1981b, 1984), estimates from retrapped individuals may also underestimate the maximum rate of deposition. Most studies of the rate of premigratory fat deposition have been based on increases in body weight. This is a reasonable approximation, as changes in fat content are by far the largest component of such weight changes (Johnston and McFarlane, 1967; Davidson, 1981a,b). However, smaller parts of weight gain may be due to other body components (see also below).

Rates of weight increase measured in retrapped birds have varied from less than 1% to about 7% of lean weight per day, the two extremes being recorded in Curlew Sandpipers (lean weight about 55 g), namely about 0.3 g/day in western Africa in autumn (Wilson *et al.*, 1980) and a mean of 3.9 g/day in westward-drifted autumn migrants in Britain (Stanley and Minton, 1972). Migrant Dunlins (lean weights about 50 g), retrapped while depositing fat in Britain, increased weight at mean rates of 0.7 g/day in autumn and 1 g/day in spring (Pienkowski *et al.*, 1979); in autumn in Sweden an average rate of 1 g/day with a peak of 3 g/day were recorded (Mascher, 1966). In spring in Iceland, Ruddy Turnstones (lean weights about 105 g) increased weight at an average of 2.5 g/day with some individuals reaching 4.6 g/day (Morrison and Wilson, 1972). Also in spring, Red Knots (lean weights about 135 g) increased weight in April and early May at Morecambe Bay, northwestern England, and in late May in Iceland, at up to about 3 g/day in the latter area (Morrison and Wilson, 1972; Prater and Wilson, 1972).

Fat accumulation appears to be achieved largely by feeding for longer than would otherwise be done to satisfy basic requirements. For example, in May in northeastern England, resident Ringed Plovers (which were already nesting) could not have fed for more than 50% of each 24 hr (because incubation is shared fairly equally between the sexes); in fact, they fed for rather less than this: birds were seen not feeding for considerable periods when feeding areas were available. In the same area and month, migrant birds en route to Greenland fed for up to 95% of the daylight hours and for some of the night also (Pienkowski, 1973, 1984c).

The quantity of fat deposited by shorebirds before migration is highly variable, depending on the length of the migratory flight, the size of bird, and possibly the weather conditions likely to be encountered en route and/or on arrival. Fat reserves of up to 90% of lean weight (which itself had also increased by 13% over winter levels to carry protein reserves) have been recorded in Sanderling (Davidson, 1981b). There are some indications that fat reserves deposited in spring tend to be larger than those in autumn (reviewed by Davidson, 1981b), one suggestion being that this allows an extra reserve for egg production or survival in the first days after arrival on the breeding grounds if spring there is late [as is now well established for some waterfowl—Snow and Barnacle Geese *Chen caerulescens* and *Branta leucopsis* (Ankney, 1977; Drent *et al.*, 1981)]. There is undoubtedly some risk to birds in such conditions, and heavy mortality has been reported occasionally (e.g., Morrison, 1975; Marcström and Mascher, 1979). However, there is, as yet, no firm evidence that shorebirds reach Arctic breeding grounds with a fat reserve intended to assist survival there, or egg production. It is possible that the apparently higher fat deposition in spring is an artifact of the location of the study sites in relation to the distance to the next destinations, or of the time of capture in relation to timing of migration. If, however, the seasonal difference is real, it could be associated with the tighter migration schedule in spring (see below), possibly allowing less opportunity of waiting for favorable winds, or less opportunity to use all intermediate staging areas (see below).

Estimations of possible flight ranges, based on the amount of stored fat, rely on various aerodynamic or metabolic models, sometimes incorporating empirical features (e.g., McNeil and Cadieux, 1972; Pennycuick, 1975; Greenewalt, 1975; Summers and Waltner, 1979). These various models can give very different estimates of maximum flight-range, in some conditions. For example, the flight ranges of Dunlins leaving the Wash in spring were estimated by Pienkowski *et al.* (1979) as 2000 km (according to Pennycuick), 2600 km (Greenewalt), or 2900 km (McNeil and Cadieux). For shorebirds, Pennycuick's model gives consistently the shortest ranges, and Summers and Waltner's the longest (Davidson, 1984).

The few data against which to compare these estimates were reviewed by Summers and Waltner (1979). A banded Ruddy Turnstone from the Pribilof islands recovered in Hawaii 3.7 days later tends to favor McNeil and Cadieux's equations over those of Pennycuick (which gave a shorter range). Greenewalt's was not used. The timing of spring migration, and location of possible staging areas of Arctic breeding shorebirds "wintering" in South Africa also tend to favor McNeil and Cadieux's estimations or longer ones (i.e., Summers and Waltner's and, sometimes, Greenewalt's). Davidson (1984) reviewed further evidence which indicates that Pennycuick's equations give too low a flight range estimate, but that further data are required to test the other models. To date, however, these estimations have not been critically field-tested in shorebirds, perhaps not surprisingly in view of the practical difficulties. However, the migrations of some populations in the western Palearctic are now sufficiently well understood to make such tests feasible.

Protein reserves are also required during long migrations, to replace normal protein turnover in organs such as the liver. These can be stored in the muscles and therefore affect lean weight. Muscle size may also be increased before migration to provide extra power to assist flight when large loads of fat are carried. At Teesmouth, the lean weights of the winter resident Dunlins and Sanderlings are higher in May than earlier. For birds migrating from "wintering" areas well to the south, there are indications that muscle size is sometimes enlarged at staging posts (Davidson, 1981b, 1983, in preparation).

VII. TIMING OF MIGRATION

Due to the complexity of shorebird migration patterns (Section IV), migration seasons for most species are prolonged. In Greenland, birds that had failed to breed could be seen, as early as July, gathering on the shore prior to migration (Ferns, 1978). The migration of Arctic shorebirds through Europe is noted from late June (Table I). In some years in some Arctic regions, breeding conditions may be very poor over a wide area. This may presumably lead to the marked differences in timing of migration at some sites between years (e.g., references in opening paragraph of Section IV). Migration may begin as early as June also for populations breeding in more temperate areas, where the season starts and may finish much earlier than in the Arctic (Table I).

In several species it has been reported that one sex (usually the male) arrives on the breeding grounds earlier than the other, although in Red Phalaropes *Phalaropus fulicarius* (which show reversed sex roles) the sexes arrived together (Myers, 1981a). Available data, reviewed by Myers (1981a), demonstrated that these early arrivals were consistent with the theory that this was due to intrasexual competition, e.g., for territories (see Hildén, 1979), rather than a consequence of any difference in wintering distributions.

For several species, differences in the timing of autumn migration are apparent between the sexes (e.g., Edelstam, 1972). These are usually related to the breeding systems of the populations (see Hildén, 1975; Pitelka *et al.*, 1974). Various reasons have been put forward for the early departure of one sex. Pitelka (1959), Recher (1966), and Pitelka *et al.* (1974) suggested that this reduced competition for food resources for the remaining parent and the young. Others (e.g., Ashkenazie and Safriel, 1979; Myers, 1981b) have suggested that the early desertion is to allow a better chance of survival for the departing adult (by allowing more time for fattening, or for reaching migration stopovers early to forestall possible prey depletion (Schneider and Harrington, 1981), or for completing molt and reaching winter sites before weather conditions lead to reduced prey availability (see Sections III and IV)), rather than to increase the chance of survival of its young. (The question of which sex departs early is complex and has been discussed by many of the authors cited above.) The two ideas are not necessarily incompatible, and there appears to be no strong evidence in favor of one rather than the other, although Myers (1981b) reviews available information that tends to favor the second. Schneider and Harrington (1981) reported substantial depletions of densities of prey animals of some shorebirds at a staging area in Massachusetts over the autumn migration period. These depletions were apparently due to predation by shorebirds. However, these workers did not investigate whether the depletion in density was sufficient to depress rate of food intake by the shorebirds.

In some populations the early departure of one sex is associated with a markedly different migration and molting schedule and, in some cases, with differences in wintering areas. For example, male Curlew Sandpipers [which leave their mates to incubate and tend young (Portenko, 1959)] migrate earlier in autumn than females and, in some areas, have an earlier and slower autumn molt (Pienkowski *et al.*, 1976; Wilson *et al.*, 1980). In Ruff *Philomachus pugnax*, a lekking species, males play no parental role beyond fertilization. They tend to winter further north (some even as far north as Britain) than females (termed "Reeves"), which reach South Africa. Their schedules of migration and molt differ correspondingly (Pienkowski *et al.*, 1976; Schmitt and Whitehouse, 1976). Myers (1981a) reviews other examples.

As discussed earlier (Section V), the autumn migration of adults, at least as far as their molting grounds, tends to precede that of juveniles.

The timing of autumn migration tends to be more spread than that in spring, this clearly being the result of conditions and performance on the breeding grounds. There are constraints on the timing of autumn migration, such as the need to fatten for migration while feeding conditions remain suitable and, if molting in the northern temperate zone, the need to complete this before winter conditions set in. The latter constraint may lead to adjustments in molting schedules and a faster molt by individuals late in the season, or by most of the population if migration is late (Pienkowski *et al.*, 1976; Johnson and Minton, 1980). The constraints on spring migration may be tighter because, to maximize breeding success, arrival as soon as conditions are adequate at the start of the season is essential.

Arctic breeding shorebirds require at least 4 weeks to lay and incubate eggs, and about 3 weeks to rear young to fledging (Green *et al.*, 1977; Green and Greenwood, 1978). Further periods are required for territory establishment, possibly pairing, egg formation, and postfledging development of the young. Therefore, a minimum of approximately $2\frac{1}{2}$ months is required for the small species (Ringed Plover, Red Knot, Sanderling, Dunlin, Ruddy Turnstone); larger species tend to take slightly longer (see, e.g., Cramp and Simmons, 1983), and these do not occur in as High Arctic regions as those mentioned. Some High Arctic areas are snow-free for less time than this, and shorebirds do not breed there (see, e.g., Green *et al.*, 1977). In many areas, snow-free patches first appear in June, and the first autumn snows occur in late August and early September. It is, therefore, essential for shorebirds to reach their breeding grounds as soon as they become snow-free. This means a departure from temperate staging areas in late May and early June (see Ferns, 1980–81; Dick, 1979). If wintering areas are further south, the schedule of departure is probably determined by the time to make the flights and, more importantly, the time to deposit reserves at staging areas. Dick (1979), basing his estimates on a coordinated set of counting, banding, and weight studies, considered that Siberian Red Knots needed to leave South Africa in mid-April (which he found they do) or western Africa (their main wintering area) in early May to travel (as he found they do) via western Europe to arrive at the Taimyr Peninsula, northern Siberia, in early to mid-June.

As mentioned earlier (Section IV), Red Knots seem to use rather few staging areas. It is not clear whether this is due to unsuitability of other areas or a preference for longer flights. Wilson (1981) has reviewed evidence that some shorebirds, including Red Knots, choose not to use some potential staging posts, which are suitable (as evidenced by their use by conspecifics moving to other areas), if a direct overflight or bypass is within their capabilities. Thus, many Ruddy Turnstones, Ringed Plovers, Dunlins, and Sanderlings breeding in northeastern Greenland overfly Iceland in spring when migrating from western Europe. In contrast, Ruddy Turnstones, Red Knots, and probably Ringed Plovers moving to northwestern Greenland and northeastern Canada in May stop in Iceland before crossing the Greenland ice cap (Wilson, 1981).

Schneider (1981) argued that the timing of northward movement follows a spring bloom of littoral benthos production. This is, however, unlikely as, in many estuaries in western Europe, densities of the main prey species of shorebirds do not start to increase until June or later, and peak in the autumn (Dugan, 1981a). It is possible that increased prey activities, and consequent availabilities, in the warmer spring conditions may be implicated. The surface activity, and possibly depth distribution, of many intertidal invertebrates is very temperature-sensitive (Pienkowski, 1981a,b; 1983b), and this leads to an increase in food intake rate in spring in some species (e.g., Goss-Custard, 1969; Smith, 1975; Pienkowski, 1982, 1983c).

Little is known of the proximate factors timing shorebird movements. Although increasing daylength in spring at temperate latitudes could act as a trigger for both the start of molt into breeding plumage and fat deposition, this has not been confirmed experimentally. It is much less likely that daylength changes are adequate to time the departures of the same species from western Africa in spring, for the changes in daylength involved are so small. More probably, as in many Arctic breeding passerines (Gwinner, 1977), a circannual clock is involved. The date of departure may be modified in temperate regions by a cold or warm spring, and even more finely adjusted by meteorological conditions, notably favorability of wind direction (Richardson, 1979).

ACKNOWLEDGMENTS

We thank our colleagues at Durham for discussions and information on many points: Miss Rowena Cooper, Dr. N. C. Davidson, Dr. P. J. Dugan, Mrs. Ann Pienkowski, Dr. D. J. Townshend. We are also grateful to the many participants in the project on Movements of Wader Populations in Western Europe organized, with the Wader Study Group, by our team at Durham, funded by the Nature Conservancy Council of Great Britain and the EEC Environment Programme. Other parts of the studies at Durham referred to in this chapter were supported by the British Ornithologists' Union, the Natural Environment Research Council, the Nuffield Foundation, and the Science Research Council.

REFERENCES

Able, K. P., 1908, Mechanisms of orientation, navigation and homing, in: *Animal Migration, Orientation, and Navigation* (S. A. Gauthreaux, Jr., ed.), pp. 283–373, Academic Press, New York.

Alerstam, T., and Högstedt, G., 1980, Spring predictability and leap-frog migration, *Ornis Scand.* **11**:196–200.

Ali, S., and Hussain, S. A., 1981, Studies on the movement and population structure of Indian avifauna, Annual Report 1 (June 1980–July 1981), Bombay Natural History Society.

Ankney, C. D., 1977, The use of nutrient reserves by breeding male Lesser Snow Geese *Chen caerulescens caerulescens, Can. J. Zool.* **55**:1984–1987.

Argyle, F. B., 1975, Report Bird-Ringing in Iran 1970 to 1974, Ornithology Unit, Division of Parks and Wildlife, Iran Department of the Environment.

Ashkenazie, S., and Safriel, U. N., 1979, Time–energy budget of the Semipalmated Sandpiper *Calidris pusilla* at Barrow, Alaska, *Ecology* **60**:783–799.

Baillie, S., 1980, The effect of the hard winter of 1978/79 on the wader populations of the Ythan Estuary, *Wader Study Group Bull.* **28**:16–17.

Baker, M. C., and Baker, A. E., 1973, Niche relationships among six species of shorebirds on their wintering and breeding ranges, *Ecol. Monogr.* **43**:183–212.

Boere, G. C., 1976, The significance of the Dutch Waddenzee in the annual life cycle of Arctic, Subarctic and boreal waders. Part 1. The functions as a moulting area, *Ardea* **64:**210–291.

Brady, F., 1943, The distribution of the fauna of some intertidal sands and muds on the Northumberland coast, *J. Anim. Ecol.* **12:**27–41.

Branson, N. J. B. A., and Minton, C. D. T., 1976, Moult, measurements and migrations of the Grey Plover, *Bird Study* **23:**257–266.

Branson, N. J. B. A., Ponting, E. D., and Minton, C. D. T., 1978, Turnstone migrations in Britain and Europe, *Bird Study* **25:**181–187.

Burger, J., 1980, Age differences in foraging Black-necked Stilts in Texas, *Auk* **97:**633–636.

Burger, J., 1981, Movements of juvenile Herring Gulls hatched at Jamaica Bay Refuge, New York, *J. Field Ornithol.* **52:**285–290.

Burger, J., Hahn, D. C., and Chase, J., 1979, Aggressive interactions in mixed-species flocks of migrating shorebirds, *Anim. Behav.* **27:**459–469.

Byrkjedal, I., 1980, Nest predation in relation to snow-cover—A possible factor influencing the start of breeding in shorebirds, *Ornis Scand.* **11:**249–252.

Clark, N. A., 1982, The effects of the severe weather in December 1981 and January 1982 on waders in Britain, *Wader Study Group Bull.* **34:**5–7.

Cramp, S., and Simmons, K. E. L., 1983, *Birds of the Western Palearctic*, Vol. 3, Oxford University Press, Oxford.

Davidson, N. C., 1981a, Survival of shorebirds (Charadrii) during severe weather: The role of nutritional reserves, in: *Feeding and Survival Strategies of Estuarine Organisms* (N. V. Jones and W. J. Wolff, eds.), pp. 231–249, Plenum Press, New York.

Davidson, N. C., 1981b, Seasonal changes in the nutritional condition of shorebirds (Charadrii) during the non-breeding season, Ph.D. thesis, University of Durham, U.K.

Davidson, N. C., 1982, Changes in body-condition of redshanks during mild winters: An inability in regulate reserves?, *Ringing Migration* **4:**51–62.

Davidson, N. C., 1983, Formulae for estimating the lean weights and fat reserves of live shorebirds, *Ringing Migration* **4:**159–166.

Davidson, N. C., 1984, How valid are flight range estimates for waders?, *Ringing Migration* **5:**49–64.

Davidson, N. C., and Evans, P. R., 1982, Mortality of redshanks and oystercatchers from starvation during severe weather, *Bird Study* **29:**183–188.

Davidson, P. E., 1971, Some foods taken by waders in Morecambe Bay, Lancashire, *Bird Study* **18:**177–186.

Dick, W. J. A. (ed.), 1976, *Oxford & Cambridge Mauritanian Expedition 1973 Report*, Cambridge.

Dick, W. J. A., 1979, Results of the WSG project on the spring migration of Siberian Knot *Calidris canutus* in 1979, *Wader Study Group Bull.* **27:**8–13.

Dick, W. J. A., and Pienkowski, M. W., 1979, Autumn and early winter weights of waders in north-west Africa, *Ornis Scand.* **10:**117–123.

Dick, W. J. A., Pienkowski, M. W., Waltner, M. A., and Minton, C. D. T., 1976, Distribution and geographical origins of knots *Calidris canutus* wintering in Europe and Africa, *Ardea* **64:**22–47.

Dobinson, H. M., and Richards, A. J., 1964, The effects of the severe winter of 1962/3 on birds in Britain, *Br. Birds* **57:**373–434.

Dorst, J., 1974, *The Life of Birds*, Vol. 2, Weidenfeld & Nicholson, London.

Drent, R., Ebbinge, B., and Weijand, B., 1981, Balancing the energy budgets of Arctic-breeding geese throughout the annual cycle: A progress report, *Verh. Ornithol. Ges. Bayern* **23:**239–264.

Duffy, D. C., Atkins, N., and Schneider, D. C., 1981, Do shorebirds compete on their wintering grounds?, *Auk* **98**:215–229.

Dugan, P. J., 1981a, Seasonal movements of shorebirds in relation to spacing behaviour and prey availability, Ph.D. thesis, University of Durham, U.K.

Dugan, P. J., 1981b, The importance of nocturnal foraging in shorebirds: A consequence of increased invertebrate prey activity, in: *Feeding and Survival Strategies of Estuarine Organisms* (N. V. Jones and W. J. Wolff, eds.), pp. 251–260, Plenum Press, New York.

Dugan, P. J., Evans, P. R., Goodyer, L. R., and Davidson, N. C., 1981, Winter fat reserves in shorebirds: Disturbance of regulated levels by severe weather conditions, *Ibis* **123**:359–363.

Ebbinge, B., Canters, K., and Drent, R., 1975, Foraging routines and estimated daily food intake in Barnacle Geese wintering in northern Netherlands, *Wildfowl* **26**:5–19.

Edelstam, C. (ed.), 1972, *The Visible Migration of Birds at Ottenby, Sweden, Vår Fågelvärld* Suppl. 7.

Ehlert, W., 1964, Zur Ökologie und Biologie der Ernährung einiger Limikolen-Arten, *J. Ornithol.* **105**:1–53.

Elliott, C. C. H., Waltner, M., Underhill, L. G., Pringle, J. S., and Dick, W. J. A., 1976, The migration system of the Curlew Sandpiper *Calidris ferruginea* in Africa, *Ostrich* **47**:191–213.

Ens, B., 1979, Territoriality in Curlews *Numenius arquata*, *Wader Study Group Bull.* **26**:28–29.

Evans, P. R., 1968, Autumn movements and orientation of waders in north-east England and southern Scotland, studied by radar, *Bird Study* **15**:53–64.

Evans, P. R., 1976, Energy balance and optimal foraging strategies: Some implications for their distributions and movements in the non-breeding season, *Ardea* **64**:117–139.

Evans, P. R., 1979a, Adaptations shown by foraging shorebirds to cyclical variations in the activity and availability of their invertebrate prey, in: *Cyclic Phenomena in Marine Plants and Animals* (E. J. Naylor and R. G. Hartnoll, eds.), pp. 357–366, Pergamon Press, Elmsford, N.Y.

Evans, P. R., 1979b, Some questions and hypotheses concerning the timing of migration in shorebirds, *Wader Study Group Bull.* **26**:30.

Evans, P. R., 1981a, Reclamation of intertidal land: Some effects on Shelduck and wader populations in the Tees Estuary, *Verh. Ornithol. Ges. Bayern* **23**:147–168.

Evans, P. R., 1981b, Migration and dispersal of shorebirds as a survival strategy, in: *Feeding and Survival Strategies of Estuarine Organisms* (N. V. Jones and W. J. Wolff, eds.), pp. 275–290, Plenum Press, New York.

Evans, P. R., and Dugan, P. J., 1984, Coastal birds: Numbers in relation to food resources, in: *Coastal Waders and Wildfowl in Winter* (P. R. Evans, J. D. Goss-Custard, and W. G., Hale, eds.), pp. 8–28, Cambridge University Press, Cambridge.

Evans, P. R., and Pienkowski, M. W., 1982, Behaviour of Shelducks *Tadorna tadorna* in a winter flock: Does regulation occur?, *J. Anim. Ecol.* **51**:241–262.

Evans, P. R., and Smith, P. C., 1975, Studies of shorebirds at Lindisfarne, Northumberland. 2. Fat and pectoral muscles as indicators of body condition in the Bar-tailed Godwit, *Wildfowl* **26**:64–74.

Evans, P. R., Herdson, D. M., Knights, P. J., and Pienkowski, M. W., 1979, Short-term effects of reclamation of part of Seal Sands, Teesmouth, on wintering waders and Shelduck, *Oecologia (Berlin)* **41**:183–206.

Ferns, P. N., 1978, General ornithological notes, in: *Joint Biological Expedition to North East Greenland 1974* (G. H. Green and J. J. D. Greenwood, eds.), pp. 152–164, Dundee University NE Greenland Expedition, Dundee.

Ferns, P. N., 1980–81, The spring migrations through Britain in 1979, *Wader Study Group Bull.* **29**:10–13; **30**:22–25; **31**:36–40; **32**:14–19; **33**:6–10.

Folkestad, A. O., 1975, Wetland bird migration in central Norway, *Ornis Fenn.* **52**:49–56.

Fournier, O., 1969, Recherches sur les Barges à queue noir *Limosa limosa* et les Combattants *Philomachus pugnax* stationnant en Camargue au printemps 1966, *Nos Oiseaux Bull. Soc. Romande Etude Prot. Oiseaux* **30**:87–102.

Fournier, O., and Dick, W. J. A., 1981, Preliminary survey of the Archipel des Bijagos, Guinea-Bissau, *Wader Study Group Bull.* **31**:24–25.

Fretwell, S. D., and Lucas, H. L., 1970, On territorial behaviour and other factors influencing habitat distribution in birds, *Acta Biotheor.* **19**:16–36.

Fuchs, E., 1973, Durchzug und Uberwinterung des Alpenstrandläufers *Calidris alpina* in der Camargue, *Ornithol. Beob.* **70**:113–134.

Goss-Custard, J. D., 1969, The winter feeding ecology of the redshank *Tringa totanus*, *Ibis* **111**:338–356.

Goss-Custard, J. D., 1970, Dispersion in some overwintering wading birds, in: *Social Behaviour in Birds and Mammals* (J. H. Crook, ed.), pp. 3–35, Academic Press, New York.

Goss-Custard, J. D., 1976, Variation in the dispersion of redshank *Tringa totanus* on their winter feeding grounds, *Ibis* **118**:257–263.

Goss-Custard, J. D., 1977, The ecology of the Wash. III. Density-related behaviour and the possible effects of a loss of feeding grounds on wading birds (Charadrii), *J. Appl. Ecol.* **14**:721–739.

Goss-Custard, J. D., 1980, Competition for food and interference among waders, *Ardea* **68**:31–52.

Goss-Custard, J. D., Jones, R. E., and Newbery, P. E., 1977a, The ecology of the Wash. I. Distribution and diet of wading birds (Charadrii), *J. Appl. Ecol.* **14**:681–700.

Goss-Custard, J. D., Jenyon, R. A., Jones, R. E., Newbery, P. E., and Williams, R. le B., 1977b, The ecology of the Wash. II. Seasonal variation in the feeding conditions of wading birds (Charadrii), *J. Appl. Ecol.* **14**:701–719.

Goss-Custard, J. D., Durell, S. E. A., McGrorty, S., Reading, C. J., and Clarke, R. T., 1981, Factors affecting the occupation of mussel (*Mytilus edulis*) beds by oystercatchers (*Haematopus ostralegus*) on the Exe Estuary, Devon, in: *Feeding and Survival Strategies of Estuarine Organisms* (N. V. Jones and W. J. Wolff, eds.), pp. 217–229, Plenum Press, New York.

Goss-Custard, J. D., Durell, S. E. A., McGrorty, S., and Reading, C. J., 1982, Use of mussel *Mytilus edulis* beds by oystercatchers *Haematopus ostralegus* according to age and population size. *J. Anim. Ecol.* **51**:543–554.

Green, G. H., and Greenwood, J. J. D. (eds.), 1978, *Joint Biological Expedition to North East Greenland 1974,* Dundee University NE Greenland Expedition 1974, Dundee.

Green, G. H., Greenwood, J. J. D., and Lloyd, C. S., 1977, The influence of snow conditions on the date of breeding of wading birds in north-east Greenland, *J. Zool.* **183**:311–328.

Green, J., 1968, *The Biology of Estuarine Animals*, Sidgwick & Jackson, London.

Greenewalt, C. H., 1975, The flight of birds, *Trans. Am. Philos. Soc. N. S.* **65**(4).

Grimes, L. G., 1974, Radar tracks of Palearctic waders departing from the coast of Ghana in spring, *Ibis* **116**:165–171.

Gromadzka, J., 1983, Results of bird ringing in Poland: Migrations of Dunlin *Calidris alpina*, *Acta Ornithol.* (Warsz.) **19**:113–136.

Groves, S., 1978, Age-related differences in Ruddy Turnstone foraging and aggressive behaviour, *Auk* **95**:95–103.

Gwinner, E., 1977, Circannual rhythms in bird migration, *Annu. Rev. Ecol. Syst.* **8**:381–405.

Gwinner, E., and Wiltschko, W., 1978, Endogenously controlled changes in migratory direction of the Garden Warbler *Sylvia borin, J. Comp. Physiol.* **125**:267–273.

Håland, A., and Kålås, J. A., 1980, Spring migration of the Siberian Knot *Calidris canutus*: Additional information, *Wader Study Group Bull.* **29**:22–23.

Hale, W. G., 1973, The distribution of the redshank *Tringa totanus* in the winter range, *Zool. J. Linn. Soc.* **50**:199–268.

Hale, W. G., 1980, *Waders*, Collins, London.

Harengerd, M., Prünte, W., and Speckmann, M., 1973, Zugphänologie und Status der Limikolen in den Rieselfeldern der Stadt Münster, *Vogelwelt* **94**:81–118, 121–146.

Harrington, B. A., and Morrison, R. I. G., 1979, Semipalmated sandpiper migration in North America, in: *Studies in Avian Biology No. 2* (F. A. Pitelka, ed.), pp. 83–100, Cooper Ornithological Society, Allen Press, Lawrence, Kansas.

Harris, M. P., 1967, The biology of oystercatchers *Haematopus ostralegus*, on Skokholm Island, S. Wales, *Ibis* **109**:180–193.

Harris, M. P., 1975, Skokholm oystercatchers and the Burry Inlet, *Rep. Skokholm Bird Observatory for 1974* pp. 17–19.

Heppleston, P. B., 1971, The feeding ecology of oystercatchers (*Haematopus ostralegus* L.) in winter in northern Scotland, *J. Anim. Ecol.* **40**:651–672.

Hicklin, P. W., and Smith, P. C., 1979, The diets of five species of migrant shorebirds in the Bay of Fundy, *Proc. N. S. Inst. Sci.* **29**:483–488.

Hildén, O., 1975, Breeding system of Temminck's Stint *Calidris temminckii, Ornis Fenn.* **52**:117–146.

Hildén, O., 1979, Territoriality and site tenacity of Temminck's Stint *Calidris temminckii, Ornis Fenn.* **56**:56–74.

Hoffmann, L., 1957, Le passage d'automne du Chevalier sylvain *Tringa glareola* en France méditerranéenne, *Alauda Rev. Int. Ornithol.* **25**:30–42.

Hofmann, H., and Hoerschelmann, H., 1969, Nahrungsuntersuchungen bei Limikolen durch Mageninhaltsenalysen, *Corax* **3**:7–22.

Holmes, R. T., 1966a, Breeding ecology and annual cycle adaptations of the Red-backed Sandpiper (*Calidris alpina*) in northern Alaska, *Condor* **68**:3–46.

Holmes, R. T., 1966b, Feeding ecology of the Red-backed Sandpiper (*Calidris alpina*) in Arctic Alaska, *Ecology* **47**:32–45.

Holmes, R. T., 1970, Differences in population density, territoriality, and food supply of Dunlin on Arctic and subarctic tundra, in: *Animal Populations in Relation to Their Food Resources* (A. Watson ed.), pp. 303–319, Blackwell, Oxford.

Holmes, R. T., 1971, Latitudinal differences in the breeding and molt schedules of Alaskan Red-backed Sandpipers, *Condor* **73**:93–99.

Holmes, R. T., and Pitelka, F. A., 1968, Food overlap among coexisting sandpipers on northern Alaskan tundra, *Syst. Zool.* **17**:305–318.

Järvinen, O., and Väisänen, R. A., 1978, Ecological zoogeography of North European waders; or why do so many waders breed in the North?, *Oikos* **30**:495–507.

Jenkins, D., Murray, M. G., and Hall, P., 1975, Structure and regulation of a Shelduck (*Tadorna tadorna* (L.)) population, *J. Anim. Ecol.* **44**:201–231.

Johnson, A. R., 1974, Wader research in the Camargue, *Proc. IWRB Wader Symp.* Warsaw 1973, pp. 63–82.

Johnson, C., and Minton, C. D. T., 1980, The primary moult of the Dunlin *Calidris alpina* at the Wash, *Ornis Scand.* **11**:190–195.

Johnson, O. W., 1979, Biology of shorebirds summering on Enewetak Atoll, in: *Studies in Avian Biology No. 2* (F. A. Pitelka, ed.), pp. 193–205, Cooper Ornithological Society, Allen Press, Lawrence, Kans.

Johnston, D. W., and McFarlane, R. W., 1967, Migration and bioenergetics of flight in the Pacific Golden Plover, *Condor* **69:**156–168.

Kaukola, A., and Lilja, I., 1972, Sirrien (*Calidris* spp.) ja Jänkäsirriäisen (*Limicola falcinellus*) muutto Yyterissä 1961–69, *Porin Lintutiet. Yhd. Vuosik.* **3:**17–23.

Kendeigh, S. C., 1970, Energy requirements for existence in relation to size of bird, *Condor* **72:**60–65.

Lack, D., 1962, Radar evidence on migratory orientation, *Br. Birds* **55:**139–158.

Lack, D., 1962–63, Migration across the southern North Sea studied by radar, *Ibis* **104:**74–85; **105:**1–54, 461–492.

Lack, D., 1968, *Ecological Adaptations for Breeding in Birds*, Methuen, London.

Larson, S., 1957, The suborder Charadrii in Arctic and boreal areas during the Tertiary and Pleistocene, *Acta Vertebr.* **1:**1–84.

Larson, S., 1960, On the influence of the Arctic Fox *Alopex lagopus* on the distribution of Arctic birds, *Oikos* **11:**276–305.

Lasiewski, R. C., and Dawson, W. R., 1967, A re-examination of the relationship between standard metabolic rate and body weight in birds, *Condor* **69:**12–23.

Le Maho, Y., Vu Van Kha, H., Koubi, H., Dewasmes, G., Girard, J., Ferré, P., and Cagnard, M., 1981, Body composition, energy expenditure, and plasma metabolites in long-term fasting geese, *Am. J. Physiol.* **241:**E342–E354.

McClure, H. E., 1974, *Migration and survival of the birds of Asia*, U. S. Army Medical Component, SEATO Medical Project, Bangkok, Thailand.

McLennon, J. A., 1979, Formation and function of mixed species wader flocks in fields, Ph.D. thesis, University of Aberdeen, U. K.

McNeil, R., 1970, Hivernage et estivage d'oiseaux aquatiques Nord-Americains dans le nord-est du Vénézuela (mue, accumulation de graisse, capacité de vol et routes de migration), *Oiseau Rev. Fr. Ornithol.* **40:**185–303.

McNeil, R., and Cadieux, F., 1972, Numerical formulae to estimate flight range of some North American shorebirds from fresh weight and wing length, *Bird-Banding* **43:**107–113.

Marcström, V., and Mascher, J. W., 1979, Weights and fat in Lapwings and Oystercatchers starved to death during a cold spell in spring, *Ornis Scand.* **10:**235–240.

Mascher, J. W., 1966, Weight variations in resting Dunlins (*Calidris alpina alpina*) on autumn migration in Sweden, *Bird-Banding* **37:**1–34.

Mason, C. F., 1969, Waders and terns in Leicestershire and an index of relative abundance, *Br. Birds* **62:**523–533.

Meltofte, H., Pihl, S., and Sørensen, B. M., 1972, Efterårstraekket af vadefugle (Charadrii) ved Blåvandshak 1963–1971, *Dan. Ornithol. Foren. Tidsskr.* **66:**63–69.

Melville, D. S., 1981, Spring measurements, weights and plumage status of *Calidris ruficollis* and *C. ferruginea* in Hong Kong, *Wader Study Group Bull.* **33:**18–21.

Middlemiss, E., 1961, Biological aspects of *Calidris minuta* while wintering in south-west Cape, *Ostrich* **32:**107–121.

Minton, C. D. T., 1975, The waders of the Wash—Ringing and biometric studies, Report of Scientific Study G, Wash Water Storage Scheme Feasibility Study to the Natural Environment Research Council.

Moreau, R. E., 1967, Water-birds over the Sahara, *Ibis* **109:**232–253.

Moreau, R. E., 1972, *The Palaearctic–African Bird Migration Systems*, Academic Press, New York.

Morrison, R. I. G., 1975, Migration and morphometrics of European knot and turnstone on Ellesmere Island, Canada, *Bird-Banding* **46**:290–301.

Morrison, R. I. G., and Wilson, J. R. (eds.), 1972, *Cambridge–Iceland Expedition 1971 Report*, Cambridge.

Muus, B. J., 1967, The fauna of Danish estuaries and lagoons: Distribution and ecology of dominating species in the shallow reaches of the mesohaline zone, *Medd. Dan. Fisk. Havunders. N.S.* **5**:1–316.

Myers, J. P., 1980, Sanderlings *Calidris alba* at Bodega Bay: Facts, inferences and shameless speculations, *Wader Study Group Bull.* **30**:26–32.

Myers, J. P., 1981a, A test of three hypotheses for latitudinal segregation of the sexes in wintering birds, *Can. J. Zool.* **59**:1527–1534.

Myers, J. P., 1981b, Cross-seasonal interactions in the evolution of sandpiper social systems, *Behav. Ecol. Sociobiol.* **8**:195–202.

Myers, J. P., Connors, P. G., and Pitelka, F. A., 1979a, Territoriality in nonbreeding shorebirds, in: *Studies in Avian Biology No. 2* (F. A. Pitelka, ed.), pp. 231–246, Cooper Ornithological Society, Allen Press, Lawrence, Kans.

Myers, J. P., Connors, P. G., and Pitelka, F. A., 1979b, Territory size in wintering Sanderlings: The effects of prey abundance and intruder density, *Auk* **96**:551–561.

Naylor, E., 1963, Temperature relationships of the locomotor rhythm of *Carcinus*, *J. Exp. Biol.* **40**:669–679.

Netterstrøm, B., 1970, Efterarstraekket af Islandsk Ryle (*Calidris canutus*) i Vestjylland, *Dan. Ornithol. Foren. Tidsskr.* **64**:223–228.

Nettleship, D. N., 1973, Breeding ecology of turnstones *Arenaria interpres* at Hazen Camp, Ellesmere Island, N. W. T., *Ibis* **115**:202–217.

Nettleship, D. N., 1974, The breeding of the knot *Calidris canutus* at Hazen Camp, Ellesmere Island, N. W. T., *Polarforschung* **44**:8–26.

NOME, 1982, Wintering waders on the Banc d'Arguin, Mauritania, Report of the Netherlands Ornithological Mauritanian Expedition 1980, Communication No. 6 of the Wadden Sea Working Group, Groningen.

Nørrevang, A., 1959, The migration patterns of some waders in Europe based on ringing results, *Vidensk. Medd. Dan. Naturhist. Foren. Kobenhavn* **121**:181–214.

Norton-Griffiths, M., 1967, Some ecological aspects of the feeding behaviour of the oystercatcher on the Edible Mussel, *Ibis* **109**:412–424.

Norton-Griffiths, M., 1969, The organisation, control and development of parental feeding in the oystercatcher (*Haematopus ostralegus*), *Behaviour* **34**:55–114.

O'Connor, R. J., and Cawthorne, R. A., 1982, How Britain's birds survived the winter, *New Sci.* **93**:786–788.

Page, G., and Middleton, A. L. A., 1972, Fat deposition during autumn migration in the Semi-palmated Sandpiper, *Bird-Banding* **43**:85–96.

Page, G., and Whitacre, D. F., 1975, Raptor predation on wintering shorebirds, *Condor* **77**:73–83.

Panov, E. N., 1963, On territorial relationships of shorebirds during migration, *Ornitologiya* **1963**(6):418–423 (in Russian).

Pearson, D. J., Phillips, J. M., and Backhurst, G. C., 1970, Weights of some Palearctic waders wintering in Kenya, *Ibis* **112**:199–208.

Pennycuick, C. J., 1975, Mechanics of flight, in: *Avian Biology*, Vol. 5 (D. S. Farmer and J. R. King, eds.), pp. 1–75, Academic Press, New York.

Pienkowski, M. W., 1973, Feeding activities of wading birds and Shelducks at Teesmouth and some possible effects of further loss of habitat, Report to Coastal Ecology Research Station, Institute of Terrestrial Ecology (formerly the Nature Conservancy).

Pienkowski, M. W. (ed.), 1975, *Studies on Coastal Birds and Wetlands in Morocco 1972*, Norwich.

Pienkowski, M. W., 1976, Recurrence of waders on autumn migration at sites in Morocco, *Vogelwarte* **28**:293–297.

Pienkowski, M. W., 1980, Aspects of the ecology and behaviour of Ringed and Grey Plovers *Charadrius hiaticula* and *Pluvialis squatarola*, Ph.D. thesis, University of Durham, U.K.

Pienkowski, M. W., 1981a, Differences in habitat requirements and distribution patterns of plovers and sandpipers as investigated by studies of feeding behaviour, *Verh. Ornithol. Ges. Bayern* **23**:105–124.

Pienkowski, M. W., 1981b, How foraging plovers cope with environmental effects on invertebrate behaviour and availability, in: *Feeding and Survival Strategies of Estuarine Organisms* (N. V. Jones and W. J. Wolff,eds.), pp. 179–192, Plenum Press, New York.

Pienkowski, M. W., 1982, Diet and energy intake of Grey and Ringed Plovers, *Pluvialis squatarola* and *Charadrius hiaticula*, in the non-breeding season, *J. Zool.* **197**:511–549.

Pienkowski, M. W., 1983a, Changes in the foraging pattern of plovers in relation to environmental factors, *Anim. Behav.* **31**:244–264.

Pienkowski, M. W., 1983b, Surface activity of some intertidal invertebrates in relation to temperature and the foraging behaviour of their shorebird predators, *Mar. Ecol. Progr. Ser.* **11**:141–150.

Pienkowski, M. W., 1983c, The effects of environmental conditions on feeding rates and prey selection of shore plovers, *Ornis Scand.* **14**:227–238.

Pienkowski, M. W., 1984a, Behaviour of young Ringed Plovers *Charadrius hiaticula* and its relationship to growth and survival to reproductive age, *Ibis*. **126**: in press.

Pienkowski, M. W., 1984b, Development of feeding and foraging behaviour in young Ringed Plovers *Charadrius hiaticula* in Greenland and Britain, *Dan. Ornithol. Foren. Tidsskr.* **77**:133–147.

Pienkowski, M. W., 1984c, Breeding biology and population dynamics of Ringed Plovers *Charadrius hiaticula* in Britain and Greenland: Nest predation as a possible factor limiting distribution and timing of breeding, *J. Zool., Lond.* **202**:83–114.

Pienkowski, M. W., and Dick, W. J. A., 1975, The migration and wintering of Dunlin *Calidris alpina* in north-west Africa, *Ornis Scand.* **6**:151–167.

Pienkowski, M. W., and Knight, P. J., 1977, La migration postnuptiale des limicoles sur la côte atlantique du Maroc, *Alauda Rev. Int. Ornithol.* **45**:165–190.

Pienkowski, M. W., and Clark, H., 1979, Preliminary results of winter dye-marking on the Firth of Forth, Scotland, *Wader Study Group Bull.* **27**:16–18.

Pienkowski, M. W., and Pienkowski, A. E., 1980, WSG project on movements of wader populations in western Europe: First progress report, *Wader Study Group Bull.* **30**:7–9.

Pienkowski, M. W., and Pienkowski, A., 1981, WSG project on movements of wader populations in western Europe: Second progress report, *Wader Study Group Bull.* **31**:16–17.

Pienkowski, M. W., and Pienkowski, A. E., 1983, WSG project on the movements of wader populations in western Europe: Eighth progress report, *Wader Study Group Bull.* **38**:13–22.

Pienkowski, M. W., Knight, P. J., Stanyard, D. J., and Argyle, F. B., 1976, The primary moult of waders on the Atlantic coast of Morocco, *Ibis* **118**:347–365.

Pienkowski, M. W., Lloyd, C. S., and Minton, C. D. T., 1979, Seasonal and migrational weight changes in Dunlins, *Bird Study* **26**:134–148.

Pienkowski, M. W., Evans, P. R., and Townshend, D. J., 1984, Leap-frog and other migration patterns of waders: A critique of the Alerstam and Högstedt hypothesis, and some alternatives, *Ornis Scand.* (in press).

Pienkowski, M. W., Ferns, P. N., Davidson, N. C., and Worrall, D. H., 1984, Balancing the budget: Problems in measuring the energy intake and requirements of shorebirds in the field, in: *Coastal Waders and Wildfowl in Winter* (P. R. Evans, J. D. Goss-Custard, and W. G. Hale, eds.), pp. 29–56, Cambridge University Press, Cambridge.

Pilcher, R. E. M., 1964, Effects of the cold weather of 1962/3 on birds of the north coast of the Wash, *Wildfowl Trust Annu. Rep.* **15**:23–26.

Pilcher, R. E. M., Beer, J. V., and Cook, W. A., 1974, Ten years of intensive late-winter surveys for waterfowl corpses on the north-west shore of the Wash, England, *Wildfowl* **25**:149–154.

Pitelka, F. A., 1959, Numbers, breeding schedule, and territoriality in Pectoral Sandpipers of northern Alaska, *Condor* **61**:233–264.

Pitelka, F. A., (ed.), 1979, *Studies in Avian Biology No. 2*, Cooper Ornithological Society, Allen Press, Lawrence, Kans.

Pitelka, F. A., Holmes, R. T., and MacLean, S. F., Jr., 1974, Ecology and evolution of social organization in Arctic sandpipers, *Am. Zool.* **14**:185–204.

Plath, L., 1976, Die Vögel der Stadt Rostock (Nonpasseres), Kulturbund der DDR, Bezirksleitung Rostok, Berzirkskommission Natur and Heimat.

Portenko, L. A., 1959, Studien an einigen seltener Limikolen aus dem nördlichen und östlichen Siberien. II. Der Sichelstrandläufer *Erolia ferruginea*, *J. Ornithol.* **98**:454–466.

Prater, A. J., 1972, The ecology of Morecambe Bay. III. The food and feeding habits of knot (*Calidris canutus* L.) in Morecambe Bay, *J. Appl. Ecol.* **9**:179–194.

Prater, A. J., 1982, Wader Research Group report, Debrecen, *Int. Waterfowl Res. Group Bull.* **47**:74–78.

Prater, A. J., and Wilson, J., 1972, Aspects of spring migration of knot in Morecambe Bay, *Wader Study Group Bull.* **5**:9–11.

Reading, C. J., and McGrorty, S., 1978, Seasonal variations in the burying depth of *Macoma balthica* (L.) and its accessibility to wading birds, *Estuarine Coastal Mar. Sci.* **6**:135–144.

Recher, H. F., 1966, Some aspects of the ecology of migrant shorebirds, *Ecology* **47**:393–407.

Recher, H. F., and Recher, J. A., 1969, Comparative foraging efficiency of adult and immature Little Blue Herons (*Florida caerulea*), *Anim. Behav.* **17**:320–322.

Richardson, W. J., 1976, Autumn migration over Puerto Rico and the western Atlantic: A radar study, *Ibis* **118**:309–332.

Richardson, W. J., 1978, Timing and amount of bird migration in relation to weather: A review, *Oikos* **30**:224–272.

Richardson, W. J., 1979, Southeastward shorebird migration over Nova Scotia and New Brunswick in autumn: A radar study, *Can. J. Zool.* **57**:109–124.

Safriel, U., 1975, On the significance of clutch size in nidifugous birds, *Ecology* **56**:703–708.

Salomonsen, F., 1954, The migration of the European redshank (*Tringa totanus* (L.)), *Dan. Ornithol. Foren. Tidsskr.* **49**:149–181.

Salomonsen, F., 1955, The evolutionary significance of bird migration, *Biol. Meddr.* **22**:1–62.

Schmitt, M. B., and Whitehouse, P. H., 1976, Moult and mensural data of Ruff on the Witwatersrand, *Ostrich* **47**:176–190.

Schneider, D. C., 1981, Food supplies and the phenology of migratory shorebirds: A hypothesis, *Wader Study Group Bull.* **33**:43–45.

Schneider, D. C., and Harrington, B. A., 1981, Timing of shorebird migration in relation to prey depletion, *Auk* **98**:801–811.

Slagsvold, T., 1982, Spring predictability and bird migration and breeding times: A comment on the phenomenon of leap-frog migration, *Ornis Scand.* **13**:145–148.

Smith, C. J., and Wolff, W. J., 1981, Birds of the Wadden Sea, Wadden Sea Working Group Report 6, Balkema, Rotterdam.

Smith, P. C., 1975, A study of the winter feeding ecology and behaviour of the Bar-tailed Godwit (*Limosa lapponica*), Ph.D. thesis, University of Durham, U.K.

Smith, P. C., and Evans, P. R., 1973, Studies of shorebirds at Lindisfarne, Northumberland. I. Feeding ecology and behaviour of the Bar-tailed Godwit, *Wildfowl* **24**:135–139.

Snow, D. W., 1968, Movements and mortality of British Kestrels *Falco tinnunculus*, *Bird Study* **15**:65–83.

Soikkeli, M., 1970, Dispersal of Dunlin *Calidris alpina* in relation to sites of birth and breeding, *Ornis Fenn.* **47**:1–9.

Stanley, P. I., and Minton, C. D. T., 1972, The unprecedented westward migration of Curlew Sandpipers in autumn 1969, *Br. Birds* **65**:365–380.

Stinson, C. H., 1977, The spatial distribution of wintering Black-bellied Plovers, *Wilson Bull.* **89**:470–472.

Summers, R. W., and Waltner, M., 1979, Seasonal variations in the mass of waders in southern Africa, with special reference to migration, *Ostrich* **50**:21–37.

Sutherland, W. J., 1982, Food supply and dispersal in the determination of wintering population levels of oystercatchers, *Haematopus ostralegus*, *Estuarine Coastal Shelf Sci.* **14**:223–229.

Swennen, C., 1971, Het voedsel van de Groenpootruiter *Tringa nebularia* tijdens het verblijt in het Nederlandse Waddengebied, *Limosa* **44**:71–83.

Swennen. C., 1984, Differences in quality of roosting flocks of Oystercatchers, in: *Coastal Waders and Wildfowl in Winter* (P. R. Evans, J. D. Goss-Custard, and W. G. Hale, eds.), pp. 177–189, Cambridge University Press, Cambridge.

Symonds, F. L., Langslow, D. R., and Pienkowski, M. W., 1984, Movements of wintering shorebirds within the Firth of Forth: Species differences in usage of an intertidal complex, *Biol. Conserv.* **28**:187–215.

Taylor, R. C., 1980, Migration of the Ringed Plover *Charadrius hiaticula*, *Ornis Scand.* **11**:30–42.

Thomas, D. G., and Dartnall, A. J., 1970, Pre-migratory deposition of fat in the Red-necked Stint, *Emu* **70**:87.

Thomas, D. G., and Dartnall, A. J., 1971, Moult of the Curlew Sandpiper in relation to its annual cycle, *Emu* **71**:153–158.

Thomson, A. L. (ed.), 1964, *A New Dictionary of Birds*, Nelson, London.

Townshend, D. J., 1981, The importance of field feeding to the survival of wintering male and female curlews *Numenius arquata* on the Tees Estuary, in: *Feeding and Survival Strategies of Estuarine Organisms* (N. V. Jones and W. J. Wolff, eds.), pp. 261–274, Plenum Press, New York.

Townshend, D. J., 1982a, The use of intertidal habitats by shorebird populations, with special reference to Grey Plover (*Pluvialis squatarola*) and Curlew (*Numenius arquata*), Ph.D. thesis, University of Durham, U.K.

Townshend, D. J., 1982b, The Lazarus syndrome in Grey Plovers, *Wader Study Group Bull.* **34**:11–12.

Townshend, D. J., 1984, Decisions for a lifetime: Establishment of spatial defence and movement patterns by juvenile shorebirds, *Ibis* (submitted).

Townshend, D. J., Dugan, P. J., and Pienkowski, M. W., 1984, The unsociable plover: Use of intertidal areas by Grey Plovers, in: *Coastal Waders and Wildfowl in Winter* (P. R. Evans, J. D. Goss-Custard, and W. G. Hale, eds.), pp. 140–159, Cambridge University Press, Cambridge.

van der Have, T., and Nieboer, E., 1984, Age-related distribution of Dunlin in the Dutch Wadden Sea in: *Coastal Waders and Wildfowl in Winter* (P. R. Evans, J. D. Goss-Custard, and W. G. Hale, eds.), pp. 160–176, Cambridge University Press, Cambridge.

Veitch, C. R., 1978, Waders of the Manukau Harbour and the Firth of Thames, *Notornis* **25:**1–24.

Vines, G., 1980, Spatial consequences of aggressive behaviour in flocks of oystercatchers *Haematopus ostralegus* L., *Anim. Behav.* **28:**1175–1185.

Welty, J. C., 1962, *The Life of Birds*, Saunders, Philadelphia.

Williams, T. C., Williams, J. M., Ireland, L. C., and Teal, J. M., 1977, Autumnal bird migration over the western North Atlantic, *Am. Birds.* **31:**251–267.

Wilson, J. R., 1981, The migration of High Arctic shorebirds through Iceland, *Bird Study* **28:**21–32.

Wilson, J. R., Czajkowski, M. A., and Pienkowski, M. W., 1980, The migration through Europe and wintering in west Africa of Curlew Sandpipers, *Wildfowl* **31:**107–122.

Wolff, W. J., 1969, Distribution of non-breeding waders in an estuarine area in relation to the distribution of their food organisms, *Ardea* **57:**1–28.

Zwarts, L., 1974, Vogels van het brakke getijgebied, ecologische onder-zoekingen op de ventsagerplaten, Amsterdam.

Zwarts, L., 1976, Density-related processes in feeding dispersion and feeding activity of teal (*Anas crecca*), *Ardea* **64:**192–209.

Zwarts, L., 1980, Intra- and inter-specific competition for space in estuarine bird species in a one-prey situation, *Proc. 17th Int. Ornithol. Congr., Berlin, 1978*, pp. 145–150.

Zwarts, L., and Drent, R. H., 1981, Prey depletion and the regulation of predator density: Oystercatchers (*Haematopus ostralegus*) feeding on mussels (*Mytilus edulis*), in: *Feeding and Survival Strategies of Estuarine Organisms* (N. V. Jones and W. J. Wolff, eds.), pp. 193–216, Plenum Press, New York.

Chapter 3

MIGRATION SYSTEMS OF SOME NEW WORLD SHOREBIRDS

R. I. G. Morrison

Canadian Wildlife Service
Ottawa, Ontario, Canada K1A 0E7.

I. INTRODUCTION

Each year, millions of shorebirds (Charadriiformes: Charadriidae, Scolopacidae, Phalaropodidae) make their way north to breeding grounds across the Arctic and subarctic regions of the New World to lay their eggs and rear their young. The group forms one of the most prominent and important components of the Arctic avifauna. To be able to use these Arctic breeding grounds, which provide the habitats, food, and other resources required for successful reproduction, the birds must make long and spectacular migrations to wintering areas where climate and food supply are suitable for subsistence, though not for breeding. Migration would clearly appear to play a central role in the annual cycle of the birds—from an energetic point of view, their migrations involve the accumulation and use of several times an individual's weight in fat reserves required to enable long flights to be made during the journey.

Most species of shorebirds breeding in the New World, particularly those occupying the central Arctic, migrate southwards to wintering areas extending from North America through South America, some as far south as the tip of Tierra del Fuego (Morrison, 1983a). Other species, however, particularly those at the eastern and western extremities of the North American Arctic, make their way to wintering areas on the European seaboard and parts of Asia and the Pacific, respectively (Morrison, 1975, 1976b, 1977a; Johnson and MacFarlane, 1967; Thompson, 1973). For many years, knowledge of the migrations and distributions of shorebird species in the New World has remained obscure and information difficult to obtain, owing to the remoteness not only of the Arctic breeding areas but also of the South American wintering grounds. In recent years, increased interest in shorebirds has led to a rapid advance in our understanding of the migrations of certain species, through the development of internationally coordinated observer networks, extensive aerial surveys on the wintering grounds, research on breeding areas, and increased banding and migration studies. As the details of different species' migration patterns emerge, it is becoming clear that they have evolved a wide variety of strategies to deal successfully with their long journeys and that the major energy-demanding events of the annual cycle (breeding, migration, molt) are elegantly interrelated in the birds' evolutionary response to the ecological factors they encounter. It appears that events during one part of the year will influence events elsewhere: for instance, migration distance appears to influence the type of breeding systems adopted by different species (Myers, 1981a). Thus, although an integrative approach may be necessary to understand the interplay of ecological factors operating during shorebirds' migration systems, this can only be achieved through a reductionist approach of understanding the system of one species at a time in as much detail as possible (e.g., Myers *et al.*, 1983).

The purpose of this chapter, therefore, will not be to attempt to provide a review of all that is known about the migration patterns of the complete spectrum of North American shorebird species. An introduction outlining the general migration systems in the New World will be followed by an account of recent advances in our knowledge of the migration systems of particular species with differing strategies, using each species to illustrate various aspects of the biology of shorebird migration. Some topics, such as the influence of weather, radar studies, orientation mechanisms, etc., will receive little coverage, the emphasis being on recent advances in our knowledge of distribution and routes. This review concentrates mostly on sandpipers, members of the genus *Calidris*.

II. MIGRATION SYSTEMS IN THE NEW WORLD

Three main groups of species may be distinguished among New World shorebirds: (1) those breeding in the northeastern Canadian High Arctic and wintering in Europe, (2) those breeding across the central North American Arctic and wintering in North, Central, and South America, forming the majority of species, and (3) those breeding in northern Alaska and migrating to wintering areas in Asia and the Pacific. Some major migration pathways for shorebirds in the New World during southward and northward migration are illustrated in Figure 1, and are discussed further below.

A. New World/Europe

It is now well established that a small number of species in the eastern High Arctic migrate to wintering areas in Europe, mostly on the northwestern European seaboard. This pattern appears to have evolved in response to the occurrence of glacial refuges in previous Ice Ages (Prater, 1981). Early taxonomic considerations regarding plumage and size indicated that populations of Red Knot (for scientific names, see the Appendix) breeding on Ellesmere Island could be referred to populations wintering in the Old World (Salomonsen, 1950–51; Godfrey, 1953, 1966) and subsequent band recoveries and research expeditions to Iceland and Ellesmere Island have confirmed these conclusions (Morrison, 1971, 1975, 1976b, 1977a; Morrison *et al.* 1971; Witts and Morrison, 1980; Witts, 1982). In the spring, Iceland is used as a stopover area for many Red Knot en route to northern Canada and Greenland from Europe (Morrison *et al.*, 1971; Morrison 1971, 1977a; Wilson, 1981). The return passage through Iceland is much lighter, many birds presumably flying directly back across the northern Atlantic Ocean to Europe (see Section III.B).

Other species that breed on Ellesmere Island and that follow a similar route include Ruddy Turnstone and Ringed Plover and probably the Sanderling (see Morrison, 1975, 1976b, 1977a; Godfrey, 1953; Parmelee and MacDonald, 1960). Although band recovery data are fewer, the early arrival dates of these species also indicate a European wintering origin (Morrison, unpublished).

Further south in the eastern Arctic, populations of the Purple Sandpiper breeding on Baffin Island may also migrate to Europe via Iceland (see Section III.E.2). It should be noted that some species whose ranges extend into central Ellesmere Island, especially Baird's Sandpiper, winter in the New World, so that there is some overlap of populations from the two wintering regions.

B. New World: North/South America

1. North America

The majority of species breeding across the Canadian Arctic and northern Alaska migrate southwards through North America to wintering

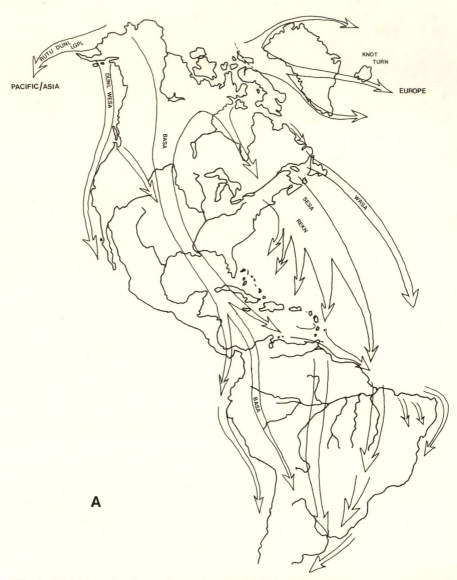

Fig. 1. (A) Major migration pathways for shorebirds in the New World during southward migration. Routes for a few selected species are indicated (see Appendix for abbreviations). (B) Major migration pathways for shorebirds in the New World during northward migration. Routes for a few selected species are indicated.

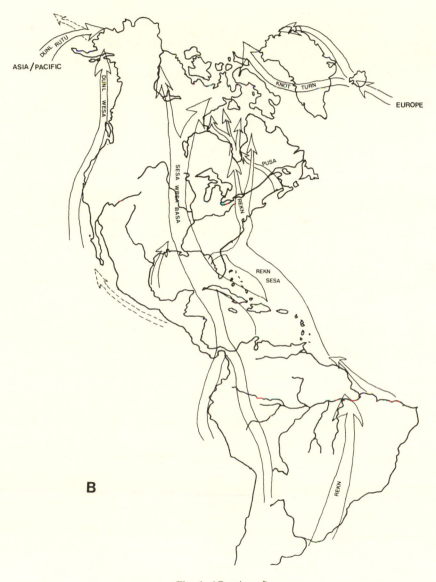

Fig. 1. (*Continued*)

areas in the United States, Central America, and South America. Within this group, however, a number of different major pathways appear to be used. Species such as the White-rumped Sandpiper migrate through the most northeasterly parts of the Atlantic seaboard of Canada and the United States, being abundant in Labrador (Austin, 1932) and the Gulf of St. Lawrence (Morrison, 1976c), but less common in the Maritime Provinces of Canada and relatively uncommon on the New England coastline (Hagar, 1956; Morrison, 1976c). Other species such as the Golden Plover, Whimbrel, and Eskimo Curlew appear to adopt this route. A transoceanic flight is undertaken direct to the north coast of South America or Caribbean islands.

The Semipalmated Sandpiper, on the other hand, though common along much of the Atlantic seaboard, is most numerous in the upper Bay of Fundy. Many individuals appear to launch a direct flight from such areas across the Atlantic Ocean to the north coast of South America. Species such as the Red Knot are found in largest concentrations further south on the Atlantic coast, in Massachusetts and New Jersey, which are again used as refueling points before a direct flight to South America (see below).

Populations of some species breeding further west in the Arctic (e.g., Semipalmated Sandpiper) migrate through central areas of the United States, stopping at concentration points such as Cheyenne Bottoms, Kansas. These populations also appear capable of making long flights, and may migrate to western parts of the wintering range on the north coast of South America, e.g., Venezuela, or cross Central America to winter on the Pacific coast of South America.

Other species, such as Baird's Sandpiper, are more alpine, following the central High Plains southwards and migrating through South America along the Andes chain (Jehl, 1979). Baird's Sandpiper and the White-rumped Sandpiper provide an interesting contrast, in that they are generally superficially similar in size and morphology, overlap in much of their breeding range in the Arctic and wintering grounds in southern South America, but take widely differing routes southwards.

To the extreme west, populations of species such as Western Sandpipers and Dunlins breeding in western Alaska appear to follow the Pacific coast southwards (Senner, 1979; Senner et al., 1981; Gill, 1979; Gill and Jorgensen, 1979).

In the spring, northward migration routes are often different from those followed in the autumn. A major constraint is that many of the areas on the northeastern seaboard, especially in the Canadian Maritime Provinces, which are of enormous importance in the autumn, have only recently thawed out from winter ice, and food resources are at a seasonal low. Many species or populations (e.g., Semipalmated Sandpiper, White-rumped Sandpiper) take a more central route northwards than southwards, presumably in response to food and climatic factors (Harrington and Morrison, 1979). However, parts of the Atlantic seaboard are of extreme importance in the spring; Delaware Bay has recently been identified as a major staging area for enormous numbers of shorebirds [especially Red Knot (Wander and Dunne, 1981; Dunne *et al.*, 1982; Harrington, 1982a)]. The birds appear to make use of the spring peak of Horseshoe crab (*Limulus polyphemus*) eggs, laid during the peak lunar cycle late in May (Harrington, 1982a).

2. South America

Migration routes within South America have remained relatively undocumented. Migrants making landfall on the north coast of the continent after a transatlantic flight or flight through the Gulf of Mexico/Caribbean, appear to move eastwards toward their ultimate wintering grounds (e.g., Semipalmated Sandpiper) or before heading further south (e.g., Red Knot), as indicated by passage dates of species through Venezuela and Surinam (McNeil, 1970; Spaans, 1978) and observations of movements in Surinam (Spaans, 1978; Harrington and Leddy, 1982b; Morrison and Spaans, unpublished observations). Recent aerial surveys have shown Surinam and French Guiana to be the principal wintering grounds for a variety of species on the north coast (Morrison, 1983a; Morrison and Ross, 1983a,b; see species accounts below).

For species wintering further south, many appear to head directly across Amazonia rather than following the coastline of Brazil. P. Antas (personal communication) has suggested that there are a number of distinct flyways used by birds during these flights. To the west, birds turning southwards around the mouth of the Orinoco River in Venezuela would pick up river systems oriented approximately north/southwards in west/central Amazonia (e.g., Caroni River, Rio Branco, Rio Negro, Rio Jiparana, Rio Guapore), bringing them to interior wetland areas in southwestern Brazil (e.g., the Pantanal), Bolivia, and Paraguay. Birds turning south around the mouth of the Amazon River in Brazil would follow other river systems running approximately north/south (e.g., Xingu, Araguaia, Tocantins) and leading either to the coast of southern Brazil, or to major river systems such as the Parana, Paraguay, and Uruguay, which would bring them to Buenos Aires Province in Argentina. Birds moving further east on the north coast of Brazil past the mouth of the Amazon River may turn southeastwards across land from areas such as Sao Luis or Fortaleza to the Salvador (Bahia) coastline, and a few may follow the coast all the way around. Recent aerial surveys, however, have shown relatively few shorebirds using estuaries on the latter part of the coast during the northern winter (unpublished results). There is some evidence that autumn and spring routes may differ in southern Brazil, with autumn routes being further inland. This is suggested by the observations of W. Belton that Red Knot are found in large numbers during northward but not southward migration along the Rio Grande do Sul coast in southern Brazil, and by the recovery of a banded knot inland from this coast during southward migration (Harrington, 1982a; Lara-Resende and Leal, 1982). It appears likely that these seasonal patterns are driven by differing but stable patterns of habitat availability (P. Antas, personal communication). For instance, water levels in the extensive Pantanal wetland in southwestern Brazil are low during September, providing much suitable habitat for shorebirds during southward migration. During March/April, however, water levels are higher and suitable habitats are not available for northward migrants. Much further work remains to be done to establish migration routes in South America and to identify ecological factors driving the system.

Species such as Baird's Sandpiper (Jehl, 1979) and Wilson's Phalarope migrate further west along the Andes, making use of the chains of high lakes (e.g., in Peru) and occurring in large numbers in areas such as Laguna Mar Chicita in Argentina (Nores and Yzurieta, 1980).

Although extensive winter surveys have not yet been carried out on the Pacific coast of South America, it would appear that some species such as the Sanderling and Whimbrel are more common on the west side of the continent than on the Atlantic side. Recent evidence (see below) indicates that highest wintering concentrations of Sanderlings are linked to coastlines with oceanic upwellings, e.g., coastal Peru and Chile (Myers et al., 1983). Whereas some species may migrate from the breeding grounds southwards along the Pacific coast (e.g., some Western Sandpipers, some Sanderling, and species generally restricted to the west coast, e.g., Surfbird, Wandering Tattler), others probably cross Central America after migration through the central United States (e.g., Semipalmated Sandpiper). With coastal deserts or unsuitable habitats along long stretches of the Pacific coast, certain coastal wetlands probably assume considerable importance to shorebirds on migration. Areas such as the Pucchun and Mejia lagoons, in southern coastal Peru, though moderate in size, are separated from the nearest suitable stopover areas by some 300 miles to the north and 800 miles to the south (Hughes, 1979).

Senner and Howe (Vol. 5 of this series) list estuaries holding large numbers (20,000 or more) of shorebirds on the west coast of the United States (see also Jurek, 1974), and it appears unlikely there are major stopover areas between the Fraser River Delta and the Copper River Delta in Alaska (Senner *et al.*, 1981), making estuaries in Washington and southern British Columbia of considerable significance (see also Widrig, 1979). It has been estimated that some 20 million shorebirds pass through the Copper River Delta in spring (Isleib, 1979), making this an area of exceptional importance for birds en route to breeding grounds elsewhere in Alaska and, probably, Siberia.

C. New World/Pacific

A number of species either moving through or breeding in Alaska and the western Arctic appear to winter in the Pacific area or along the Asian coastline. Species such as the Golden Plover, Ruddy Turnstone, Whimbrel, Bristle-thighed Curlew, and Wandering Tattler winter on Pacific islands, and some nonbreeding birds remain during the summer (Johnson, 1979). Extensive banding of turnstones passing southwards through the Pribiloff Islands showed dispersal to winter areas throughout the Pacific and a possible return passage along the coastline of Asia (Thompson, 1973). Dunlin breeding along the north coast of Alaska are known to migrate to wintering areas on the coast of Asia (MacLean and Holmes, 1971; Norton, 1971; Browning, 1977). The wintering origin of the large number of Red Knot passing through the Copper River Delta in southern Alaska in spring is unknown (Isleib, 1979). The species does pass through estuaries such as Leadbetter Point, Willapa Bay, Washington, in moderate to large numbers (Widrig, 1979, 1980), though whether the wintering area is within the New World or may perhaps be in Australasia remains to be determined.

III. MIGRATION OF INDIVIDUAL SPECIES

It is when one comes to consider the details of the migration patterns of individual species that the biological and ecological factors driving the system may be revealed. This section will consider what is known of the migration strategies of a number of different species in terms of the birds' distributions, movements, morphometrics, and biology. The Semipalmated Sandpiper will be considered in some detail, while other species are given a much briefer treatment to provide a contrast in migration patterns and strategies.

A. Semipalmated Sandpiper

The Semipalmated Sandpiper is the most numerous small shorebird in eastern North America (Harrington and Morrison, 1979). The species has been relatively intensively studied recently, with the result that details of its migration are becoming fairly well known, various aspects of its annual cycle can be related, and questions posed concerning differences between populations from different parts of the breeding range.

1. Breeding Range and Morphometrics

The breeding grounds of the Semipalmated Sandpiper stretch from Alaska eastwards across the Canadian Arctic to northern Quebec, central Baffin Island, and northern Labrador (AOU, 1957; Godfrey, 1966) and it has been suggested that there may be three relatively distinct populations (Manning *et al.*, 1956; Palmer, 1967; Harrington and Morrison, 1979), which may verge on subspecific status (H. Ouellet, personal communication). The largest birds occur in the eastern Arctic, with both sexes becoming progressively smaller as one moves westwards toward Alaska. There is an apparent step in the cline in the area of Southampton Island (see Manning *et al.*, 1956; Harrington and Morrison, 1979), and possibly a further one toward the extreme west of the range in Alaska (H. Ouellet, personal communication), though this was not evident in data presented by Harrington and Morrison (1979), possibly because the Alaskan specimens were principally from Barrow on the north coast. The species is sexually dimorphic, with females averaging 1.5–2 mm longer in bill length than males; eastern birds are up to 2–2.5 mm longer in bill length than those from Alaska, so that the "large" eastern males overlap considerably in bill size with the "small" females from the western Arctic. These trends are illustrated in Fig. 2, which shows plots of mean bill/mean wing versus mean bill length for birds from different parts of the breeding range (Harrington and Morrison, 1979). The reasons for this variation in bill size across the breeding range are not known, though it is interesting to note that mean bill lengths for males and females from different parts of the Arctic are correlated with mean June temperatures from the region (Fig. 3), a possible example of Allen's rule, which states that exposed or protuberant parts of the body of warm-blooded animals tend to be reduced in cooler climates. Hale (1980) has assembled evidence suggesting that selection for size in many species of shorebirds tends to occur on the breeding rather than wintering range. Spaans and Swennen (1982), however, reported that several shorebird species, including the Semipalmated Sandpiper, which uses very soft mudflats while wintering in Surinam, had

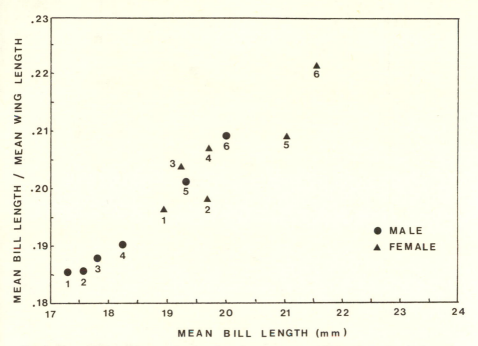

Fig. 2. Mean bill length/mean wing length ratio versus mean bill length for Semipalmated Sandpipers from different parts of the breeding range. Circles = males, triangles = females; numbers indicate breeding area: Alaskan (area 1), central (2–4), and eastern (5, 6) parts of the breeding range, respectively. Alaskan/central and eastern birds form two groups, with eastern males and Alaskan/central females overlapping.

larger toes and larger palmations than those species that did not frequent this habitat, and suggested that these morphological features had evolved to facilitate the exploitation of the soft, otherwise unaccessible mudflats. Morrison (unpublished results) found a similar result with regard to length of tarsus. The degree of morphometric variation across the breeding range is of great use in helping identify populations of Semipalmated Sandpipers at migration stopovers and on the wintering grounds, especially where sexed samples are available (see below).

2. Distribution during Migration and on Wintering Areas

Migration. The relative abundance of Semipalmated Sandpipers in North and South America during autumn migration and on the wintering grounds is illustrated in Fig. 4. Features that are immediately apparent are the large concentrations of birds on the eastern seaboard of North

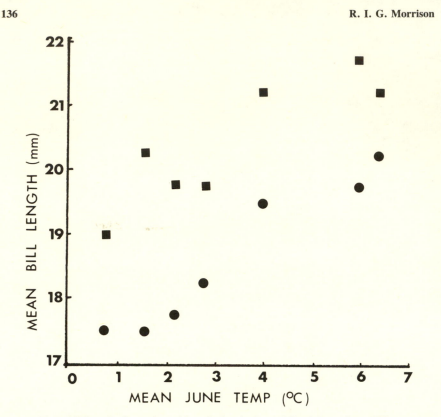

Fig. 3. Mean bill lengths for male and female Semipalmated Sandpipers from different parts of the breeding range, plotted against mean daily June temperature (T). Males = circles: regression line is bill = $0.51T + 16.80$, $r = 0.96$, $p < 0.001$. Females = squares: regression line is bill = $0.40T + 18.96$, $p < 0.002$.

America during migration, the large concentrations along the north coast of South America during the winter, and more scattered large concentrations inland during migration, either at coastal sites (e.g., along James Bay) or in the interior (e.g., Cheyenne Bottoms, Kans.). In recent years, a series of internationally coordinated survey schemes has provided much new information concerning the details of these distribution patterns and has enabled the identification of many areas of international importance for the species. Such schemes include the International Shorebird Survey/Maritimes Shorebird Survey schemes, involving a network of volunteer observers in the eastern United States (Leddy and Harrington, 1978) and Canada (Morrison, 1976c,d, 1978b, 1983b; Morrison and Campbell, 1983a,b; Morrison and Gratto, 1979b), ground surveys by the Canadian Wildlife Service along the St. Lawrence River (Broussard, 1981),

Fig. 4. Relative abundance of Semipalmated Sandpipers in North and South America during autumn migration and on the wintering grounds. Each point indicates the maximum abundance reported in a given area or a given location. Adapted from Cadieux (1970), Burton (1974), Lank (1983), and Morrison (1983a). Increasing sizes of circles represent: open circles—rare, occasional; closed circles—regular or fairly common, common, very common or numerous, abundant. Breeding (hatched) and wintering (cross-hatched) ranges are adapted from AOU (1957), Godfrey (1966), and Phillips (1975).

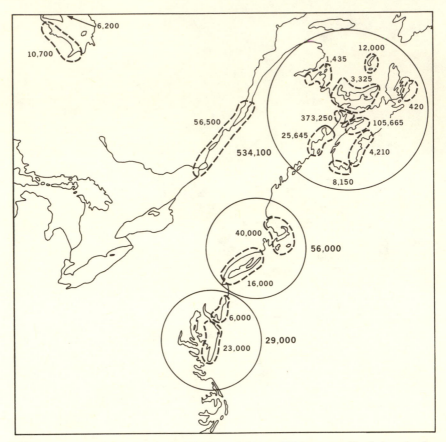

Fig. 5. Counts of Semipalmated Sandpipers (*Calidris pusilla*) in eastern North America from Canadian Wildlife Service aerial and ground surveys and from the International Shorebird Survey/Maritimes Shorebird Survey schemes. [From Morrison (1983a).]

aerial surveys by the Canadian Wildlife Service in James Bay and Hudson Bay (see Morrison, 1983a), and the extensive program of aerial surveys of wintering areas in South America undertaken by the Canadian Wildlife Service and counterpart agencies under its Latin American Program (see Morrison and Harrington, 1979; Morrison, 1983a; Morrison and Ross, 1983a,b).

On the east coast of Canada and the United States, the upper Bay of Fundy is clearly the focal point of the migration of Semipalmated Sandpipers (Fig. 5). Concentrations of several hundred thousands occur regularly at Mary's Point, New Brunswick, and numbers there are typically up to ten times greater than those in areas further south along the coasts of New England and the mid-Atlantic states—or along the St. Lawrence estuary. Survey results have demonstrated clearly that the birds tend to concentrate heavily at a relatively limited number of sites or in a relatively small percentage of the available habitat. For instance, data from the Canadian Maritime Provinces, derived from maximum counts of Semipalmated Sandpipers at 61 sites covered by Maritimes Shorebird Survey participants during the 5-year period 1974–1978 (Fig. 5), show a total of over half a million birds: of these, the northwest arm of the Bay of Fundy held over 373,000, or some 70%, and the southeast arm held over 105,000, or a further 20%, so that the upper Bay of Fundy accounted for about 90% of the total for the Maritime Provinces. Within these areas, the bulk of the birds occurred at a restricted number of sites: although the species was practically ubiquitous, occurring at 60 of the 61 sites, the top five sites nevertheless held 81% of the birds (Morrison, 1983a).

The same concentration effect is observed in other areas during migration, as well as on the wintering grounds (see below). For instance, aerial surveys covering the southern half of James Bay in July 1977 revealed that 71% of the peeps were observed in the most heavily used part of the coast comprising some 10% of the coastline covered (Table I). The top three sections accounted for 88% of the birds in 18% of the coast. Similarly, along the St. Lawrence River 45% of the Semipalmated Sandpipers occurred around Quebec City in 11.5% of the available habitat covered between Cornwall, Ontario, and La Pocatière, Quebec (Table II) (Broussard, 1981; Morrison, 1983a). Leddy and Harrington (1978) reported similar results from counts at 20 International Shorebird Survey sites on the Atlantic coast of the United States in 1976. Although Semipalmated Sandpipers were again widespread, occurring at 19 of the 20 sites, the top site held 46% of the total and the top three sites 85% of the birds.

The most important wintering grounds for the species appear to be on the north coast of South America (see Phillips, 1975), with particularly large concentrations occurring in Surinam and, to a slightly lesser extent, in French Guiana (Morrison, 1983a). Aerial surveys in January–February 1982, covering the north coast from Lake Maracaibo in Venezuela, eastwards through Trinidad, Guyana, Surinam, French Guiana, and Brazil to Sao Luis east of the mouth of the Amazon River, resulted in a count of over 1.9 million peeps, presumed mostly to be Semipalmated Sandpipers. Surinam held a full 70% of the total, with a further 20% in French Guiana, so that these two small countries alone held 90% of the peeps counted during the surveys (Fig. 6). The importance of these areas is related to the occurrence of very extensive mudbanks, especially along the coast of Surinam: the mudbanks are formed from sediments deposited after

Table I. Distribution of Small Sandpipers (Mostly Semipalmated Sandpipers *Calidris pusilla*) during High-Tide Aerial Surveys in Southern James Bay, 26 July 1977[a]

Survey zone	Distance surveyed (km)	Small sandpipers
A–1	16.25	—
B–1	68.0	410
A–2	11.0	15
C–1	26.25	25
D–1	43.5 (9.7%)	12,915 (70.8%)
E–1	16.25 (3.6%)	1,830 (10.1%)
F–1	28.75	190
E–2	11.0	270
F–2	24.25	50
C–2	33.75	290
B–2	29.75	—
G–1	4.5	130
D–2	8.5	380
G–2	19.0 (4.3%)	1,295 (7.1%)
H–1	9.25	95
G–3	41.5	310
C–3	16.75	—
I–1	31.5	30
J–1	7.25	—
Total	447.0	18,235

[a] Survey zones extend from the Ontario/Québec border (A–1), along the Ontario coastline northwards to Ekwan Point.

Table II. Distribution of Semipalmated Sandpipers (*Calidris pusilla*) along the St. Lawrence River[a]

Area surveyed	Area of habitat (ha)	Semipalmated Sandpipers
Montreal	2461	469
Montreal–Lac St.-Pierre	1090	792
Lac St.-Pierre	423	85
Lac St.-Pierre–Québec	2220 (16.2%)	6,378 (10.9%)
Québec	1569 (11.5%)	26,440 (45.2%)
Cap St.-Ignace–La Pocatière	2799	5,683
Cap Tourment–La Malbaie	1007	66
Iles de l'estuaire	2118 (15.5%)	18,609 (31.8%)

[a] From Broussard (1981).

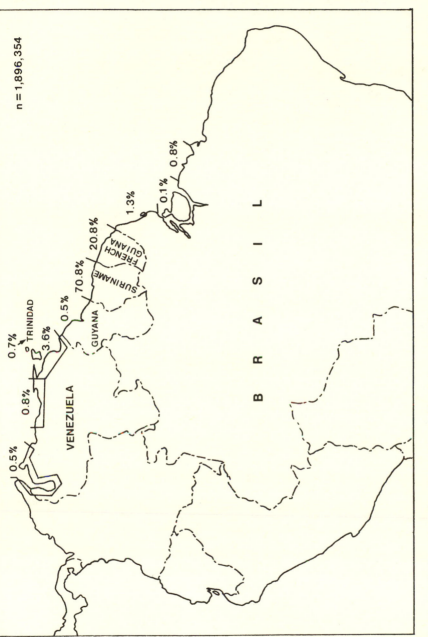

Fig. 6. Distribution of small sandpipers (mostly Semipalmated Sandpipers *Calidris pusilla*) on the north coast of South America during aerial surveys in January/February 1982. [From Morrison (1983a).]

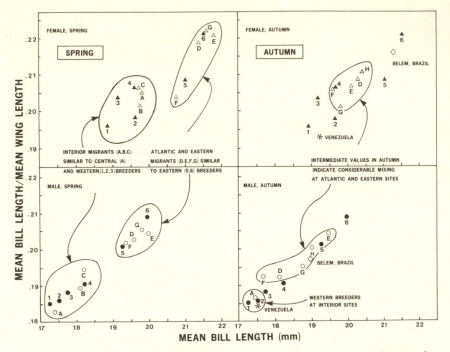

Fig. 7. Analysis of Semipalmated Sandpiper (*Calidris pusilla*) migration through comparison of measurements of birds from breeding areas (1 = Alaska; 2, 3, 4 = central Arctic; 5, 6 = eastern Arctic) with those from migration areas (A, B, C = interior/prairie Canada and U.S. sites; D, E, H = eastern seaboard U. S., Canada; F, G = St. Lawrence River estuary). Note that in spring, Atlantic seaboard and eastern migrants in North America (D, E, F, G) are similar to eastern Arctic populations, whereas interior/prairie migrants (A, B, C) apparently derive from central and western breeders (1, 2, 3, 4). In the autumn, intermediate values of bill/wing ratios indicate considerable mixing of populations at Atlantic and eastern sites, with western breeders migrating south through the interior/prairies. Samples from Belem, Brazil, and from Venezuela indicate wintering/migrating birds in those areas are from eastern and western breeding populations, respectively. [From Harrington and Morrison (1979), and Morrison (1983a).]

being discharged from the Amazon River and being carried northwards and westwards by the Guiana current. Further surveys in Brazil in January 1983 indicated very few birds migrated to estuarine areas on the eastern Atlantic coast of Brazil (e.g., Salvador, Rio de Janeiro), few being found even in areas of suitable-looking habitat (unpublished results), though the inclusion of these areas in the wintering range is substantiated by a variety of specimens in Brazilian museums (Morrison, unpublished data).

Less is known about the distribution of the species in Central America or on the Pacific coast of South America, though it does occur there in moderate abundance (see below), especially along the Pacific coast of Central America (Phillips, 1975; Smith and Stiles, 1979; Schneider and Mallory, 1982).

3. Migration Patterns

It has proved possible to trace the movements of Semipalmated Sand-pipers from different parts of the breeding grounds, using a combination of morphometric analysis of birds at migration areas and banding results (Harrington and Morrison, 1979; Morrison, 1983a). These techniques are also being used to identify the origins of populations using particular parts of the wintering range (Morrison, 1983a; and see below).

Morphometric Analysis. By comparing specimens of Semipalmated Sandpipers from migration areas with reference samples from different parts of the breeding range (see above), Harrington and Morrison (1979) were able to trace the northward and southward migrations of different populations using a combination of bill lengths, wing/bill ratios, and coefficients of variation of these measurements. The results are summarized in Figs. 7 and 8. During the northward spring migration, Atlantic coast migrants were clearly from eastern Arctic breeding areas, whereas migrants from interior sites involved birds from western and central areas (Fig. 7). In the autumn, western breeders returned southwards through the prairies, but measurements intermediate to eastern and western breeding populations and increased coefficients of variation at Atlantic and eastern sites, indicated that considerable mixing of eastern and central Arctic breeders takes place. Thus, many central breeders and possibly some western birds take a southeastward route through the eastern seaboard: this results in something of an elliptical pattern for central Arctic breeders, with a central spring route and an autumn passage through the east coast.

This type of analysis may clearly be extended to the wintering areas to identify the breeding origins of the birds. Such work is currently in progress, and some preliminary results are included in Figs. 7 and 8. Long-

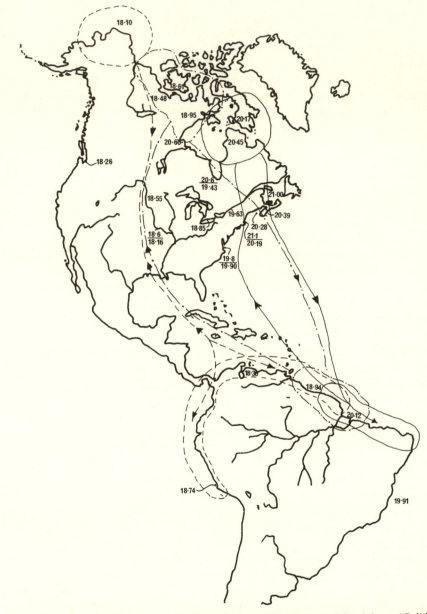

Fig. 8. Summary of migration patterns of populations of Semipalmated Sandpipers (*Calidris pusilla*). Broad outlines of migration patterns of birds from western, central, and eastern Arctic breeding areas to wintering areas in South America are based on analysis of measurements of birds and marking studies. Measurements of bill lengths are from breeding, migration, and wintering areas: values are the average of means for males and females from

billed birds wintering east of the mouth of the Amazon River in Brazil would clearly appear to be from eastern breeding populations, whereas those passing through Venezuela are predominantly from western populations. Recently obtained measurements from birds wintering in Peru indicate that birds on the Pacific coast are also from western breeding populations (Morrison and Myers, unpublished data). Intermediate measurements of samples from Surinam presumably indicate a mixture of breeding populations, as both eastern and western groups are known to reach the country (see below) and the very high percentage of birds on the north coast using this area (see above) suggests that several populations would overlap in their wintering range in Surinam (Fig. 8). Further information on this topic is needed. Measurements from various areas and a summary of the proposed migration systems for the various breeding populations are illustrated in Fig. 8.

Banding Results. Sightings and recoveries of Semipalmated Sandpipers marked during various major studies in recent years have confirmed the pattern of movements described above.

The pattern of dispersal of migrants passing through central areas is illustrated in Fig. 9, showing results obtained by Lank (1979, 1983) from marking operations at Sibley Lake, North Dakota, and Kent Island, New Brunswick. The western Arctic origin of many of the birds passing southwards through North Dakota was indicated not only by the bill length (mean of mixed sex samples of live birds 18.5 mm), but also by the capture of a bird hatched near Prudhoe Bay, Alaska, earlier the same summer. Onward dispersal was principally southeastwards, with sightings of birds in Minnesota, Tennesee, Wisconsin, North Dakota, Louisiana, Kansas, and Florida. Three birds were seen in Latin America, one in Venezuela, one on Aruba, and one in Panama, suggesting a movement to the western part of the wintering range.

Further evidence for the western origin of central migrants came from banding operations at Cheyenne Bottoms, Kansas, by E. F. Martinez. One bird banded in spring in Kansas was recovered on the Alaskan North Slope, and two birds banded at Barrow, Alaska, were subsequently trapped at Cheyenne Bottoms (Martinez, 1974, 1979). Alaskan birds have also been recovered in Florida, Cuba, and Surinam (Martinez, 1974; Hanson and Eberhardt, 1978; Lank, 1979; Spaans, 1979).

---

the respective areas. For migration areas, first/second values represent spring/autumn measurements; single values are from autumn migration. Data from Harrington and Morrison (1979), Lank (1983), Kaiser (personal communication), and Morrison (unpublished). Values shown for breeding areas, Venezuela, and Brazil are from museum specimens; other measurements are from unsexed samples of live birds caught for banding.

Fig. 9. Dispersal of Semipalmated Sandpipers from Sibley Lake, North Dakota (closed symbols), and Kent Island, New Brunswick (open symbols). Star, banding location; circle, fall record; square, spring record; triangle, marking location of birds seen or captured at Sibley Lake or Kent Island. Lines connect same-season movements. [From Lank (1979, 1983).]

Dispersal of birds marked by Lank from Kent Island, New Brunswick, on the other hand, involved sightings and recoveries along the Atlantic coast and in Guyana and Surinam (Fig. 9; Lank, 1979, 1983). Mean bill lengths (20.3 mm) and the capture and sightings of birds marked in James Bay indicated the birds derived from eastern and central Arctic populations. A similar pattern of sightings and recoveries was obtained from banding by McNeil and Burton (1973, 1977) on the Magdalen islands in the Gulf of St. Lawrence (see Table III).

Both McNeil and Burton (1973) and Lank (1983) concluded that most Semipalmated Sandpipers reached South America by a direct flight from the eastern seaboard of Canada and the United States, on the basis of the relatively small numbers of sightings received from east coast locations when compared to the much less densely populated areas of the Caribbean and of South America. Radar studies have shown shorebird tracks departing southeastwards on a transoceanic route from the Maritime Provinces of Canada and the New England states (Drury and Keith, 1962; Nisbet, 1963; Richardson, 1972, 1979; Williams *et al.*, 1977), usually under conditions of following winds, rising barometric pressure, and clearing conditions (Richardson, 1979). Stoddard *et al.* (1983) have recently reported computer simulations of the proposed routes based on radar data indicating that flights typically would take up to 50 to 60 hr. Cadieux (1970), McNeil and Cadieux (1972), and Lank (1983) present data indicating that Semipalmated Sandpipers departing from east coast locations can accumulate adequate fat reserves to undertake such flights.

Between 1974 and 1982, the Canadian Wildlife Service captured some 40,000 Semipalmated Sandpipers on the southwest coast of James Bay. Measurements of live birds and specimens indicated a breeding origin in the central Arctic (see below), and two birds marked in James Bay were found on the breeding grounds near Churchill, Manitoba (Morrison and Gratto, 1979a). Dispersal of marked birds occurs in a broad fan to the eastern seaboard of North America (Fig. 10; Morrison, 1978a; Morrison and Gratto, 1979a), with many birds passing through important stopover areas in the upper Bay of Fundy (New Brunswick/Nova Scotia), Maine, Massachusetts, New York, New Jersey, and Pennsylvania: sightings and band recoveries on the wintering grounds have come from Guyana, Surinam, French Guiana, and Brazil (see Table III). The patterns of band recoveries and of sightings of color-marked birds captured at North Point, James Bay, are compared in Table III. Band recoveries and sightings were distributed very differently among the four areas considered ($\chi^2 = 18.72$, 3 *df*, $p < 0.001$). By far the most sightings came from the eastern seaboard of Canada and the United States, reflecting the presence of an active bird-watching public and lack of hunting of shorebirds in North America. In contrast, the majority of band recoveries came from the Caribbean and South America, where there are very few bird-watchers and where shorebirds are hunted. This is dramatically illustrated by the band recoveries: some 87% of the 109 recoveries came from the Caribbean and South America, with 60 out of these being reported from Guyana alone.

Table III. Summary of Band Recoveries and Sightings of Color-Marked Semipalmated Sandpipers Captured in James Bay and in the Magdalen Islands

| | Magdalen Islands[a] | | | | | James Bay[b] | | | |
| | Band recoveries | | | Sightings (total) | 1970–71 total | Band recoveries to 1980 | | | Sightings (1978) (total) |
	Ad	Imm	Total			Ad	Imm	Total	
Canada	3	2	5	17	22	1	3[c]	4	192[d]
USA	—	—	—	6	6	8	2	10	227
Caribbean	3	6	9	1	10	12	16	28	3
South America	1	1	2	—	2	(55)	(12)	(67)	13
Guyana						51	9	60	
Surinam						3	1	4	
French Guiana						—	2	2	
Brazil						1	—	1	
Total	7	9	16	24	40	76	33	109	435

[a] Data from McNeil and Burton (1973).
[b] Data from Morrison and Gratto (1979a and unpublished).
[c] Includes 2 from Ontario/Québec.
[d] Includes 29 from Ontario/Québec.

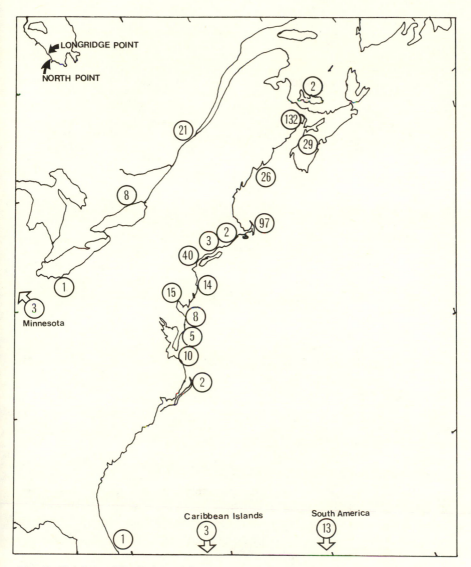

Fig. 10. Sightings of Semipalmated Sandpipers color-marked in James Bay during July and August 1978. One "bird-day" was counted for each day a color-marked bird was reported in a given locality. [From Morrison and Gratto (1979a).]

Table IV. Months of Recoveries of Adult and Juvenile Semipalmated Sandpipers Banded at North Point, James Bay, in the Caribbean and South America (Reports Received to 24 October, 1980)

	Guyana		Surinam		French Guiana		Brazil		Caribbean	
	Ad	Juv	Ad	Juv	Ad	Juv	Ad	Juv	Ad	Juv
July	1	1	—	—	—	—	—	—	—	—
August	8	—	—	—	—	—	—	—	1	—
September	33	2	1	1	—	—	—	—	11	10
October	3	3	—	—	—	—	1	—	—	6
November	2	—	—	—	—	1	—	—	—	—
December	—	—	—	—	—	1	—	—	—	—
January	—	1	—	—	—	—	—	—	—	—
February	1	1	—	—	—	—	—	—	—	—
March	—	—	2	—	—	—	—	—	—	—
April	1	—	—	—	—	—	—	—	—	—
May	2	1	—	—	—	—	—	—	—	—
June	—	—	—	—	—	—	—	—	—	—
Total	51	9	3	1	—	2	1	—	12	16
	60		4		2		1		28	

Ten times the number of birds were recovered in Guyana (60) compared with Surinam and French Guiana together (6), despite the fact that winter aerial surveys indicated that only 0.5% of the wintering population surveyed (total 1.9 million) occurred in Guyana compared with 91.6% of the population in Surinam and French Guiana! As might be anticipated, most of the birds appear to be obtained during migration through Guyana, mostly in the autumn, especially September, and to a lesser extent in the spring (Table IV). There are two major factors affecting these results: first, the distribution of human population on the coast, and second, hunting practices. In Surinam and French Guiana, access to the coast is relatively restricted and few villages are situated right on the coast. In Guyana, on the other hand, large numbers of people live in fishing villages all along the eastern section of the country, and it is known that shorebirds are hunted methodically as a source of food, both for private use and for the market (D. R. Osbourne, personal communication, and others). The very small number of recoveries of adults during the winter period (approx. November–March) presumably reflects the fact that the bulk of the birds winter along uninhabited coastlines further to the east. Eastward movements are suggested by a passage of the species through eastern Venezuela starting in late July (McNeil, 1970), followed by a substantial movement of Semipalmated Sandpipers eastwards through Surinam in late August and early September (Spaans, 1978). Many shorebirds may be observed passing eastwards along the coast of Surinam in September (Spaans, 1978; personal observation). There is a smaller return passage in Surinam in the spring in May (Spaans, 1978). These results are consistent with the pattern of band recoveries described above (see Table IV).

Most sightings and band recoveries of birds marked by Spaans in Surinam occurred in the central United States on northward migration, and in eastern Canada and the United States during the autumn southward migration, reflecting the elliptical migration pattern of migrants from central and western breeding areas. The involvement of western breeders was demonstrated by the recovery of a bird banded on its nest in eastern Alaska in June (Spaans, 1978). Spring sightings in the United States of birds marked in Surinam occurred in the latter part of May and early June, and Spaans suggested that those birds moving northwards through the central United States between mid-March and mid-May were from populations wintering in northwestern South America and Central America (Spaans, 1978).

Pattern of Adult and Juvenile Recoveries/Distribution. Band recoveries of adults and juveniles captured at North Point appeared to be differently distributed between the four regions considered in Table III (χ^2 = 18.72, 3 df, $p < 0.001$). Whereas 72% of adult recoveries came from South America, only 36% of juvenile recoveries came from this region. The other major difference was the greater proportion of juveniles (48%) recovered in the Caribbean compared to adults (16%). The higher incidence of recovery of juveniles on Caribbean islands could be caused by several factors: e.g., different flight range capability, leading to "shortfall" of juveniles on Caribbean islands, with a higher proportion of adults successfully completing a direct flight to South America; different departure points on the United States seaboard, leading to different likely arrival locations (see Stoddard et al., 1983); different navigational capabilities, with more juveniles being deflected westwards by prevailing wind patterns, etc. All but 1 of the 12 band recoveries of adults in the Caribbean were from September, with juvenile recoveries being in September (10) and October (6), reflecting the later passage of immature birds. The length of the hunting season must also be taken into account in assessing how representative the resulting temporal and spatial pattern of band recoveries is in relation to the migration pattern.

Little information is available on the geographical distribution patterns of adult and juvenile Semipalmated Sandpipers in the New World on their wintering grounds. Myers (1981b) found no differences in the age distribution patterns of specimens in museum collections for several species of shorebirds, with the exception of first-winter male Sanderling, where a more northerly winter distribution was thought to be related to the advantages of an early arrival on the breeding grounds.

4. Strategies at Stopover Areas

The question of how shorebirds use stopover areas, the purposes for which these areas are used, and what makes such areas attractive has received study in recent years, and some selected aspects of this topic will now be considered.

Factors Affecting Distribution. A number of studies have shown that shorebird abundance on wintering areas is related to the abundance of preferred food items at different locations (e.g., Goss-Custard, 1977a,b; Goss-Custard et al., 1977; Bryant, 1979; see O'Connor, 1981), and one might anticipate a similar phenomenon to occur during migration. Studies in James Bay indicate that Semipalmated Sandpipers do, in fact, respond spatially to abundance of their prey, and that changes in prey availability may affect their distribution during the season. For instance, at North Point, James Bay, the distribution of Semipalmated Sandpipers on a transect running across the marsh perpendicular to the coast was related to the distribution of their preferred prey in marsh habitats, dipteran larvae (and oligochaete worms) (Fig. 11, Table V). The preferred feeding zone in late July for adults was along the edge of the short-grass *Puccinellia phryganodes* marsh, at the junction of the vegetation and tidal flats. When Semipalmated Sandpipers were censused along this habitat over an approximately 15-km stretch of coast, their distribution was again related to the relative abundance of their prey, dipteran larvae (Morrison, 1983a).

Censuses at North Point at a second transect covering both marsh and intertidal areas indicated a habitat shift occurred between late July and mid/late August, from the edge of the marsh to lower intertidal areas on the flats, apparently in response to a dramatic drop in the availability of dipteran larvae/oligochaete worms along the edge of the marsh, and the increased availability of the bivalve *Macoma balthica* and the gastropod *Hydrobia minuta* on the flats (Fig. 12). It is not known to what extent predation may cause reduction of dipteran larvae/oligochaete worms, but studies elsewhere indicate that predation substantially reduces available food supplies during migration (see Lank, 1983; Boates and Smith, 1979; Boates, 1980). Studies on the wintering grounds, where birds are, of course, present much longer, also indicate predation is significant in reducing food supplies (Drinnan, 1957; Brown and O'Connor, 1974; Goss-Custard, 1969, 1977b, 1980; Evans et al., 1979; see O'Connor,

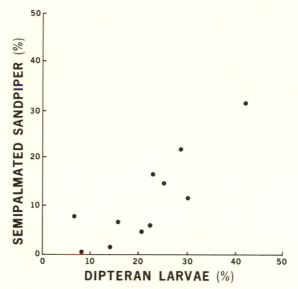

Fig. 11. Relative distribution of Semipalmated Sandpipers (*Calidris pusilla*) and dipteran larvae on a transect across a coastal marsh at North Point on the southwest coast of James Bay. [From Morrison (1983a).]

1981). It may be noted that the habitat shift in James Bay between late July and mid/late August coincides roughly with the passage of adult and juvenile birds, respectively (see below). Whether adult males and females, which differ in their mean bill lengths (by about 2 mm, i.e., approx. 10%), may show consistent differences in habitat use related to their sexual size dimorphism remains to be established, though Harrington (1982b) claimed that there was a difference in habitat use related to bill length among Semipalmated Sandpipers on migration at Plymouth, Massachusetts: this occurred only when birds were feeding by tactile methods, and long-billed

Table V. Food Items Taken by Semipalmated Sandpipers Feeding in Different Habitats at North Point, James Bay, 1978[a]

Habitat	Sample size	Diptera			Oligochaetes			Macoma			Hydrobia		
		n	%	$\bar{x}$	n	%	$\bar{x}$	n	%	$\bar{x}$	n	%	$\bar{x}$
Marsh	30	30	100	30.9	2	7	21	—	—	—	—	—	—
Flats	10	—	—	—	—	—	—	8	80	6.3	8	80	42.1

[a] Sample size = No. of stomachs examined; n = No. of stomachs containing item; % = % of stomachs containing item; $\bar{x}$ = mean No. items per stomach containing item.

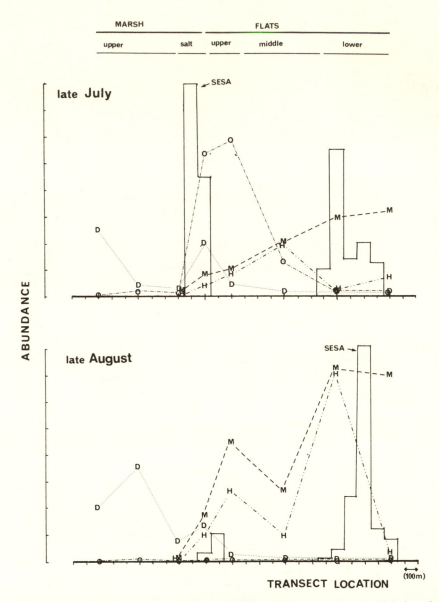

Fig. 12. Seasonal changes in the distribution of Semipalmated Sandpipers and their major prey items on marsh (oligochaetes, dipteran larvae) and intertidal flats (*Macoma balthica*, *Hydrobia minuta*) on a transect at North Point, James Bay, during July and August, 1982.

birds tended to be found feeding in soft muddy substrates whereas shorter-billed birds were found more often in sandier areas of the flats.

Aggressive interactions are often observed among flocks of shore-birds during migration (e.g., Recher and Recher, 1969; Burger *et al.*, 1979; Harrington and Groves, 1977), and it seems likely that some degree of competition may occur for food resources. Studies in James Bay suggest that interspecific competition would tend to be reduced through use of different habitats by different species, and use of different size classes of prey or different prey items where food items overlap, as suggested by Recher (1966). Lank (1983) noted territorial behavior of Semipalmated Sandpipers on migration at Kent Island, New Brunswick, and such behavior has also been seen in James Bay, suggesting competition is operative. One would anticipate territoriality would occur at intermediate food and predator densities, for low food resources might not be worth defending and high densities would attract large numbers of predators, making defense impractical when food was in any case abundant, as was convincingly demonstrated by Myers *et al.* (1979) for Sanderling wintering in California. Harrington and Groves (1977) reported aggression among Semipalmated Sandpipers on migration in Massachusetts and noted that juvenile birds were more frequently aggressive than conspecific adults under the conditions observed.

Migration Sequence. For the majority of shorebirds, including the Semipalmated Sandpiper, adults precede juveniles on southward migration, owing to the earlier departure of the adults from the breeding grounds. Exceptions may occur with some species, for instance where wing molt occurs before migration to the main wintering areas (e.g., Dunlin in North America). The passage of adults and juveniles through James Bay in 1982 is illustrated in Fig. 13: adult migration peaked in the last week of July, whereas juveniles peaked in the latter half of August. The juvenile migration was considered to be up to a week later than in most years owing to a delayed breeding season in 1982 (Morrison *et al.*, 1982). Similar patterns occur in other areas further south, though at a later date, of course (e.g., Page and Bradstreet, 1968).

Recent analyses have yielded further information on the sequence of ages and sexes within this pattern. For instance, the first wave of migrants in James Bay appears to contain a high proportion of 1-year-old (subadult) birds and probably failed breeders and other nonbreeders. One-year-old Semipalmated Sandpipers may be identified by their having undergone a partial primary wing molt during the previous winter (involving molt of some of the outer primaries and inner secondaries only) or heavily worn feathers, the combination of these factors being the only currently known reliable method of determining the age of 1-year-old birds at this time of

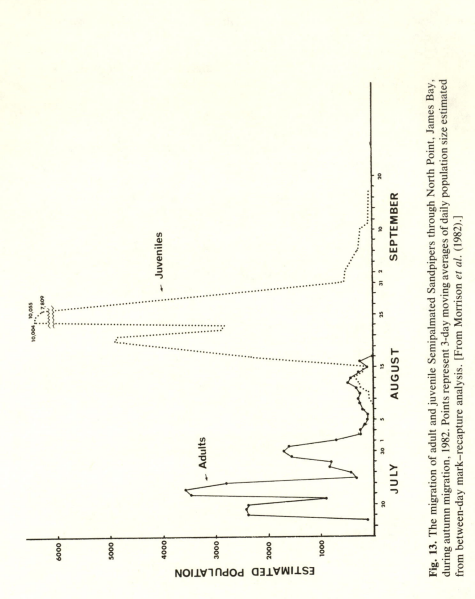

Fig. 13. The migration of adult and juvenile Semipalmated Sandpipers through North Point, James Bay, during autumn migration, 1982. Points represent 3-day moving averages of daily population size estimated from between-day mark–recapture analysis. [From Morrison *et al.* (1982).]

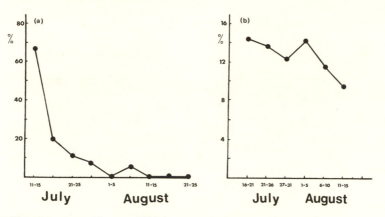

Fig. 14. The percentage of 1-year-old birds (a) in specimens obtained during autumn migration of Semipalmated Sandpipers at North Point, James Bay, 1979–1981, and (b) among live birds caught at North Point in 1982 showing that 1-year-old birds tend to form a high proportion of the earliest migrants.

the year (Gratto and Morrison, 1981; Gratto, 1983; see Prater *et al.*, 1977). Gratto showed that from 3 to 10% of the breeding population at Churchill, Manitoba, from 1980 to 1982 consisted of 1-year-old birds, and that in one year (1980) such birds had a significantly lower breeding success than adults (Gratto *et al.*, 1981, 1983; Gratto, 1983). At North Point, the first wave of migrants contains a higher proportion of 1-year-old birds than later adults, as may be seen both from specimens collected over the years 1979–1981 and from live birds, captured in 1982 (Fig. 14). It is reasonable to suppose that 1-year-old birds, which may have lowered breeding success at least some years and/or difficulty entering the breeding population, would be among the first migrants to head south.

During the main adult migration that follows, females generally migrate somewhat earlier than males. This is consistent with considerations of breeding biology, many studies of which have shown that females depart from the breeding grounds earlier than males (e.g., Pitelka *et al.*, 1974). Myers (1981a) has shown that the type of breeding system and parental care adopted by a species is influenced by the migration distance to the wintering quarters, with those species migrating the farthest tending to show an increasing tendency for early departure of one of the adults, usually the female, and an increasing tendency toward a promiscuous breeding system. The early departure of one adult was thought to be concerned more with decreasing the risk of the subsequent long-distance migration, as suggested for Semipalmated Sandpipers by Ashkenazie and Safriel (1979), than, for instance, with reducing competitive pressure on

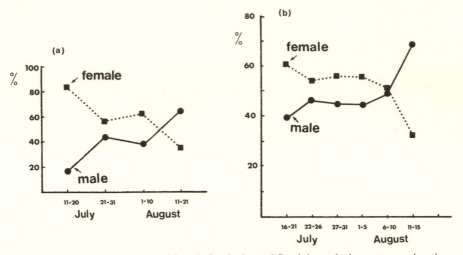

Fig. 15. The passage of male and female Semipalmated Sandpipers during autumn migration at North Point, James Bay, (a) among specimens of known sex obtained during the period 1979–1981, (b) among live birds sexed on the basis of measurements from known-sex specimens during 1982.

food resources between adults and young, as had been previously suggested (Pitelka, 1959; Pitelka *et al.*, 1974).

Data showing the earlier migration of females than males at North Point, James Bay, are illustrated in Fig. 15, both from sexed specimens obtained between 1979 and 1981, and from live birds captured in 1982 and whose sex was determined from mensural characteristics derived from the specimens of known sex. Over 85% of the specimens could be sexed correctly on the basis of bill length, although this may not apply elsewhere (see Harrington and Taylor, 1982). Analysis of the specimen data using percent cumulative frequency methods described by Harding (1949) and Griffiths (1968) indicated that the midpoints of female and male migration were separated by about 4–5 days (female midpoint 21–22 July, male midpoint 25–26 July) and that 80% of each sex migrated through the area over a period of about 23 days (females 15 July–7 August, males 19 July–11 August) (unpublished results).

Interestingly, a similar preponderance of females over males early in the season was observed in 1-year-old birds (Fig. 16), an early ratio of approximately 70:30 falling to around 50:50 by the end of the migration: the overall proportion of females (167, 62%) to males (102, 38%) departed significantly from an even sex ratio ($\chi^2 = 157$, 1 *df*, $p < 0.001$). While the proportion of 1-year-old females (8) to males (3) in the specimens was

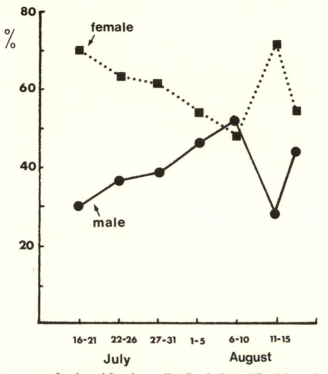

Fig. 16. The passage of male and female yearling Semipalmated Sandpipers during autumn migration at North Point, James Bay, 1982. One-year-old birds were aged on the presence of partial postjuvenile molt (see text) and sexed on the basis of measurements derived from specimens of known age.

not statistically significantly different from being even, it was suggestively similar. It may be noted that Gratto (1983) found a slight overall preponderance of female (13) to male (9) yearlings over 3 years in the breeding population at Churchill, Manitoba, which, although again not significantly different statistically from an even ratio, represents a similar trend. One might speculate that it would be easier for a yearling female to enter the breeding population, as a yearling male would not only have to find an available bird to pair with but also establish a territory.

    Morphometric Aspects of Migration. The mean bill lengths of Semipalmated Sandpipers caught during migration fall during the course of the autumn season, as shown by studies in James Bay, Massachusetts, North Dakota, and New Brunswick (Harrington and Morrison, 1979; Morrison, unpublished results; Lank, 1983). Two possible factors may cause this decline. First, the earlier migration of larger females and later migration

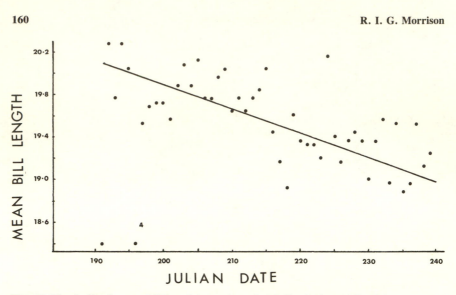

Fig. 17. The decline in mean bill length in catches of adult Semipalmated Sandpipers trapped during autumn migration at North Point, James Bay, in 1979. The regression line is bill = -0.0224(Julian date $+$ 79,000) $+$ 1796.47, $r = 0.16$, $p < 0.001$.

of smaller males would produce a decrease in observed mean bill lengths, but a geographical component caused by the later migration of smaller birds from the western Arctic could contribute to the decline at some sites. To what extent do the two components contribute to the observed pattern at any given site? This question is of relevance in deciding whether birds in a given area come from a restricted area in the Arctic or whether they derive from a wide geographical zone.

The decline in bill lengths in adult Semipalmated Sandpipers trapped in James Bay in 1979 is shown in Fig. 17. The overall mean bill length was 19.56 (S.D. 1.26) mm, the regression line having a slope of -0.0224 mm/day, indicating a decline of approximately 0.90 mm during a 40-day migration period from mid-July to the latter part of August. Mean bill measurements for specimens of adult Semipalmated Sandpipers from North Point obtained during autumn migration in the years 1979–1981 are given in Table VI. Using these measurements, one may calculate that if the sex ratio falls from approximately 80:20 females to males to 30:70 during the season, as suggested by data from the specimens and banding (see Fig. 15), this would lead to a decrease in observed mean bill length of 19.81 mm to 18.96 mm, or 0.85 mm, which is remarkably similar to that seen in the banding samples (Fig. 17).

Bill lengths and bill/wing ratios of autumn specimens from James Bay, both regarded as a useful indication of breeding origin by Harrington and

Table VI. Bill Measurements of Specimens of Adult
Semipalmated Sandpipers from James Bay on Autumn
Migration, 1979–1981

	Males	Females
n	44	67
Mean bill length (mm)	18.45	20.15
S.D.	0.75	1.06
C.V.	4.07	5.27
Bill/wing ratio	0.1877	0.1984

Morrison (1979), indicate that Semipalmated Sandpipers found at North Point originate from the central Arctic, being somewhat larger than reference measurements from the Keewatin district, Northwest Territories, suggesting a breeding origin west of Hudson Bay: this fits well with the sighting of two birds from North Point on the breeding grounds at Churchill, Manitoba.

The bill/wing ratio of banding samples falls during the migration period (Fig. 18), but again by an amount (2.76% of the mean value over 40 days) very similar to that predicted by a change in female:male ratio of 80:20 to 30:70 during the season (2.75%).

Examination of the coefficients of variation of specimens from James Bay (Table VI) indicates that the value for males was within the range expected for a single-sex sample from one breeding zone, with the value for females a little higher, but not significantly different from that of the males. The coefficient of variation did not change significantly during the season for either the specimens (unpublished results) or among mixed-sex banding samples (Fig. 19), the overall mean for specimens and banding samples (6.48%) lying on the upper end of the range expected for mixed-sex samples from one breeding zone (cf. Harrington and Morrison, 1979). The scatter of individual points suggests that catches on some days contained birds from more than one area, though large coefficients of variation tended to be associated with small samples, usually early and late in the season. It would thus appear that the bulk of variability in catches, especially during the main migration period, can be accounted for by the differential migration of sexes from a reasonably discrete part of the breeding grounds. Examination of data from other years will be required to assess the interyear variation in this phenomenon.

Weight Gain. One of the most, if not the most, important purposes that staging areas are used for by shorebirds is as an area in which fat reserves are laid down for a subsequent long flight. Where the staging

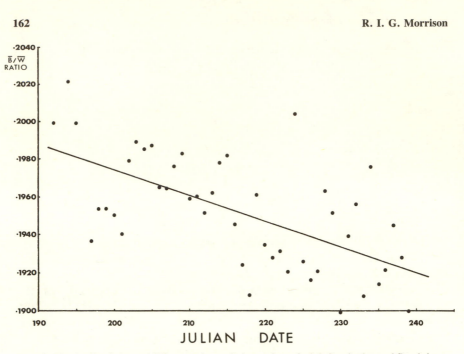

Fig. 18. The decline in mean bill/mean wing ratio in catches of adult Semipalmated Sandpipers trapped during autumn migration at North Point, James Bay, in 1979. The regression line is mean bill/mean wing ratio = -0.0001349(Julian date) + 0.2244, r = 0.62, p < 0.001.

area is situated before a major "ecological barrier" (e.g., the ocean, the boreal forest), then a considerable amount of fat may need to be accumulated to enable the bird to reach its destination.

The pattern of weights shown in catches of adult Semipalmated Sandpipers in James Bay during autumn migration in 1979 is shown in Fig. 20. Mean weights of catches varied little between mid-July and early August, and because many birds arrived during this period if would appear that mean weights remained fairly constant either because weight gains and departures of birds that had arrived previously were balanced by arrivals of new, lighter birds, or that little weight gain took place in the local population. After about 4 August 1979, mean weights of catches rose from approximately 25.5 g to about 32.5 g by 23 August (Fig. 20), with individual birds up to 41 g. The slope of the line indicated mean weights of catches increased by 0.35 g/day, excluding the four points that appeared to lie below the line (slope = 0.29 g/day for all points). Mean weights of juveniles did not increase during the period of trapping, which ended on 27 August 1979.

The pattern of mean weight changes shown by individual adult Semipalmated Sandpipers retrapped in James Bay in 1979 is shown in Fig. 21.

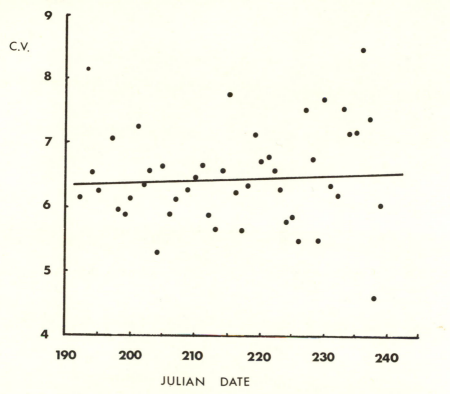

Fig. 19. Coefficient of variation of bill lengths in catches of Semipalmated Sandpipers during autumn migration in James Bay, 1979. The regression line [C.V. = 0.00407(Julian date) + 5.562] is not significant, $r = 0.08$, n.s.

Typically, birds lost weight for several days, the mean value not returning to the initial level until about 7 days after first capture: the regression line for retrap periods of 7 days or more indicated adults gained weight at a rate of 0.36 g/day, a value similar to that derived above from mean weights of catches after 4 August. Juveniles showed a similar pattern of weight loss after capture, returning to initial values after about 7 days and gaining weight thereafter at a rate of 0.30 g/day.

A similar pattern was observed at Long Point, Ontario, on Lake Erie (Page and Salvadori, 1969; Page, 1970; Page and Middleton, 1972), where adults and juveniles both lost weight after capture, but only for a period of about 2 days, weights increasing above initial levels after 3 days. The average weight gains for adults and juveniles were 0.76 and 0.50 g/day, respectively, and if allowance was made for the period of initial weight loss, corrected values of 1.23 and 0.76 g/day were obtained, respectively.

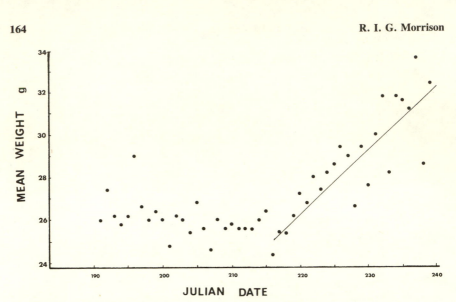

Fig. 20. Mean weights of catches of adult Semipalmated Sandpipers trapped during autumn migration at North Point, James Bay, in 1979. Regression line for points after 4 August (Julian date 215) is weight = 0.2902(Julian date) − 37.34, $r = 0.84$, $p < 0.001$.

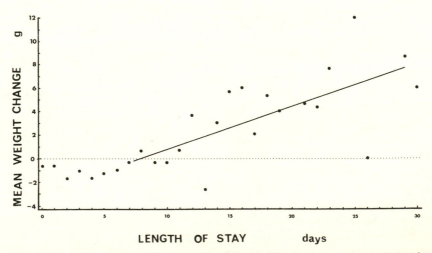

Fig. 21. Mean weight changes of individual Semipalmated Sandpipers retrapped after various intervals during autumn migration at North Point, James Bay, in 1979. The regression line for points after 7 days is weight change = 0.364(stay length) − 2.83, $r = 0.50$, $p < 0.05$.

Lank (1983) reported that Hicklin and Sherman estimated that gains of 1.5–1.8 g/day were normal for Semipalmated Sandpipers in the upper Bay of Fundy, with a maximum value of 2.5 g/day having been observed. Lank (1983) reported that juveniles were capable of fattening at a rate of 1.3 g/day at Sibley Lake, North Dakota, and juveniles and adults at a rate of up to 2.5 g/day, during restricted periods in early September and late August at the two locations, respectively; deriving these figures from mean weights of catches over periods of only a few days may not necessarily represent the performances by individual birds.

It would thus appear that rate of gain of weight may vary widely, both with location as well as seasonal date, and as preliminary analysis of James Bay data indicates, between years.

Less is known about the relative performance of adults and juveniles or of the two sexes. Some studies have indicated that adults are able to gain weight more rapidly than juveniles (e.g., Page and Salvadori, 1969) and this occurred in James Bay in some years but not others (see Gratto, 1983). Preliminary analysis of retrap data in James Bay suggested that adult females gained weight more rapidly than males, using initial recapture records only and excluding subsequent captures of individuals (Gratto, 1983), though inclusion of multiple retrap records suggested little difference occurred (Morrison, unpublished). Female Ruff gain weight more rapidly than males during autumn migration in Germany (M. Hargengerd, personal communication), and Puttick (1981) reported that female Curlew Sandpipers foraged both faster and more successfully than males in South Africa, while Smith and Evans (1973) reported a similar finding for female Bar-tailed Godwits.

The amount of fat accumulated by the bird at a stopover area appears to be related to the distance of the subsequent flight. McNeil and Cadieux (1972) and McNeil and Burton (1973) have shown that a number of species of shorebirds have higher fat reserves in the autumn on the Magdalen Islands when preparing for a transoceanic flight from the eastern seaboard of North America to South America than in Venezuela in the spring, when a shorter flight across the Caribbean to a route through the interior of North America is undertaken.

Lank (1983) noted that Semipalmated Sandpipers on fall migration at Kent Island, New Brunswick, on the east coast arrived with greater far stores than birds at Sibley Lake, North Dakota, in the interior of the continent. Similar results are found when weights are compared between the Bay of Fundy and James Bay (Table VII). Mean weights on any given date were considerably higher in the Bay of Fundy (by up to 10 g). Not only were the minimum weights greater than in James Bay, suggesting that birds arrive on the east coast with some fat reserves left, but max-

Table VII. Comparison of Daily Mean Weights of Semipalmated Sandpipers on Migration in James Bay and the Bay of Fundy, and of Overall Mean Weights during Autumn Migration in Various Areas

	Daily mean weights(g) in James Bay and the Bay of Fundy					
	James Bay (1979)			Starr's Point, N.S. (1978 to 1979)[a]		
Date	n	$\bar{x} \pm$ S.D.	Min–max	n	$\bar{x} \pm$ S.D.	Min–max
16 July	28	26.64 ± 1.80	22.5–30.0	15	28.79 ± 3.42	24.0–37.4
18 July	30	26.45 ± 2.67	23.0–32.5	13	33.02 ± 7.16	23.5–46.0
26 July	148	24.52 ± 2.57	19.5–34.5	26	34.42 ± 6.18	22.0–44.4
27 July	85	25.98 ± 2.43	19.0–31.0	3	34.60 ± 5.80	28.5–40.0
8 August	192	27.15 ± 3.97	18.5–38.0	46	32.06 ± 5.38	23.5–44.0
16 August	16	26.59 ± 3.56	21.5–35.0	8	36.91 ± 6.64	24.5–47.0
23 August	15	31.53 ± 3.02	26.5–37.5	31	38.71 ± 3.51	31.0–44.5

Overall mean weights (g) during autumn migration at various locations			
	n	Overall mean ± S.D.	Source
Coastal			
James Bay, 1976–1979	22,875	28.2 ± 4.3	Unpublished results, see Gratto (1983)
Sackville, N.B., Bay of Fundy	—	32.5 ± 3.2	Hicklin (unpublished), see Gratto (1983)
Kent Island, N.B.	1,456	33.0 ± 5.8	Lank (1983)
New Jersey	102	28.1 ± 4.0	Murray and Jehl (1964)
North Carolina	27	29.0 ± 5.5	Post and Browne (1976)
Inland			
Long Point, Ont.	229	29.8 ± 5.3	Page and Salvadori (1969)
Sibley Lake, N.D.	1,476	29.6 ± 5.6	Lank (1983)
Postmigration			
Surinam	56	24.8 ± 3.1	Spaans (unpublished), see Gratto (1983)

[a] From Boates (1980).

imum weights rose to much higher levels, reflecting the longer subsequent flight faced by birds taking the direct, transoceanic route to South America. Lank (1983) suggested that the low arrival weights of birds in the interior may have resulted from depletion of reserves during a long, direct flight between breeding areas along the Arctic coast and staging areas on the great plains. Birds do have the option of making short flights from inland locations, as shown by short-range dispersal of marked birds from both Long Point, Ontario (Page and Middleton, 1972) and Sibley Lake (Lank, 1983), but many may make a long, nonstop flight overland to wintering or subsequent staging areas in South America, as suggested by Jehl (1979) for Baird's Sandpipers and Senner and Martinez (1982) for Western Sandpipers. Highest mean weights, however, are found in catches of birds from major staging areas around the Bay of Fundy (Table VII).

The simple hypothesis that birds arrive in a stopover area, gain weight (at an optimal rate?) to a threshold level determined by the length of the subsequent flight and then depart, while attractive, does not appear to be valid. A number of studies, for instance, have concluded that the amount of fat a bird is carrying is a poor predictor of the subsequent length of stay (Page and Middleton, 1972; Lank, 1983): results in James Bay were similar (Morrison, unpublished). Lank (1983) found that the length of stay was best predicted by seasonal date for Kent Island birds, and results from James Bay were again similar (unpublished results). However, even though fat levels were not good predictors of length of stay, Lank (1983) showed that fat levels did affect the probability of departing. It would thus appear that there is a complicated interplay among the various factors involved in weight gain and migration—age and foraging efficiency, available food resources and their change and/or depletion during the season, number of competitors and population levels, date during the season, amount of fat required to complete the subsequent flight successfully, and weather patterns. Further studies and detailed seasonal analysis of pattern of weight changes in relation to geographical area and available food resources are needed to clarify these questions.

B. Red Knot

As mentioned in Section I, three main groups of knot occur in North America: (1) those breeding in the eastern High Arctic and wintering in Europe, (2) those breeding throughout the central Arctic and wintering in the New World, and (3) those passing through southern Alaska in the spring whose breeding and wintering areas are currently uncertain. Knot may be classified as long-distance migrants that occur in large numbers in relatively few locations; they have fairly specialized food requirements and appear to be susceptible to disturbance. As such, they have been considered of conservational concern in both the Old World and the New World. This section will review some of the basic features of the migration patterns of populations visiting the New World, and review new information on the North American race *Calidris canutus rufa*.

1. European Wintering Population

Taxonomic considerations indicated that Red Knot breeding on Ellesmere Island and adjacent areas of the eastern High Arctic belonged to

Table VIII. Occurrence of Red Knot at Sites on the Atlantic Coast of North America from International Shorebird Survey Autumn (July–October) Counts, 1975–1978, Using Maximum Recorded Counts[a]

	1975	1976	1977	1978
Total	6630	7763	9046	7730
% in Massachusetts and New Jersey	87.3%	87.2%	60.4%	76.8%
% at top five sites	79.5%	77.3%	75.9%	84.3%
% of sites at which knot were observed (n)	13.9% (36)	12.8% (39)	10.2% (49)	11.6% (43)

[a] Data from Morrison (1983a) and Morrison *et al.* (1980).

the nominate race *Calidris canutus canutus* (Salomonsen, 1950–51; Godfrey, 1953, 1966), a conclusion that has been confirmed by a variety of band recoveries (Morrison, 1975, 1976b, 1977a; Witts and Morrison, 1980; Witts, 1982). Distances from northern Ellesmere Island to the Canadian provincial capitals, stretched roughly along the Canada/United States border and to London, England, are all about 2500 miles, so that the European wintering grounds are considerably closer than wintering areas used by the New World race *Calidris canutus rufa*. The very dry climate and relative lack of snow on Ellesmere Island compared to areas in the central Arctic (Parmelee and MacDonald, 1960) also result in earlier arrival times on the breeding grounds (late May/early June).

In the spring, many knot use Iceland as a stopover area in May en route from European wintering areas to breeding grounds in Arctic Canada and Greenland (Morrison *et al.*, 1971; Morrison, 1971, 1977a; Wilson, 1981). In Iceland, knot may accumulate up to 50% of their arrival weight in fat, increasing from approximately 135 g to 200 g between early and late May (Morrison, 1977a), enabling them to make the long flight across the Greenland ice cap. Arrival weights on Ellesmere Island (ca. 135 g) are similar to spring arrival weights in Iceland, and to early winter weights in the United Kingdom (ca. 135 g) (see Morrison, 1975, 1977a), showing that most of the fat acquired before migration is used during the journey. Some reserves are retained, however, and these may be important in ensuring survival during poor weather early in the breeding season, when food may be relatively scarce, and/or in enabling the birds to breed. Severe weather in June has been shown to cause death due to starvation on Ellesmere Island (Morrison, 1975).

In the autumn, details of return routes are less clear. Birds leaving Ellesmere Island may pass to the north, over, or to the south of Greenland (Salomonsen, 1950–51; Morrison, 1975, 1977a; and see Wilson, 1981), and many take a direct route across the north Atlantic from Greenland to Europe, as the autumn passage in Iceland is much lighter (Morrison, 1977a; Wilson, 1981). Migration weights for knot in Iceland are much lighter in the autumn than in the spring, as might be anticipated from the shorter subsequent flight and lowered requirement for retaining fat reserves. Adult knot arrive in Iceland from mid-July onwards at weights of 110–120 g and depart after they attain weights of around 150 g (Morrison, 1977a). In this context, it is interesting to note that McNeil and Cadieux (1972) reported that several species of shorebirds in North America show the opposite pattern to knot in Iceland, with higher fat reserves during the autumn southward migration than during the spring northward migration. In both cases, the fat reserves are related to the length of the subsequent flight.

2. New World Populations

A separate race, *Calidris canutus rufa*, occupies breeding grounds in the central Canadian Arctic. *Rufa* is somewhat paler in plumage and slightly smaller than the nominate race (Conover, 1943; Salomonsen, 1950–51; Godfrey, 1953, 1966), and migrates to wintering areas located principally in Argentina (Harrington and Morrison, 1980a; Harrington, 1982a; Morrison, 1983a).

From the breeding grounds, *rufa* heads first to the coasts of Hudson Bay and James Bay, where it is common on migration. Aerial surveys of the Ontario coastline of James and Hudson bays in 1976 and 1977 revealed approximately 7000 Red Knot, with areas in the north and south of James Bay characterized by extensive sandflats appearing to be most important for the species (Morrison and Harrington, 1979; Morrison et al., 1980; Morrison, 1983a). Knot concentrated to a considerable extent in favored areas: in southern James Bay some 40% of the birds were found in only 11% of the coast, rising to 70% in 22% of the coast (Morrison, 1983a).

Knot are also highly concentrated at their principal stopover areas on the eastern seaboard of North America during the autumn migration. Of the 6000–9000 birds recorded on International Shorebird Survey censuses in the period 1975–1978, 60–90% of the birds occurred in two states alone (Massachusetts and New Jersey) and the top five sites in any given year accounted for a full 75–85% of the total: in fact, knot occurred at only 10–14% of the sites censused (Morrison and Harrington, 1979; Morrison et al., 1980; Morrison, 1983a) (Table VIII).

From the eastern United States, the birds make a transoceanic flight that takes them through the Guianas on the north coast of South America.

Morrison and Spaans (1979) reported finding 600–900 knot in Surinam in September 1978, including three birds that had been color-marked together only 23 days previously in James Bay. Harrington and Leddy (1982b) reported finding approximately 2450 Red Knot during aerial surveys in the Guianas in late August/early September 1982. Highest numbers of knot occur in Surinam during August/September (Spaans, 1978), with few remaining through the winter; no birds were found in wing molt among adult birds captured in September 1978 (unpublished results) and no knot were identified on winter aerial surveys in February 1982 (Morrison and Ross, 1983a,b). Observations have been made of many shorebirds, including knots, passing eastwards along the coast of Surinam (Spaans, 1978; Morrison and Spaans, unpublished observations; Harrington and Leddy, 1982b), and it seems likely that knot move on to staging grounds, possibly in northern Brazil, before heading south across Amazonia. Few knot are observed during southward migration on the coast of Rio Grande do Sul in southern Brazil (W. Belton in litt., see Harrington, 1982a) and the recovery of a knot banded in Massachusetts inland in southern Brazil (Lara-Resende and Leal, 1982) indicates one of the inland flyways described in the introduction may be used, bringing birds to the coast of Buenos Aires Province in Argentina, where they occur on passage and winter in small numbers (Myers and Myers, 1979; M. Norres, personal communication; Harrington and Morrison, 1980a,b).

The majority of Red Knot found on aerial surveys in Argentina in January 1982 occurred on the central and southern sections of the coast (Fig. 22) (Morrison and Ross, 1983a,b; Morrison, 1983a). Of the total of about 14,500 knot counted on the surveys, the largest concentration (51% of the total) occurred in the Bahia Bustamente area on the northern side of Golfo San Jorge, and other important areas included the Valdez Peninsula, also known to be important during migration (Harrington, 1982a), the restinga on the southern part of Golfo San Jorge from Caleta Olivia to Cabo Blanco, the coast south to Puerto Deseado, and the Atlantic coast of Tierra del Fuego. Ground surveys in Tierra del Fuego revealed totals of 5000 birds at Rio Grande in 1979 (Harrington and Morrison, 1980a,b) and 1500 knot north and south of Rio Grande in 1982 (Morrison and Ross, 1983a,b), and with a similar report of 5000 birds at Rio Grande in 1976 by Devillers and Terschuren (1976), Tierra del Fuego is clearly an important wintering area.

The northward migration may be traced through the Valdez Peninsula, the Rio Grande do Sul coastline of southern Brazil, Surinam and the southeastern coast of the United States (Harrington, 1982a; Morrison, 1983a).

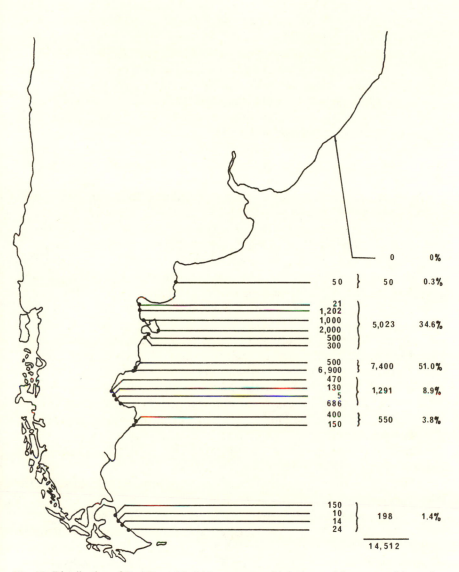

Fig. 22. Distribution of Red Knot (*Calidris canutus rufa*) during aerial surveys of Argentina and southern Brazil, January 1982. [From Morrison (1983a).]

Migration proceeds at a steady pace up the southeastern coast of the United States, with a huge buildup of knot in Delaware Bay in late May. In May 1981, some 61,000 Red Knot were found on the Cape May county shores of Delaware Bay (Harrington and Leddy, 1982a), and aerial surveys of both sides of Delaware Bay in 1982 found over 95,500 knot (Dunne *et al.*, 1982), clearly representing a substantial proportion of the New World population. Here the migration is timed to coincide with an apparent superabundance of horseshoe crab (*Limulus polyphemus*) eggs. From Delaware Bay, the main flight heads inland toward James Bay, where a large passage occurs in late May.

A major goal of current research is to determine the population size of *rufa*. Initial guesstimates of about 30,000 were based mainly on the observation that both aerial surveys in James Bay and I.S.S. surveys on the Atlantic seaboard of North America produced totals of some 7000–10,000 Red Knot. However, the discovery of 61,000 and over 95,000 birds in Delaware Bay on spring migration in 1981 and 1982, respectively, shows that the population must be much larger, and current estimates based on resightings of marked birds are in the region of 150,000 for the population wintering in Argentina (Harrington, 1982a). Clearly, these findings raise questions concerning the detection efficiency of knot on ground and aerial surveys, and a critical assessment of population sizes based on resightings of marked birds needs to be made, given the facts that marked birds do not appear to disperse randomly between flocks (Harrington and Leddy, 1982c) and that they appear to be highly traditional in their use of stopover areas (Morrison, 1977a; see Harrington *et al.*, 1981).

Research by personnel of the Manomet Bird Observatory, Massachusetts, has revealed the existence of a second population of knot, which winters in Florida and the Gulf of Mexico: this group appears to be considerably smaller than the population wintering in Argentina, and current estimates for the group in west Florida are about 10,000 birds (Harrington, 1982a).

Band recoveries and sightings of marked birds have confirmed the migration patterns described above, with records of movements between Massachusetts, the Guianas, southern Brazil, Argentina, New Jersey, and James Bay (Harrington, 1982a; Harrington and Leddy, 1982a; Harrington *et al.*, 1981). Although the Argentina and Florida wintering populations appear to use the same stopover areas in North America, no exchanges between wintering groups have been recorded and they appear to remain separated (Harrington, 1982a). Molt schedules of the two groups are also different (B. A. Harrington, personal communication). Further work is needed to establish the breeding origins and status of these two apparently different populations.

C. Sanderling

Rather little is known about the details of the migration pattern of Sanderlings of the New World, although considerable work has been carried out on wintering populations in California in recent years (Myers *et al.*, 1979, 1980a,b; Connors *et al.*, 1981; Myers, 1980b). This brief section will therefore focus on several interesting demographic and ecological factors that influence Sanderling migration, which have been pointed out by Myers *et al.* (1983).

Comparison of the breeding and wintering areas of Sanderling (see Fig. 23) shows that whereas the breeding grounds cover a rather restricted latitudinal range in the Arctic, mostly north of 73°N, the wintering areas cover an enormous latitudinal range, spanning some 80–100° latitude from about 50°N on the Pacific coast of North America to around 50°S in the southern parts of South America. This implies that different parts of the population migrate enormously different distances, which, although not necessarily unusual among shorebird populations (see e.g., Hale, 1980), immediately raises the question of what the advantages and disadvantages are of migrating shorter or longer distances. Implicit in the assumptions generally made are that increased migration distance will increase mortality associated with migration, and that this will be balanced by increased winter survivorship because of a greater or more predictable food supply, less stressful climate, lowered predation, lowered competition, etc. As yet, no quantitative studies are available to demonstrate differences in mortality resulting from such factors, and studies partitioning mortality between the different phases of the annual cycle are currently a major requirement in our understanding of where selection occurs. Data currently available on the winter distribution of Sanderling are summarized in Fig. 23 (Myers *et al.*, 1983), and represent one of the spectacular early successes of current efforts to coordinate shorebird work on an international scale. Information was drawn together by Myers *et al.* (1983) for the Pacific coasts of North and South America from surveys by Myers and 31 collaborators, from the north and Atlantic coasts of South America from aerial surveys carried out under the Canadian Wildlife Service Latin American Program (R. I. G. Morrison and R. K. Ross), and on the east coast of North America by J. P. Walters. Figure 23 clearly shows that Sanderling winter at much higher densities on the Pacific coast of both continents, and that winter densities are generally higher in South America than in North America. It also shows clearly that major wintering areas are closely associated with areas of major coastal upwellings (see e.g. Fairbridge, 1966), which would thus appear to be the major factor

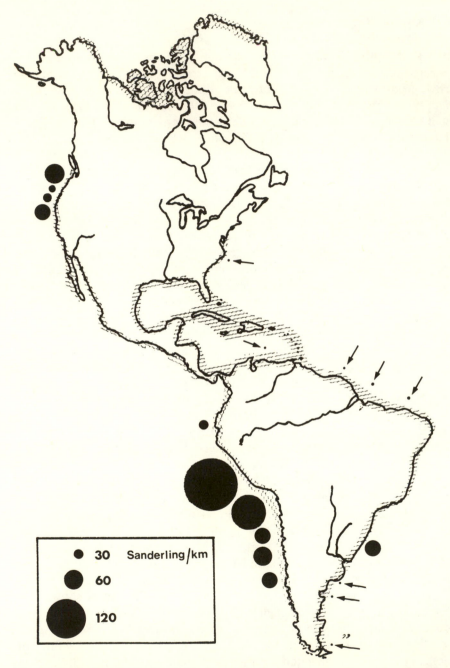

Fig. 23. Sanderling distribution in North and South America. Breeding and nonbreeding distributions are indicated by hatching in Arctic and more southerly latitudes, respectively. [From Myers *et al.* (1983)].

influencing overall distribution, presumably through the food supply available to the birds. Myers *et al.* (1983) further show that Sanderling wintering in California are generally lighter than those in Chile, and that the latter spend far less time feeding to maintain a higher weight, suggesting that resource conditions are more favorable in Chile than in California. Detailed, comparative studies, such as those being undertaken by Myers, will be necessary to identify factors contributing to the annual cycle of survivorship and reproduction, and ecological parameters driving the birds' demography—and of the role migration plays in these processes.

D. White-rumped Sandpiper and Baird's Sandpiper

These two species, though superficially similar in size and form, have rather different patterns of habitat use, different breeding systems, and considerably different migration routes. This brief section is intended principally to draw attention to the latter. Little detailed information has been gathered on either species, though the autumn migration pattern of Baird's Sandpiper has been reviewed by Jehl (1979).

Baird's Sandpiper breeds across the Arctic, from Alaska through the Canadian Arctic islands and into northwestern Greenland (see Godfrey, 1966; Jehl, 1979). White-rumped Sandpipers are not as widely distributed, being restricted to the central parts of the Arctic and Arctic islands, and absent or uncommon in eastern Baffin Island, Ellesmere Island (Godfrey, 1966), and Alaska (Gabrielson and Lincoln, 1959). Adult Baird's Sandpipers migrate very early through North America [generally late July–mid-August (Jehl, 1979)], while adult White-rumped Sandpipers move through eastern Canada somewhat later [early August—early September (Morrison, 1976c; Tufts, 1973; McNeil and Burton, 1973; McLaren, 1981)]. On the wintering grounds in Argentina, this difference appears to be maintained, with Baird's Sandpipers arriving by mid-August and White-rumped Sandpipers by late August in Buenos Aires Province (Myers and Myers, 1979). The wintering grounds overlap broadly, with Baird's Sandpiper ranging from the Andes of Ecuador, Central Peru, Chile, and southern Bolivia through Argentina to Tierra del Fuego, and the White-rumped Sandpiper chiefly east of the Andes from southern Brazil and Paraguay to Tierra del Fuego (AOU, 1957; Jehl, 1979; Meyer de Schauensee, 1970; Jehl and Rumboll, 1976; Blake, 1977). The two species tend to be associated with one another on the wintering grounds (Myers and Myers, 1979; Jehl and Rumboll, 1976), though the White-rumped Sandpiper tends to use littoral habitats to a greater extent than

Baird's Sandpiper, which makes more extensive use of inland wetlands and grassland (Myers and Myers, 1979; Myers, 1980a).

Jehl (1979) has assembled evidence from museum specimens and the literature that adult Baird's Sandpipers move rapidly through the central High Plains of North America, in an area bounded by eastern Alberta and western Manitoba in southern Canada, and central Colorado and central Kansas in the United States. The species is very uncommon in Central America, and Jehl (1979) concludes that most adults depart from the central staging grounds on a direct flight of some 4000 miles to the Andes in northern South America. In contrast to the adults, the migration of juveniles was much more widely dispersed across North America. Of general interest also was that adult females preceded males on migration, as might be anticipated from a species with a "conservative" breeding system [which Baird's Sandpiper appears to have (Norton, 1972; Morrison and Witts, unpublished results; see Pitelka et al., 1974; Jehl, 1979)], and that the molt schedule of the birds and their weight levels were consistent with this hypothesis.

White-rumped Sandpipers, on the other hand, appear to migrate southwards through the northeasternmost parts of the Atlantic seaboard, making a direct flight across the north Atlantic to the north coast of South America. The species is one of the commonest shorebirds in Labrador (Austin, 1932; Todd, 1963), it is very common on the Magdalen Islands (Hagar, 1956), moderately common in the Gulf of St. Lawrence in New Brunswick and Prince Edward Island, as well as around the Bay of Fundy in New Brunswick and Nova Scotia, and apparently progressively less common further south on the eastern seaboard of the United States (Hagar, 1956; Palmer, 1949; Urner and Storer, 1949; Morrison, 1976c). The lack of sightings of color-marked birds both from James Bay (Morrison and Gratto, 1979a,b) and from the Magdalen Islands (McNeil and Burton, 1973) is consistent with the species taking a northerly and easterly route through eastern Canada and making a direct flight to northern South America.

There is a marked and rapid passage through Surinam in late August and September, many birds passing eastwards along the coast (Spaans, 1978). A few birds may occasionally winter in Surinam, but most move on, probably taking a trans-Amazonian route to wintering grounds extending southwards from southern Brazil.

The spring migration of White-rumped Sandpipers is principally through the central United States, as shown by their abundance in central regions and scarcity on the east coast (see Cooke, 1910; Bent, 1927; AOU, 1957; Palmer, 1967) and as illustrated by recoveries of birds banded during spring migration in the central United States and recovered during autumn migration on the eastern seaboard (Burton, 1974; McNeil and Burton, 1973, 1977; Morrison, unpublished).

Weight levels of birds moving through the Magdalen Islands on southward migration in the fall were considerably greater than those of birds on spring migration moving northwards through Venezuela, as would be expected from the longer transoceanic flight undertaken in the autumn (McNeil and Burton, 1973, 1977; McNeil and Cadieux, 1972).

White-rumped Sandpipers have a polygynous mating system (see Parmelee *et al.*, 1967; Pitelka *et al.*, 1974), with males setting up territories, mating with more than one female, but taking no part in incubation. The species shows relatively little sexual dimorphism (Pitelka *et al.*, 1974), and males depart earlier than females from the breeding grounds (Parmelee *et al.*, 1967). Data from the Magdalen Islands (McNeil and Cadieux, 1972) suggested an earlier migration of males than females in August, but few details on this or other aspects of the migration of this interesting species are available.

E. Dunlin and Purple Sandpiper

This section will consider two species with considerably shorter migration patterns than those discussed so far.

1. Dunlin

A succession of papers has considered the racial status of Dunlin in North America (Todd, 1953; MacLean and Holmes, 1971; Browning, 1977) and it is currently considered that three separable races occur (Browning, 1977), though the AOU checklist (1957) recognizes only one (*C. a. pacifica*). The three races are *Calidris alpina hudsonia* breeding in north-central Canada (principally to the north and west of Hudson Bay), *C. a. pacifica* breeding in western Alaska, and *C. a. arcticola* breeding in northern Alaska, which may be separated from one another by a combination of size and plumage characteristics (Browning, 1977). Browning (1977) considered the birds breeding in northern Alaska sufficiently different from those breeding in Siberia (*C. a. sakhalina*) to be given separate racial status [*C. a. arcticola*, as proposed by Todd (1953)].

The migration patterns of the three races are quite distinct. Band recoveries and taxonomic considerations indicate that *C. a. arcticola* migrates from northern Aalska to the Asian side of the Pacific, *C. a. pacifica* moves from western Alaska to wintering quarters along the Pacific coast of North America, and *C. a. hudsonia* migrates to the Gulf and eastern

Atlantic coasts of the United States (see MacLean and Holmes, 1971; Norton, 1971; Morrison and Gratto, 1979a; Gill, 1979; Senner, 1979; Senner *et al*, 1981; Banding Laboratory, unpublished records).

All three North American races of Dunlin undergo a full postnuptial molt, including flight feathers, before migration to the wintering grounds (Holmes, 1966, 1971; Morrison *et al*., 1982, and unpublished results), in contrast to the European races, which either undergo a suspended molt or delay molt until arrival on the wintering grounds (see Salomonsen, 1950–51; Soikkeli, 1967; Nieboer, 1972; Boere, 1976; Hardy and Minton, 1980). Compared with other New World shorebirds, the short migration of North American Dunlin to wintering grounds within the Northern Hemisphere is thought to be significant in enabling the birds to molt before migration.

2. Purple Sandpiper

Rather little information is currently available on Purple Sandpipers breeding and wintering in the New World. The Purple Sandpiper has one of the most northerly wintering distributions of all shorebirds and is the only shorebird commonly found in eastern Canada during the winter (Godfrey, 1966; Morrison, 1976c, and unpublished data): its winter distribution extends southwards on the Atlantic coast to Maryland (AOU, 1957; Snyder, 1957). Its breeding distribution is not especially northerly, however: it breeds from southern Baffin Island and Southampton Island north to Devon Island and northern Bylot Island, as well as on the Belcher Islands (Freeman, 1970) and other islands in Hudson Bay (see Todd, 1963; specimens in Royal Ontario Museum), and the Twin Islands (Manning, 1981) and possibly other islands in northern James Bay (S. G. Curtis, personal communication).

It seems likely that Purple Sandpipers breeding in the Canadian Arctic may migrate to two separate wintering areas, in a manner somewhat analogous to that of the Red Knot (see above). Measurement data (see Table IX) and a single band recovery from 225 Purple Sandpipers trapped in February 1975 in Passamaquoddy Bay, New Brunswick, indicate that birds wintering in the lower Bay of Fundy migrate to breeding areas in the Belcher islands/Hudson Bay (see Fig. 24). Departure and arrival dates, as well as observations and specimens of Purple Sandpipers at inland locations (Morrison, 1972; and see Godfrey, 1966), indicate a direct overland route is used. Godfrey (1973) suggested that Arctic Terns (*Sterna paradisaea*) make a similar flight.

Measurements and the few available long-distance band recovery records indicate that some Purple Sandpipers breeding further north on

Table IX. Bill Measurements of Purple Sandpipers in Canada and Iceland[a]

	Males	Females	Notes[b]
Canadian Arctic: Baffin Island	28.9 ± 1.50 (46)	33.6 ± 1.50 (28)	1
Canadian Arctic: Baffin Island	29.3 ± 1.17 (4)	33.6 ± 0.62 (3)	2
Iceland: shore, museum specimens	28.7 ± 1.05 —	33.4 ± 1.40 —	3, 4a
Iceland: shore, spring, live	29.5 ± 1.65 —	34.1 ± 1.35 —	4b
Iceland: shore, autumn, live	29.3 ± 1.00 —	34.3 ± 1.40 —	4c
Eastern Canada (live, winter, St. Andrews, N.B.)	27.5 ± 0.90 —	32.2 ± 1.25 —	4d
Belcher islands, N.W.T.	28.2 ± 0.78 (4)	32.1 ± 1.35 (7)	2

[a] Measurements in mm, mean ± S.D., sample size.
[b] 1, specimens in National Museum of Natural Sciences, Ottawa.
2, specimens in Royal Ontario Museum, Toronto.
3, specimens in Museum of Natural History, Reykjavik.
4, live birds trapped in Iceland 1972 and Canada 1975. Bill measurements derived by percent cumulative frequency method (Harding, 1949; Griffiths, 1968). Sample sizes: (a) 83, (b) 264, (c) 174, (d) 170.

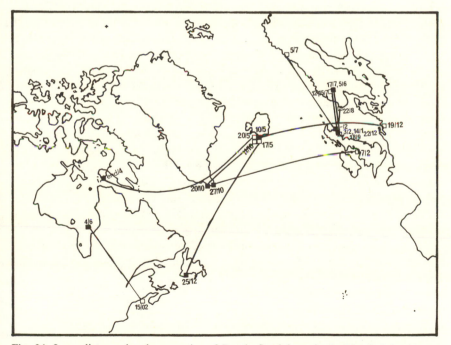

Fig. 24. Long-distance band recoveries of Purple Sandpipers in the North Atlantic area. Open symbols: place of banding; closed symbols: place of recovery or capture. Figures indicate date (day/month).

Baffin Island may migrate toward wintering areas in the Old World. A series of overlapping band recovery records (Holland to Iceland, England to Greenland, Iceland to Greenland, Iceland to Baffin Island) invite speculation as to whether Canadian populations in Baffin Island move to wintering areas in Europe where they would apparently mix with populations breeding in the Old World (see Fig. 24 and Atkinson *et al.*, 1981). The situation is complicated, however, as it would appear that some birds in Greenland and Iceland may be resident (Salomonsen, 1950–51, 1967; Morrison, 1972) and further work is needed to clarify the migration of North Atlantic wintering populations.

The Purple Sandpiper has a relatively short migration to northerly wintering quarters, and most populations [e.g., in Iceland (Morrison, 1976a), Spitsbergen (Bengtson, 1975), eastern Canada (Stresemann and Stresemann (1966)] molt before migration to the wintering area. The molt is very rapid, apparently an adaptation related to ecological conditions connected with their northerly wintering range (Morrison, 1976a). Band recoveries indicate that different populations may have widely different migration distances, however, and it is not known how and if this may influence adaptive responses, for instance in breeding system and molt pattern. Morrison (1976a) pointed out that the need for a very rapid molt in the Purple Sandpiper was not entirely obvious in populations with a short migration or which are resident, and suggested that factors connected with the northerly wintering range, such as climate, decreasing daylight, and/or deteriorating feeding conditions in the autumn, may make a rapid molt advantageous. Several factors are worth commenting on here in relation to the interaction between molt, breeding system, wintering area, and migration. First, Purple Sandpipers have a monogamous "conservative" (Pitelka *et al.*, 1974) breeding system, but there is a very pronounced tendency for the early departure of the female, the male alone usually tending the brood (both sexes share incubation) (see Bengtson, 1970). Myers (1981a) has pointed out that the parental care system of shorebirds is significantly correlated with migration distance for a variety of *Calidris* species, those with the longest migrations showing greater tendencies for the early departure of one of the sexes (usually the female). In this context, the early departure of female Purple Sandpipers would appear to go against the general trend as it has a short migration and Myers (1981a) specifically comments that the species is largely responsible for producing a poorer correlation between the variables within all shorebirds considered, compared with "Western Hemisphere" species only, in which the species was not included (and for which a much higher correlation was obtained). It is possible in this case that those pressures that select for an early departure of one bird and that are connected with migration distance for most species, are replaced by other pressures that select for an early departure, but that are connected with the ecological constraints imposed by the northerly wintering niche. In this case, one would predict that populations moving to a relatively mild wintering area might show some relaxation of selection pressures, and there is some indication that this is the case for Purple Sandpipers wintering in eastern Scotland. Atkinson *et al.* (1981) showed that Purple Sandpipers in eastern Scotland did not show the midwinter peak in weight observed in many other shorebirds wintering in Britain, and which is thought to be related to climatic stress (Pienkowski *et al.*, 1979; Davidson, 1982). They suggested that this lack of winter fattening might be related to the birds' feeding ecology and a relatively more available prey compared to other wintering species. Purple Sandpipers in Scotland are relatively far south compared to other wintering populations in Europe (Atkinson *et al.*, 1978; Harrison, 1982), and one might expect populations wintering in colder areas to have higher wintering weights, and this does appear to be the case—see Table X. The results of Atkinson *et al.* (1981) also indicated that some birds molted their flight feathers after arrival on the wintering grounds, and that this molt was completed later than birds molting on the breeding grounds in Iceland or Spitsbergen (Morrison, 1976a; Bengtson, 1975). Thus, there also appears to be some relaxation in molt schedule. No information was given as to whether early arrivals on the wintering grounds may have comprised either non- or failed breeders returning early without having reproduced. These differences appear similar to the differences in molt schedule reported for Dunlin from northern and western Alaska (see above) and to the weight differences of Sanderlings migrating to different wintering areas in California and Chile (see above).

F. Hudsonian Godwit

Perhaps the most spectacular migration pattern of any shorebird moving between North and South America is that of the Hudsonian Godwit (Morrison and Harrington, 1979; Morrison, 1983a). As recently as 40 years ago, the species was thought to be on the verge of extinction, because so few were seen on the east coast of North America. When ornithologists started visiting James Bay, however, sightings involving several thousands of birds were reported during the southward period of migration (Hope and Shortt, 1944; Hagar, 1966), and Canadian Wildlife Service surveys during the 1970s have shown them to be a common bird on the sandflats of both James Bay and Hudson Bay (unpublished results). Up

Table X. Weights and Wing Lengths of Purple Sandpipers and Mean January/February Temperatures for Populations Wintering in Different Areas

Wintering area	Month	Mean weight (g)						Source[a]	Mean Temperature (°C)		
		Males		Females		Mixed			Jan.	Feb.	Source
		(n)	$\bar{x}$	(n)	$\bar{x}$	(n)	$\bar{x}$				
E. Scotland	Dec.	—	—	—	—	(79)	63.2	(1)	2.1	2.2	(2)
E. Canada	Dec.	(18)	66.2	(14)	78.9	—	72.5	(3)	−4.4	−3.4	(4)
Russia (Murmansk)	(—)	(72)	66.4	(92)	86.2	—	81.5	(5)	−10.1	−10.5	(2)
		Mean wing length (mm)									
E. Scotland	Dec.	—	—	—	—	(79)	130.5				
E. Canada	Dec.	(18)	128.8	(14)	135.0	—	131.9				
Russia (Murmansk)	(—)	(25)	125.4	(34)	128.7	—	127.1				

[a] (1) Atkinson et al. (1981); (2) U.S. Department of Commerce (1966); (3) Specimens in Royal Ontario Museum, Toronto; (4) U.S. Department of Commerce (1965); (5) Dement'ev et al. (1951).

to 10,000 godwits were discovered north of the Albany River in 1974 (Morrison and Harrington, 1979; Morrison, 1983a).

From James Bay, the godwits essentially disappear until reaching their wintering grounds in Argentina. They are not common on the east coast of North America, counts seldom rising above a few tens in this region (Fig. 25), nor have large concentrations been discovered in northern South America, although the existence of a staging area, presumably in this region, is implied from departure and arrival dates in Canada and Argentina, respectively (Harrington and Morrison, 1980b). These observations indicate that the Hudsonian Godwit probably makes a direct flight from James Bay to South America, a distance of at least 4500 km, underlining the international importance of the coast of James Bay to the species, and eclipsing the often-quoted example of the transoceanic flight of the Golden Plover (Lincoln, 1952).

Aerial and ground surveys in Argentina have recently identified some of the most important wintering areas for the species (Fig. 26). Many birds pass through Buenos Aires Province (Myers and Myers, 1979), where up to 3000 birds have been reported (M. Norres, personal communication), but by midwinter, most birds appear to have moved further south to the coastlines of Patagonia and Tierra del Fuego (Harrington and Morrison, 1980b; Morrison, 1983a; Morrison and Ross, 1983a,b). During aerial surveys in January 1982, major concentrations were found on the very extensive flats of Bahia Union/Bahia Anegada in the estuary of the Rio Colorado, on the restinga areas in southern Bahia San Jorge and near Puerto Deseado, and especially on the mudflats of Bahia San Sebastian in Tierra del Fuego. Ground surveys at Bahia San Sebastian resulted in the discovery of an estimated 6000–8000 Hudsonian Godwits, by far the largest concentration yet observed on the wintering grounds. Williams and Pringle (1982) also reported sizable concentrations of godwits at Bahia San Sebastian.

IV. DISCUSSION

The major intention of this chapter has been to present a very selective review of the different migration patterns shown by a number of New World species of shorebirds. Emphasis has been placed on drawing attention to the diversity of migration systems and strategies adopted by a series of closely related shorebirds, and to illustrating how various aspects of the annual cycle reflect adaptive solutions to the complex of ecological conditions faced by the birds during their travels.

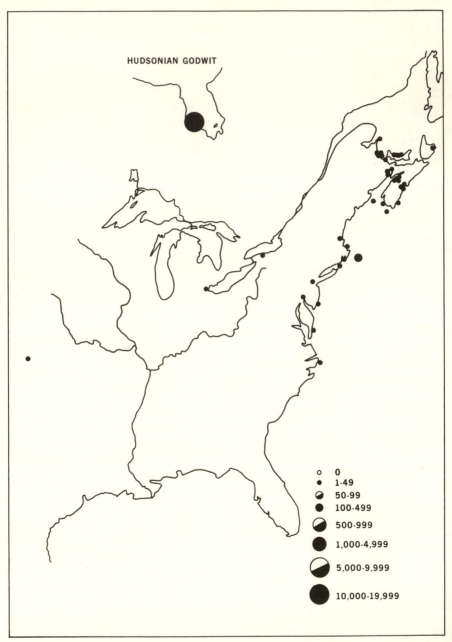

Fig. 25. Distribution of Hudsonian Godwits (*Limosa haemastica*) in eastern North America from Maritimes Shorebird Surveys, International Shorebird Surveys (ISS), and Canadian Wildlife Service surveys. [ISS information from Harrington and Leddy (1979); from Morrison (1983a).]

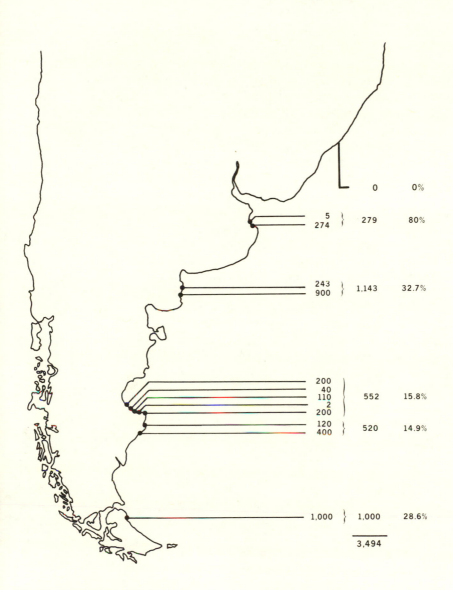

Fig. 26. Distribution of Hudsonian Godwits (*Limosa haemastica*) during aerial surveys of Argentina and southern Brazil, January 1982. [From Morrison (1983a).]

A. Migration Patterns and Strategies

The species discussed in the accounts above show widely differing migration distances, routes, and strategies. These include species such as the Purple Sandpiper, with a moderately high Arctic and subarctic breeding range and very northerly wintering range; the Dunlin, with a fairly low Arctic breeding range and moderate migration distance to wintering areas in the Northern Hemisphere; the Semipalmated Sandpiper, with a trans-Arctic breeding distribution and migration to wintering areas principally along the north coast of South America; the Baird's Sandpiper and White-rumped Sandpiper, with migrations from high Arctic breeding grounds to wintering grounds in southern South America, the former utilizing a central route in autumn and the latter a transoceanic flight across the Atlantic; and the Red Knot and Hudsonian Godwit, with migrations from high and subarctic breeding areas to wintering grounds in southern South America, the former concentrating heavily at a series of favored stopover areas and the latter making very long-distance flights, overflying much of eastern North America and probably large tracts of South America.

Various lines of evidence suggest that migration plays a major role in influencing various aspects of a species' annual cycle. Myers (1981a) demonstrated that both breeding and parental care systems were related to migration distance, increasing distance being correlated with increasingly promiscuous systems and early departure of one parent. It was suggested that behavior on the breeding grounds was influenced more by the advantages of an early arrival on migration areas (Ashkenazie and Safriel, 1979; Myers, 1981a,b) rather than, for instance, reduction of pressure on the food supply for the remaining young and other adult (Pitelka *et al.*, 1974).

Adults generally precede young birds on migration, after an initial wave of probable non- and failed breeders that may contain a high proportion of subadult birds, and there is often a slight differential in timing of migration of the sexes, related to their different departure schedules from the breeding grounds.

The difference in timing of migration of adults and young makes it likely the two age groups will encounter different feeding conditions during their journey. Food appears to be influential in determining the birds' distribution on migration (see above; Morrison, 1983a), as many studies have shown on the wintering grounds, and it appears likely that food resources are considerably depleted by shorebirds at stopover areas (e.g., Schneider, 1978; Schneider and Harrington, 1981; Boates, 1980), as well as on the wintering grounds (see O'Connor, 1981). This implies that competition occurs for resources during migration, and this has been noted in various studies (see above). Despite the central role and apparent risks of migration, very little direct information is available on the mortality associated with migration itself, and how this may differ for populations migrating different distances. Ketterson and Nolan (1982) showed that distance traveled during migration is probably closely and positively correlated with both mortality rate during migration and survival rate during the winter for Dark-eyed Juncos (*Junco h. hyemalis*), so that counterbalancing effects will tend to lead to similar overall mortality rates for different populations, ages, or sexes, where different wintering ranges are occupied by different groups. Little is known about whether there are any major differences in winter distribution of ages and/or sexes in New World shorebirds, and the studies available suggest such differences are not prevalent (Myers, 1981b): the slight tendency of first-year male Sanderling to winter further north than females was thought to be related to the advantages of an early return to the breeding grounds (Myers, 1981b). The work of Myers *et al.* (1983) with Sanderling suggests that populations wintering in Chile encounter easier feeding and climatic conditions than birds wintering in California. Recent results from the California study suggested that up to 95% of the birds present at the end of the winter returned the following autumn after breeding, suggesting that migration does not contribute a particularly large proportion of the annual mortality (J. P. Myers, personal communication); more work on determining where mortality occurs during the annual cycle is needed.

B. Migration and Molt

Molt is generally regarded as an energy-demanding process that does not overlap with migration when long-distance flights are involved (Holmes, 1966). Migration appears to be highly influential in affecting where and when molt of wing and body feathers occurs, and a remarkable variety of molt patterns have evolved among shorebirds. The majority of New World shorebirds with long-distance migrations delay wing molt until after migration, while a smaller number molt before migration (e.g., Dunlin, Purple Sandpiper, as discussed in the species accounts above). Some species undergo a suspended molt, replacing a variable number of inner primaries while on or near the breeding grounds, and resuming and completing molt of the remaining flight feathers after migration has been completed: examples include the Greater Yellowlegs and Black-bellied Plover. One molt adaptation that appears to be clearly related to migration is that of partial postjuvenile wing molt among first-winter birds (Gratto and Morrison, 1981). This involves replacement of some of the outer primaries and inner secondaries only, and has been shown to occur in a variety of species, especially among small *Calidris* and *Tringa* species (Gratto and Morrison, 1981). It appears to be an adaptation enabling 1-year-old birds to migrate north to the breeding grounds more effectively than would otherwise be possible, for without any flight feather molt, a juvenile bird would undertake three long-distance migrations before its first molt. This would result in excessive and potentially disastrous feather wear, as shown by data from Semipalmated Sandpipers originally banded as juveniles and recaptured as yearlings in James Bay: those undergoing the partial molt showed little average change in wing length after their first year (-0.8 mm, S.D. 1.3 mm, $n = 16$), whereas yearlings that had failed to molt had lost an average of 5.6 mm in wing length (S.D. $= 1.7$ mm, $n = 7$) ($P < 0.001$, Mann–Whitney test) and their feathers were very abraded indeed (see Gratto and Morrison, 1981; Gratto, 1983).

Many shorebirds begin molt of their body feathers on the breeding grounds (see review by Ferns, 1978) and body molt may be continued during migratory stopovers, though it is generally thought to be suspended before long-distance flights (Holmes, 1966; McNeil and Cadieux, 1972). Most shorebirds passing through the Magdalen Islands were in various stages of body molt (McNeil and Cadieux, 1972) and the same phenomenon was observed in James Bay (Table XI).

Not only scheduling but duration of molt may vary both within and between species. As discussed in the species accounts above, Dunlin in Europe molt either after migration or undergo a suspended molt after starting on or near the breeding grounds, whereas populations in North America all molt before migration. Dunlin in northern Alaska not only have a shorter duration of molt, but molt overlaps with breeding activities, when compared to populations in western Alaska, apparently in response to more stressful environmental/climatic conditions in northern Alaska (Holmes, 1966, 1970, 1971). More rapid molt either between or within species appears to be attained by having more feathers in growth at once (Morrison, 1976a) and/or more rapid growth of the feathers themselves (Pienkowski *et al.*, 1976).

C. Migration and Weather/Radar Studies

It is generally agreed that weather plays an influential role both locally in determining when migratory movements occur as well as globally in the evolution and maintenance of migration patterns in the New World.

Table XI. Molt Status of Shorebirds at North Point, James Bay, 1982[a]

	Adults			Juveniles, body	First summer, wing
	Wing	Tail	Body		
Semipalmated Plover	X	X	P	P	X
Lesser Golden Plover	X	X	P	X	(PPW?)
Black-bellied Plover	X	X	P	—	X
Ruddy Turnstone	X	X	P	(P)	X
Whimbrel	X	X	P	—	X
Spotted Sandpiper	—	—	—	P	X
Solitary Sandpiper	X	X	X	X	X
Greater Yellowlegs	(P)	(P)	P	P	X
Lesser Yellowlegs	X	X	P	P	PPW
Red Knot	X	X	P	X	PPW
Pectoral Sandpiper	X	X	P	X	X
White-rumped Sandpiper	X	X	P	P	X
Least Sandpiper	X	X	(P)	P	PPW
Dunlin	C	C	C	C	X
Short-billed Dowitcher	—	—	—	P	—
Semipalmated Sandpiper	X	X	P	(P)	PPW
Hudsonian Godwit	X	X	P	—	—
Sanderling	X	X	P	P	—
Wilson's Phalarope	—	—	—	P	—
Northern Phalarope	—	—	—	(—)	—

[a] C, complete molt; PPW, partial postjuvenile wing molt of outer primaries/inner secondaries by birds during their first winter; P, partial molt; X, molt not observed.

These topics have received considerable coverage recently, often in conjunction with radar studies, and will be summarized only very briefly (see Drury and Keith, 1962; Nisbet, 1963; Richardson, 1972, 1974, 1979; Hilditch *et al.*, 1973; Williams *et al.*, 1977, 1978; Williams and Williams, 1978; Gauthreaux, 1980; Stoddard *et al.*, 1983). Shorebird migration in the autumn generally occurs after the passage of a cold front, with northerly or northwesterly winds. For shorebirds departing from the east coast of Canada, this involves leaving with a following wind, though prevailing northeasterly winds are subsequently encountered that tend to bring the birds into the Caribbean. Spring migrants taking a central route through North America will also tend to have following winds (Harrington and Morrison, 1979; Richardson, 1974; Gauthreaux, 1980). Radar studies have provided much evidence of the direct, transoceanic flight by shorebirds from Nova Scotia to the north coast of South America in the autumn, and computer simulations based on conditions under which birds appeared to make the flights indicate that a shorebird departing from Nova Scotia on a heading of 134° for 10 hr and then turning to 170° for the remainder of the flight would reach Surinam after about 70 hr (Stoddard *et al.*, 1983). Radar studies have also confirmed that spring and autumn routes differ for many birds, with a much heavier volume of offshore, transatlantic migration between eastern North America and northern South America in the fall than in the spring (Richardson, 1974).

V. MIGRATION AND CONSERVATION

Shorebirds are one of the most highly migratory groups of birds in the avifauna of the New World. Some of the most significant recent advances in our knowledge of shorebird migration discussed in this chapter have come from internationally coordinated research programs. Results of this work are not only of academic biological interest, but are essential in providing the theoretical framework and baseline knowledge required to design sound management policies for the future conservational requirements of this important group of birds. Shorebird conservation must be approached on an international level, as it would be highly ineffective to mobilize extensive political and financial resources in one area if the birds' requirements were being compromised in a separate but equally important part of their range. Morrison (1983a) has outlined some of the developmental threats facing shorebirds in the eastern half of North and South America, and Senner and Howe (Vol. 5 of this series) provide a review of conservational concerns affecting shorebirds. A broad international perspective on conservational problems is required not only to identify major threats at critical points in the birds' annual cycles, but also to ensure that small-scale, creeping erosion of habitats or other resources over a wide geographic scale does not suddenly reach a level where an irreversible collapse in shorebird populations would follow.

Shorebirds do indeed appear to face developmental pressures in many parts of their range. Oil and gas exploration and transportation in the Arctic could affect both breeding and premigration feeding areas. Offshore oil development, hydroelectric projects, lignite mining, increased traffic, and a deepwater port are among real or proposed projects that could significantly affect important migration areas in James Bay and Hudson Bay. On the east coast, the Fundy Tidal Power project would affect major shorebird sites for a variety of species (Morrison, 1977b) and a deepwater port facility for supertankers at Eastpoint, Maine, has also been proposed in an area where narrow channels and strong currents combine to attract major concentrations of phalaropes and seabirds. Important stopover sites for knot in Massachusetts and New Jersey are privately owned, being considered for development, or on major transportation corridors for tankers carrying oil and chemicals. In Surinam, pesticides and toxic chemicals form a potential threat to some species. In Argentina, though much of the coast is remote, drainage of agricultural chemicals into coastal areas such as Bahia Samborombon in Buenos Aires Province poses an unknown threat, and oil development is a potential problem in Patagonia and Tierra del Fuego—15% of Red Knot observed on ground surveys in 1979 in Golfo San Jorge were found to be oiled (Harrington and Morrison, 1980a,b; Morrison, 1983a).

On the Pacific coast, oil development and tanker traffic in the Gulf of Alaska is widely regarded as an environmental threat. Estuaries of undoubted significance to shorebirds on the coast of Washington have been considered for development (Widrig, 1979, 1980). Loss of wetlands has occurred at a dramatic rate in California during the present century (Speth, 1979). In South America, isolated but strategically placed wetlands such as those at Mejia, Peru, have been threatened with development (Myers, 1982).

For species whose migrations take them through a number of such areas, there is clearly a need for a coordinated international approach to conservational planning. There is not only a need but, as more detailed information becomes available throughout the birds' ranges, a potential to develop coordinated conservation strategies, where protection of a series of sites in different countries would protect the major areas used by a species throughout its annual cycle (Morrison, 1983a). This approach is potentially far more effective than haphazard or piecemeal protection of single sites in different countries, which may or may not meet the requirements of the birds. A network of "twin" or "sister" parks, or whatever designation of protected area would be appropriate, in a series of countries could be designed for the conservation of individual species, or groups of species. For instance, for the Hudsonian Godwit, "twin" reserves in James Bay, Canada, and at Bahia San Sebastian, Argentina, would link the two known areas of major international importance for this species. For the Red Knot, sites in James Bay, Massachusetts, New Jersey, Surinam, southern Brazil, and Argentina would form a chain of sites that would protect vital areas for the birds. Such a concept offers a clear and exciting approach, both biologically and politically, to the future conservation of shorebirds. The series of international programs now in progress, such as the Shorebird Atlas Project of the Latin American Program of the Canadian Wildlife Service, survey projects of the New World section of the Wader Study Group, banding projects of the Pan-American Shorebird Program, and projects of the Wader Research Group of the International Waterfowl Research Bureau, will provide the framework for coordinated work on a large geographical scale (see Morrison and Harrington, 1979; Morrison, 1983a; Myers, 1983). Just as the biological knowledge on which conservation programs are based must be accumulated, so must the political will be cultivated to translate such knowledge into conservational action.

ACKNOWLEDGMENTS

For studies of the geographical scale described in this chapter, it would be difficult to provide a complete list of people and organizations who have contributed to the work, and apologies go to those whose names may not have been mentioned below. For Canadian Wildlife Service surveys in South America, magnificent cooperation and assistance was received from counterpart agencies in all the countries visited. Of particular assistance were the Argentinian National Parks Service (Dr. Arturo Tarak, Pablo Canevari), the Brazilian Banding Office (Paulo Antas, Susana de M. Lara-Resende, Renato P. Leal), the Surinam Forestry Service (Ferdinand Baal, Ben de Jong, Henk Reichart), the Venezuelan Wildlife Service (Douglas Figueroa, Frank Espinosa), and the Wildlife Division of the Forestry Division in the Ministry of Agriculture, Lands, and Fisheries in Trinidad (Bheesham Ramdial). We acknowledge the considerable contribution of the Argentinian Air Force in providing an aircraft for surveys in Argentina and of the Trinidad Ministry of National Security for making a coast guard helicopter available for flights in Trinidad. Thanks also to the curators of the following museum collections for providing access to specimens and facilities during visits: Sao Paulo University Zoological Museum, Sao Paulo, Brazil (Dr. P. Vanzolini); National Museum, Rio de Janeiro, Brazil (Dr. D. Texeira); Goeldi Museum, Belem, Brazil (Dr. F. Novaes); Venezuelan Wildlife Service, Maracay, Venezuela (D. Figueroa); Royal Ontario Museum, Toronto, Canada (Dr. A. J. Baker); National Museum of Natural Sciences, Ottawa, Canada (Dr. W. E. Godfrey). Iola Price has provided much assistance as Co-ordinator of the Latin American Program. We thank the many volunteer participants in the Maritimes and International Shorebird Survey schemes for their contributions to our knowledge of shorebird migration: special thanks go to Brian Harrington, Manomet Bird Observatory, Massachusetts, who coordinates ISS participants in the United States and points south. Ken Ross and Barbara Campbell of the Canadian Wildlife Service have made particularly valuable contributions to work in South America and in James Bay, respectively. Thanks go to the many paid and unpaid participants in the shorebird banding program at North Point, James Bay, and to the Ontario Ministry of Natural Resources for providing facilities and personnel at the camp. Cheri Gratto and Jill Kearney provided invaluable assistance in analysis of banding data. We thank the World Wildlife Fund and National Geographic Society for support of work referred to in this chapter. Thanks also to J. P. Myers for the chance to experience the intellectual and physical rigors of Pacific coast shorebird studies! Sharon Bradford provided stalwart assistance with typing. Thanks once more to Hugh Boyd, whose ongoing encouragement and support have been instrumental in translating ideas into action.

APPENDIX

SCIENTIFIC NAMES AND ABBREVIATIONS OF SHOREBIRD SPECIES MENTIONED IN THE TEXT

BBPL	Black-bellied Plover	*Pluvialis squatarola*
LGPL	Lesser Golden Plover	*Pluvialis dominica*
RIPL	Ringed Plover	*Charadrius hiaticula*
SEPL	Semipalmated Plover	*Charadrius semipalmatus*
GRYE	Greater Yellowlegs	*Tringa melanoleuca*
LEYE	Lesser Yellowlegs	*Tringa flavipes*
SOSA	Solitary Sandpiper	*Tringa solitaria*
WATA	Wandering Tattler	*Heteroscelus incanus*
SPSA	Spotted Sandpiper	*Actitis macularia*
ESCU	Eskimo Curlew	*Numenius borealis*
WHIM	Whimbrel	*Numenius phaeopus*
BTCU	Bristle-thighed Curlew	*Numenius tahitiensis*
HUGO	Hudsonian Godwit	*Limosa haemastica*
BTGO	Bar-tailed Godwit	*Limosa lapponica*
RUTU	Ruddy Turnstone	*Arenaria interpres*
SURF	Surfbird	*Aphriza virgata*
REKN	Red Knot	*Calidris canutus*
SAND	Sanderling	*Calidris alba*
SESA	Semipalmated Sandpiper	*Calidris pusilla*
WESA	Western Sandpiper	*Calidris mauri*
LESA	Least Sandpiper	*Calidris minutilla*
WRSA	White-rumped Sandpiper	*Calidris fuscicollis*
BASA	Baird's Sandpiper	*Calidris bairdii*
PESA	Pectoral Sandpiper	*Calidris melanotus*
PUSA	Purple Sandpiper	*Calidris maritima*

DUNL	Dunlin	*Calidris alpina*
CUSA	Curlew Sandpiper	*Calidris ferruginea*
RUFF	Ruff	*Philomachus pugnax*
SBDO	Short-billed Dowitcher	*Limnodromus griseus*
WIPH	Wilson's Phalarope	*Phalaropus tricolor*

REFERENCES

AOU, 1957, *Check-list of North American Birds*, 5th ed., American Ornithologists' Union, Baltimore. (Including changes and corrections in the 32nd and 33rd Supplements to the Check-list, *Auk* **90**:411–419, 887; **93**:975–979.)

AOU, 1982, *Thirty-fourth Supplement to the American Ornithologists' Union Check-list of North American Birds, Auk* **99**(Suppl.).

Ashkenazie, S., and Safriel, U. N., 1979, Time–energy budget of the Semipalmated Sandpiper *Calidris pusilla* at Barrow, Alaska, *Ecology* **60**:783–799.

Atkinson, N. K., Davies, M., and Prater, A. J., 1978, The winter distribution of Purple Sandpipers in Britain, *Bird Study* **25**:223–228.

Atkinson, N. K., Summers, R. W., Nicoll, M., and Greenwood, J. J. D., 1981, Population, movements and biometrics of the Purple Sandpiper *Calidris maritima* in eastern Scotland, *Ornis Scand.* **12**:18–27.

Austin, O. L., 1932, *The Birds of Newfoundland Labrador*, Nuttall Ornithological Club, Cambridge, Mass.

Bengtson, S.-A., 1970, Breeding behaviour of the Purple Sandpiper *Calidris maritima* in west Spitsbergen, *Ornis Scand.* **1**:17–25.

Bengtson, S.-A., 1975, Timing of the moult of the Purple Sandpiper *Calidris maritima* in Spitsbergen, *Ibis* **117**:100–102.

Bent, A. C., 1927, *Life Histories of North American Shorebirds*, Part 1 (Dover Reprint, New York, 1962).

Blake, E. R., 1977, *Manual of Neotropical Birds*, Vol. 1, University of Chicago Press, Chicago.

Boates, J. S., 1980, Foraging Semipalmated Sandpipers *Calidris pusilla* L. and their major prey *Corophium volutator* (Pallas) on the Starrs Point mudflat, Minas Basin, M.Sc. thesis, Acadia University, Wolfville, N.S.

Boates, J. S., and Smith, P. C., 1979, Length–weight relationships, energy content and the effects of predation on *Corophium volutator* (Pallas) (Crustacea: Amphipoda), *Proc. N. S. Inst. Sci.* **29**:489–499.

Boere, G. C., 1976, The significance of the Dutch Waddenzee in the annual life cycle of Arctic, subarctic and boreal waders, *Ardea* **64**:210–291.

Broussard, J. P., 1981, Distribution et abondance des oiseaux de rivage le long du Saint-Laurent, Unpublished report, Canadian Wildlife Service Québec Region.

Brown, R. A., and O'Connor, R. J., 1974, Some observations on the relationships between oystercatchers *Haematopus ostralegus* L. and cockles *Cardium edule* L. in Strangford Lough, *Ir. Nat. J.* **18**:73–80.

Browning, M. R., 1977, Geographic variation in Dunlins, *Calidris alpina*, of North America, *Can. Field Nat.* **91**:391–393.

Bryant, D. M., 1979, Effects of prey density and site character on estuary usage by over-wintering waders (*Charadrii*), *Estuarine Coastal Mar. Sci.* **9**:369–384.

Burger, J., Hahn, D. C., and Chase, J., 1979, Aggressive interactions in mixed-species flocks of migrating shorebirds, *Anim. Behav.* **27**:459–469.

Burton, J., 1974, La migration des oiseaux de rivage (Charadriidae et Scolopacidae) dans l'est de l'amérique du nord, Ph.D. thesis, University of Montréal, Montréal, Québec.

Cadieux, F., 1970, Capacité de vol et routes de migration automnale de certains oiseaux de rivage nord-américains (*Charadriidae* et *Scolopacidae*), M.Sc. thesis, University of Montréal, Montréal, Québec.

Connors, P. G., Myers, J. P., Connors, C. S. W., and Pitelka, F. A., 1981, Interhabitat movements by Sanderlings in relation to foraging profitability and the tidal cycle, *Auk* **98**:49–64.

Conover, B., 1943, The races of the knot (*Calidris canutus*), *Condor* **45**:226–228.

Cooke, W. W., 1910, Distribution and migration of North American shorebirds, *U.S. Dep. Agric. Biol. Surv. Bull.* **35**.

Davidson, N.C., 1982, Changes in the body condition of redshanks during mild winters: An inability to regulate reserves? *Ringing and Migration* **4**:51–62.

Dement'ev G. P., Gladkov, N. A., and Spangenberg, E. P., 1951, Ptitsy Sovetskogo Soyuza, Vol. III, Nauka Press, Moscow. [Translated as *The Birds of the Soviet Union*, Vol. III (G. P. Dement'ev and N. A. Gladkov, eds.), Israel Program for Scientific Translations, Jerusalem, 1969.]

Devillers, P., and Terschuren, J. A., 1976, Some distributional records of migrant North American Charadriiformes in coastal South America (Continental Argentina, Falkland, Tierra del Fuego, Chile and Ecuador), *Gerfaut* **66**:107–125.

Drinnan, R. E., 1957, The winter feeding of the oystercatcher (*Haematopus ostralegus*) on the edible cockle (*Cardium edule*), *J. Anim. Ecol.* **26**:441–469.

Drury, W. H., and Keith, J. A., 1962, Radar studies of songbird migration in coastal New England, *Ibis* **104**:449–489.

Dunne, P., Sibley, D., Sutton, C., and Wander, W., 1982, Aerial surveys in Delaware Bay: Confirming an enormous spring staging area for shorebirds, *Wader Study Group Bull.* **35**:32–33.

Evans, P. R., Herdson, D. M., Knights, P. J., and Pienkowski, M. W., 1979, Short-term effects of reclamation of part of Seal Sands, Teesmouth, on wintering waders and Shelduck. I. Shorebird diets, invertebrate densities and the impact of predation on the invertebrates, *Oecologia (Berlin)* **41**:183–206.

Fairbridge, R. W. (ed.), 1966, *The Encyclopedia of Oceanography*, Van Nostrand–Reinhold, Princeton, N.J.

Ferns, P. N., 1978, The onset of prebasic body moult during the breeding season in some high-Arctic waders, *Bull. Br. Ornithol. Club* **98**:118–122.

Freeman, M. M. R., 1970, The birds of the Belcher Islands, N.W.T., Canada, *Can. Field Nat.* **84**:227–290.

Gabrielson, I. N., and Lincoln, F. C., 1959, *The Birds of Alaska*, Wildlife Management Institute, Washington, D.C.

Gauthreaux, S. A., 1980, The influence of global climatological factors on the evolution of bird migratory pathways, *Acta XVII Congressus Internationalis Ornithologici, Berlin 1978*, pp. 517–525, Verlag der Deutschen Ornithologen-Gesellschaft, Berlin.

Gill, R., 1979, Shorebird studies in western Alaska 1976–1978, *Wader Study Group Bull.* **25**:37–40.

Gill, R., and Jorgensen, P. D., 1979, A preliminary assessment of timing and migration of shorebirds along the north central Alaska peninsula, in: *Studies in Avian Biology No. 2* (F. A. Pitelka, ed.), pp. 113–123, Cooper Ornithological Society, Allen Press, Lawrence, Kans.

Godfrey, W. E., 1953, Notes on Ellesmere Island birds, *Can. Field Nat.* **67**:89–93.

Godfrey, W. E., 1966, The birds of Canada, *Natl. Mus. Can. Bull.* **203**.

Godfrey, W. E., 1973, A possible shortcut spring migration route of the Arctic Tern to James Bay, Canada, *Can. Field Nat.* **87**:51–52.

Goss-Custard, J. D., 1969, The winter feeding ecology of the redshank, *Tringa totanus*, *Ibis* **111**:338–356.

Goss-Custard, J. D., 1977a, The ecology of the Wash. III. Density-related behaviour and the possible effects of a loss of feeding grounds on wading birds (*Charadrii*), *J. Appl. Ecol.* **14**:721–739.

Goss-Custard, J. D., 1977b, Predator responses and prey mortality in redshank, *Tringa totanus* (L.), and a preferred prey, *Corophium volutator* (Pallas), *J. Anim. Ecol.* **46**:21–35.

Goss-Custard, J. D., 1980, Competition for food and interference among waders, *Ardea* **68**:31–52.

Goss-Custard, J. D., Kay, D. G., and Blindell, R. M., 1977, The density of migratory and overwintering redshank, *Tringa totanus* (L.), and curlew *Numenius arquata* (L.), in relation to the density of their prey in south-east England, *Estuarine Coastal Mar. Sci.* **5**:497–510.

Gratto, C. L., 1983, Migratory and reproductive strategies of the Semipalmated Sandpiper, M.Sc. thesis, Queen's University, Kingston, Ontario.

Gratto, C. L., and Morrison, R. I. G., 1981, Partial postjuvenile moult of the Semipalmated Sandpiper (*Calidris pusilla*), *Wader Study Group Bull.* **33**:33–37.

Gratto, C. L., Cooke, F., and Morrison, R. I. G., 1981, Hatching success of yearling and older breeders in the Semipalmated Sandpiper, *Wader Study Group Bull.* **33**:37–38.

Gratto, C. L., Cooke, F., and Morrison, R. I. G., 1983, Nesting success of yearling and older breeders in the Semipalmated Sandpiper, *Calidris pusilla, Can. J. Zool.* **61**:1133–1137.

Griffiths, J., 1968, Multi-modal frequency distributions in bird populations, *Bird Study* **15**:29–32.

Hagar, J. A., 1956, Magdalen Islands—1956, Unpublished manuscript (copy in Canadian Wildlife Service Library, Sackville, N.B.).

Hagar, J. A., 1966, Nesting of the Hudsonian Godwit at Churchill, Manitoba, *Living Bird* **5**:5–43.

Hale, W. G., 1980, *Waders*, Collins, Glasgow.

Hanson, W. C., and Eberhardt, L. E., 1978, Ecological consequences of petroleum development in northern Alaska, in: *Biomedical and Environmental Research Program of the LASL Health Division*, Jan.–Dec. 1977, pp. 17–22, Los Alamos Sci. Lab., Los Alamos.

Harding, J. P., 1949, The use of probability paper for the graphical analysis of polymodal frequency distributions, *J. Mar. Biol. Assoc. U.K.* **28**:141–153.

Hardy, A. R., and Minton, C. D. T., 1980, Dunlin migration in Britain and Ireland, *Bird Study* **27**:81–92.

Harrington, B. A., 1982a, Untying the enigma of the Red Knot, *Living Bird Q.* **1**:4–7.

Harrington, B. A., 1982b, Morphometric variation and habitat use of Semipalmated Sandpipers during a migratory stopover, *J. Field Ornithol.* **53**:258–262.

Harrington, B. A., and Groves, S., 1977, Aggression in foraging migrant Semipalmated Sandpipers, *Wilson Bull.* **89**:336–338.

Harrington, B. A., and Leddy, L. E., 1979, MBO/ISS Update—1979, Newsletter to International Shorebird Survey Participants, Manomet Bird Observatory, Mass.

Harrington, B. A., and Leddy, L. E., 1982a, Sightings of knots banded and color-marked in Massachusetts in August 1980, *J. Field Ornithol.* **53**:55–57.

Harrington, B. A., and Leddy, L. E., 1982b, A study of autumn migration of Red Knots *Calidris canutus* in the Guianas, South America, Unpublished report for ICBP—Pan American Section, August/September 1982, Manomet Bird Observatory, Mass.

Harrington, B. A., and Leddy, L. E., 1982c, Are wader flocks random groupings? A knotty problem, *Wader Study Group Bull.* **36**:20–21.

Harrington, B. A., and Morrison, R. I. G., 1979, Semipalmated Sandpiper migration in North America, in: *Studies in Avian Biology No. 2* (F. A. Pitelka, ed.), pp. 83–100, Cooper Ornithological Society, Allen Press, Lawrence, Kans.

Harrington, B. A., and Morrison, R. I. G., 1980a, Notes on the wintering areas of Red Knot *Calidris canutus rufa* in Argentina, South America, *Wader Study Group Bull.* **28**:40–42.

Harrington, B. A., and Morrison, R. I. G., 1980b, An investigation of wintering areas of Red Knots *Calidris canutus* and Hudsonian Godwits *Limosa haemastica* in Argentina, Report to the World Wildlife Fund, Washington, D.C., and Toronto, Canada.

Harrington, B. A., and Taylor, A. L., 1982, Methods for sex identification and estimation of wing area in Semipalmated Sandpipers, *J. Field Ornithol.* **53**:174–177.

Harrington, B. A., Twitchell, D., and Leddy, L. E., 1981, *Knotters Anonymous*, Newsletter to Knot Study Cooperators, Manomet Bird Observatory, Mass.

Harrison, C., 1982, *An Atlas of the Birds of the Western Palaearctic*, Collins, Glasgow.

Hilditch, C. D. M., Williams, T. C., and Nisbet, I. C. T., 1973, Autumnal bird migration over Antigua, W. I., *Bird-Banding* **44**:171–179.

Holmes, R. T., 1966, Molt cycle of the Red-backed Sandpiper *Calidris alpina* in western North America, *Auk* **83**:517–533.

Holmes, R. T., 1970, Differences in population density, territoriality, and food supply of Dunlin on Arctic and subarctic tundra, in: *Animal Populations in Relation to Their Food Resources* (A. Watson, ed.), pp. 303–319, Blackwell, Oxford.

Holmes, R. T., 1971, Latitudinal differences in the breeding and molt schedules of Alaskan Red-backed Sandpipers *Calidris alpina, Condor* **73**:93–99.

Hope, C. E., and Shortt, T. M., 1944, Southward migration of adult shorebirds on the west coast of James Bay, Ontario, *Auk* **61**:572–576.

Hughes, R. A., 1979, Notes on the Charadriiformes of the south coast of Peru, in: *Studies in Avian Biology No. 2* (F. A. Pitelka, ed.), pp. 49–53, Cooper Ornithological Society, Allen Press, Lawrence, Kans.

Isleib, M. E., 1979, Migratory shorebird populations on the Copper River Delta and eastern Prince William Sound, Alaska, in: *Studies in Avian Biology No. 2* (F. A. Pitelka, ed.), pp. 125–129, Cooper Ornithological Society, Allen Press, Lawrence, Kans.

Jehl, J. R., Jr., 1979, The autumnal migration of Baird's Sandpiper, in: *Studies in Avian Biology No. 2* (F. A. Pitelka, ed.), pp. 55–68, Cooper Ornithological Society, Allen Press, Lawrence, Kans.

Jehl, J. R., Jr., and Rumboll, M. A. E., 1976, Notes on the avifauna of Isla Grande and Patagonia, Argentina, *Trans. San Diego Soc. Nat. Hist.* **18**:145–154.

Johnson, D. W., and MacFarlane, R. W., 1967, Migration and bioenergetics of flight in the Pacific Golden Plover, *Condor* **69**:156–168.

Johnson, O. W., 1979, Biology of shorebirds summering on Enewetak Atoll, in: *Studies in Avian Biology No. 2* (F. A. Pitelka, ed.), pp. 193–205, Cooper Ornithological Society, Allen Press, Lawrence, Kans.

Jurek, R. M., 1974, California Shorebird Survey 1969–1974, California Department of Fish and Game, Resources Agency.

Ketterson, E. D., and Nolan, V., 1982, The role of migration and winter mortality in the life history of a temperate-zone migrant, the Dark-eyed Junco, as determined from demographic analyses of winter populations, *Auk* **99**:243–259.

Lank, D., 1979, Dispersal and predation rates of wing-tagged Semipalmated Sandpipers *Calidris pusilla* and an evaluation of the technique, *Wader Study Group Bull.* **27**:41–46.

Lank, D. B., 1983, Migratory behavior of Semipalmated Sandpipers at inland and coastal staging areas, Ph.D. Thesis, Cornell University, Ithaca, N.Y.

Lara-Resende, S. de M., and Leal, R. P., 1982, Recuperacao de Anilhas Estrangeiras no Brasil, *Brasil Florestal* **52**:27–53.

Leddy, L. E., and Harrington, B. A., 1978, The 1976 International Shorebird Survey, Co-operators' Report, Manomet Bird Observatory, Mass.

Lincoln, F. C., 1952, *Migration of Birds*, Doubleday, New York.

McLaren, I. A., 1981, The birds of Sable Island, Nova Scotia, *Proc. N.S. Inst. Sci.* **31**:1–84.

MacLean, S. F., and Holmes, R. T., 1971, Bill length, wintering areas, and taxonomy of North American Dunlins, *Calidris alpina, Auk* **88**:893–901.

McNeil, R., 1970, Hivernage et estivage d'oiseaux aquatiques Nord-Américains dans le nord-est du Vénézuela (mue, accumulation de graisse, capacité de vol et routes de migration), *Oiseau Rev. Fr. Ornithol.* **40**:185–302.

McNeil, R., and Burton, J., 1973, Dispersal of some southbound migrating North American shorebirds away from the Magdalen islands, Gulf of St. Lawrence, and Sable Island, Nova Scotia, *Caribb. J. Sci.* **13**:257–278.

McNeil, R., and Burton, J., 1977, Southbound migration of shorebirds from the Gulf of St. Lawrence, *Wilson Bull.* **89**:167–171.

McNeil, R., and Cadieux, F., 1972, Fat content and flight-range capabilities of some adult spring and fall migrant North American shorebirds in relation to migration routes on the Atlantic coast, *Nat. Can.* **99**:589–605.

Manning, T. H., 1981, Birds of the Twin Islands, James Bay, N.W.T., Canada, Syllogeus 30, National Museums of Canada, Ottawa.

Manning, T. H., Hohn, E. O., and Macpherson, A. H., 1956, The birds of Banks Island, *Natl. Mus. Can. Bull.* **143**.

Martinez, E. F., 1974, Recovery of a Semipalmated Sandpiper at Prudhoe Bay, Alaska, *Bird-Banding* **45**:364–365.

Martinez, E. F., 1979, Shorebird banding at the Cheyenne Bottoms Waterfowl Management Area, *Wader Study Group Bull.* **25**:40–41.

Meyer de Schauensee, R., 1970, *A Guide to the Birds of South America*, Academy of Natural Sciences of Philadelphia (reprinted by Pan American Section ICBP, 1982).

Morrison, R. I. G., 1971, Cambridge Iceland Expedition 1971, Expedition Report, Cambridge, England.

Morrison, R. I. G., 1972, Migration and biometrics of the Purple Sandpiper *Calidris maritima* in Iceland, Unpublished manuscript.

Morrison, R. I. G., 1975, Migration and morphometrics of European knot and turnstone on Ellesmere Island, Canada, *Bird-Banding* **46**:290–301.

Morrison, R. I. G., 1976a, Moult of the Purple Sandpiper *Calidris maritima* in Iceland, *Ibis* **118**:237–246.

Morrison, R. I. G., 1976b, Further records, including the first double-journey recovery, of European-banded Ruddy Turnstones on Ellesmere Island, N. W. T., *Bird-Banding* **47**:274.

Morrison, R. I. G., 1976c, Maritimes Shorebird Survey 1974, Contributors' Report, Shorebird Project Report, Canadian Wildlife Service, Ottawa.

Morrison, R. I. G., 1976d, Maritimes Shorebird Survey 1975, Preliminary Report, Shorebird Project Report, Canadian Wildlife Service, Ottawa.

Morrison, R. I. G., 1977a, Migration of Arctic waders wintering in Europe, *Polar Rec.* **18**:475–486.

Morrison, R. I. G., 1977b, Use of the Bay of Fundy by shorebirds, in: *Fundy Tidal Power and the Environment* (G. A. Daborn, ed.), pp. 187–199, The Acadia University Institute, Wolfville, N.S.

Morrison, R. I. G., 1978a, Shorebird banding and colour-making studies in James Bay, 1977, *Wader Study Group Bull.* **23**:36–43.

Morrison, R. I. G., 1978b, Maritimes Shorebird Survey 1976, Preliminary Report, Shorebird Project Report, Canadian Wildlife Service, Ottawa.

Morrison, R. I. G., 1983a, A hemispheric perspective on the distribution and migration of some shorebirds in North and South America, *First western hemisphere waterfowl and waterbird symposium* (H. Boyd, ed.), pp. 84–94, Canadian Wildlife Service, Ottawa.

Morrison, R. I. G., 1983b, Maritimes Shorebird Survey 1979, Preliminary Report, Canadian Wildlife Service, Ottawa.

Morrison, R. I. G., and Campbell, B. A., 1983a, Maritimes Shorebird Survey 1980, Preliminary Report, Shorebird Project Report, Canadian Wildlife Service, Ottawa.

Morrison, R. I. G., and Campbell, B. A., 1983b, Maritimes Shorebird Survey 1981, Preliminary Report, Shorebird Project Report, Canadian Wildlife Service, Ottawa.

Morrison, R. I. G., and Gratto, C. L., 1979a, Canadian Wildlife Service Shorebird Colour-marking Program, James Bay 1978, Contributors' Progress Report, Canadian Wildlife Service, Ottawa.

Morrison, R. I. G., and Gratto, C. L., 1979b, Maritimes Shorebird Survey 1978, Preliminary Report, Shorebird Project Report, Canadian Wildlife Service, Ottawa.

Morrison, R. I. G., and Harrington, B. A., 1979, Critical shorebird resources in James Bay and eastern North America, *Transactions of the 44th North American Wildlife and Natural Resources Conference*, pp. 498–507, Wildlife Management Institute, Washington, D.C.

Morrison, R. I. G., and Ross, R. K., 1983a, Aerial surveys of shorebirds and other wildlife in South America: Some preliminary results, *Can. Wildl. Serv. Prog. Notes, in press.*

Morrison, R. I. G., and Ross, R. K., 1983b, Aerial surveys of shorebirds in South America: Some preliminary results, *Wader Study Group Bull.* **37**:41–45.

Morrison, R. I. G., and Spaans, A. L., 1979, National Geographic mini-expedition to Suriname, 1978, *Wader Study Group Bull.* **26**:37–41.

Morrison, R. I. G., Pienkowski, M. W., and Stanley, P. I., 1971, Cambridge–London Iceland Expedition 1970, Expedition Report, Cambridge, England.

Morrison, R. I. G., Harrington, B. A., and Leddy, L. E., 1980, Migration routes and stopover areas of North American Red Knot *Calidris canutus* wintering in South America, *Wader Study Group Bull.* **28**:35–39.

Morrison, R. I. G., Rimmer, C. C., and Campbell, B. A., 1982, Shorebird Banding, Migration and Related Studies at North Point, Ontario, Shorebird Project Report, Canadian Wildlife Service, Ottawa.

Murray, B. G., and Jehl, J. R., Jr., 1964, Weights of autumn migrants from New Jersey, *Bird-Banding* **35**:253–263.

Myers, J. P., 1980a, The Pampas shorebird community: Interactions between breeding and non-breeding members, in: *Migrant Birds in the Neotropics: Ecology, Behavior, Dis-*

tribution and Conservation (A. J. Keast and E. S. Morton, eds.), pp. 37–49, Smithsonian Institution, Washington, D.C.

Myers, J. P., 1980b, Sanderlings at Bodega Bay: Facts, inferences and shameless speculations, *Wader Study Group Bull.* **30**:26–31.

Myers, J. P., 1981a, Cross-seasonal interactions in the evolution of sandpiper social systems, *Behav. Ecol. Sociobiol.* **8**:195–202.

Myers, J. P., 1981b, A test of three hypotheses for latitudinal segregation of the sexes in wintering birds, *Can. J. Zool.* **59**:1527–1534.

Myers, J. P., 1982, Mejia Lagoon: Gone but perhaps not forever, *Wader Study Group Bull.* **35**:29.

Myers, J. P., 1983, Conservation of migrating shorebirds: Staging areas, geographic bottlenecks, and regional movements, *Am. Birds* **37**:23–24.

Myers, J. P., and Myers, L. P., 1979, Shorebirds of coastal Buenos Aires Province, Argentina, *Ibis* **121**:186–200.

Myers, J. P., Connors, P. G., and Pitelka, F. A., 1979, Territory size in wintering Sanderlings: The effects of prey abundance and intruder density, *Auk* **96**:551–561.

Myers, J. P., Connors, P. G., and Pitelka, F. A., 1980a, Optimal territory size and the Sanderling: Compromises in a variable environment, in: *Foraging Behaviour* (A. C. Kamil and T. Sargent, eds.), pp. 135–158, Garland STPM Press, New York.

Myers, J. P., Williams, S. L., and Pitelka, F. A., 1980b, An experimental analysis of prey availability for Sanderlings (Aves: Scolopacidae) feeding on sandy beach crustaceans, *Can. J. Zool.* **58**:1564–1574.

Myers, J. P., Maron, J. L., and Sallaberry, M., 1983, Going to extremes: Why do Sanderlings migrate to the Neotropics?, in: *Neotropical Ornithology*. (P. A. Buckley, M. S. Foster, E. S. Morton, R. S. Ridgeley, eds.), A. O. U. Monographs, Allen Press, Lawrence, Kans.

Nieboer, E., 1972, Preliminary notes on the primary moult in Dunlins *Calidris alpina*, *Ardea* **60**:112–119.

Nisbet, I. C. T., 1963, Measurements with radar of the height of nocturnal migration over Cape Cod, Massachusetts, *Bird-Banding* **34**:57–67.

Nores, M., and Yzurieta, D., 1980, Aves de ambientes acuaticos de Cordoba y Centro de Argentina, Secretaria de Estado de Agricultura y Ganaderia, Cordoba.

Norton, D. W., 1971, Two Soviet recoveries of Dunlins banded at Point Barrow, Alaska, *Auk* **88**:927.

Norton, D. W., 1972, Incubation schedules of four species of calidridine sandpipers at Barrow, Alaska, *Condor* **74**:164–176.

O'Connor, R. J., 1981, Patterns of shorebird feeding, in: *Estuary Birds of Britain and Ireland* (A. J. Prater, ed.), Poyser, Calton, England.

Page, G., 1970, The relationship between fat deposition and migration in the Semipalmated Sandpiper, M.Sc. thesis, University of Guelph, Guelph, Ontario.

Page, G., and Bradstreet, M., 1968, Size and composition of a fall population of Least and Semipalmated Sandpipers at Long Point, Ontario, *Ont. Bird Banding* **4**:50–81.

Page, G., and Middleton, A. L. A., 1972, Fat deposition during autumn migration in the Semipalmated Sandpiper, *Bird-Banding* **43**:85–96.

Page, G., and Salvadori, A., 1969, Weight changes of Semipalmated and Least Sandpipers during autumn migration, *Ont. Bird Banding* **5**:52–58.

Palmer, R. S. 1949, Maine birds, *Bull. Mus. Comp. Zool. Harv. Univ.* **102**.

Palmer, R. S., 1967, Species accounts, in: *The Shorebirds of North America* (G. D. Stout, ed.), pp. 143–267, Viking Press, New York.

Parmelee, D. F., and MacDonald, S. D., 1960, The birds of west-central Ellesmere Island and adjacent areas, *Natl. Mus. Can. Bull.* **169**.

Parmelee, D. F., Stephens, H. A., and Schmidt, R. H., 1967, The birds of southeastern Victoria Island and adjacent small islands, *Natl. Mus. Can. Bull.* **222**.

Phillips, A. R., 1975, Semipalmated Sandpiper: Identification, migrations, summer and winter ranges, *Am. Birds* **29**:799–806.

Pienkowski, M. W., Knight, P. J., Stanyard, D. J., and Argyle, F. B., 1976, The primary moult of waders on the Atlantic coast of Morocco, *Ibis* **118**:347–365.

Pienkowski, M. W., Lloyd, C. S., and Minton, C. D. T., 1979, Seasonal and migrational weight changes in Dunlin, *Bird Study* **26**:134–148.

Pitelka, F. A., 1959, Numbers, breeding schedule, and territoriality in Pectoral Sandpipers of northern Alaska, *Condor* **61**:233–264.

Pitekla, F. A., Holmes, R. T., and MacLean, S. F., Jr., 1974, Ecology and evolution of social organization in Arctic sandpipers, *Am. Zool.* **14**:185–204.

Post, W., and Browne, M. M., 1976, Length of stay and weights of inland migrating shorebirds, *Bird-Banding* **47**:333–339.

Prater, A. J. (ed.), 1981, *Estuary Birds of Britain and Ireland*, Poyser, Calton, England.

Prater, A. J., Marchant, J. H., and Vuorinen, J., 1977, Guide to the identification and ageing of Holarctic waders, BTO Guide No. 17, British Trust for Ornithology, Tring, Herts.

Puttick, G. M., 1981, Sex-related differences in foraging behaviour of Curlew Sandpipers, *Ornis Scand.* **12**:13–17.

Recher, H. F., 1966, Some aspects of the ecology of migrant shorebirds, *Ecology* **47**:393–407.

Recher, H. F., and Recher, J. A., 1969, Some aspects of the ecology of migrant shorebirds. II. Aggression, *Wilson Bull.* **81**:140–154.

Richardson, W. J., 1972, Autumn migration and weather in eastern Canada: A radar study, *Am. Birds* **26**:10–17.

Richardson, W. J., 1974, Spring migration over Puerto Rico and the western Atlantic: A radar study, *Ibis* **116**:172–193.

Richardson, W. J., 1979, Southeastern shorebird migration over Nova Scotia and New Brunswick in autumn: A radar study, *Can. J. Zool.* **57**:107–124.

Salomonsen, F., 1950–51, *Gronlands Fugle*, Munksgaard, Copenhagen.

Salomonsen, F., 1967, *Fuglene pa Gronland*, Rhodes, Copenhagen.

Senner, S. E., 1979, An evaluation of the Copper River Delta as critical habitat for migrating shorebirds, in: *Studies in Avian Biology No. 2* (F. A. Pitelka, ed.), pp. 131–145, Cooper Ornithological Society, Allen Press, Lawrence, Kans.

Senner, S. E., and Martinez, E. F., 1982, A review of Western Sandpiper migration in interior North America, *Southwest. Nat.* **27**:149–159.

Senner, S. E., West, G. C., and Norton, D. W., 1981, The spring migration of Western Sandpipers and Dunlins in south central Alaska: Numbers, timing and sex ratios, *J. Field Ornithol.* **52**:271–284.

Smith, P. C., and Evans, P. R., 1973, Studies of shorebirds at Lindisfarne, Northumberland. I. Feeding ecology and behaviour of the Bar-tailed Godwit, *Wildfowl* **24**:135–139.

Smith, S. M., and Stiles, F. G., 1979, Banding studies of migrant shorebirds in northwestern Costa Rica, in: *Studies in Avian Biology No. 2* (F. A. Pitelka, ed.), pp. 41–47, Cooper Ornithological Society, Allen Press, Lawrence, Kans.

Snyder, L. L., 1957, *Arctic Birds of Canada*, University of Toronto Press, Toronto.

Soikkeli, M., 1967, Breeding cycle and population dynamics in the Dunlin *Calidris alpina*, *Ann. Zool. Fenn.* **4**:158–198.

Spaans, A. L., 1978, Status and numerical fluctuations of some North American waders along the Suriname coast, *Wilson Bull.* **90**:60–83.

Spaans, A. L., 1979, Wader studies in Suriname, South America, *Wader Study Group Bull.* **25**:32–37.

Spaans, A. L., and Swennen, C., 1982, The 1980 Dutch mini-expedition to Suriname, *Wader Study Group Bull.* **34**:32–34.

Speth, J., 1979, Conservation and management of coastal wetlands in California, in: *Studies in Avian Biology No. 2* (F. A. Pitelka, ed.), pp. 151–155, Cooper Ornithological Society, Allen Press, Lawrence, Kans.

Stoddard, P. K., Marsden, J. E., and Williams, T. C., 1983, Computer simulation of autumnal bird migration over the western North Atlantic, *Anim. Behav.* **31**:173–180.

Stresemann, E., and Stresemann, V., 1966, Die Mauser der Vogel, *J. Ornithol. Sonderheft* **107**:1–448.

Thompson, M. C., 1973, Migratory patterns of Ruddy Turnstones in the Central Pacific region, *Living Bird* **12**:5–23.

Todd, W. E. C., 1953, A taxonomic study of the American Dunlin *Erolia alpina* subsp., *J. Wash. Acad. Sci.* **43**:85–88.

Todd, W. E. C., 1963, *Birds of the Labrador Peninsula and Adjacent Areas*, University of Toronto Press, Toronto.

Tufts, R. W., 1973, *The Birds of Nova Scotia*, 2nd ed., Nova Scotia Museum, Halifax.

Urner, C. A., and Storer, R. W., 1949, The distribution and abundance of shorebirds on the north and central New Jersey coast, 1928–1938, *Auk* **66**:177–194.

U.S. Department of Commerce, 1965, World Weather Records 1951–1960, Vol. 1, North America, U.S. Department of Commerce, Washington, D.C.

U.S. Department of Commerce, 1966, World Weather Records 1951–1960, Vol. 2, Europe, U.S. Department of Commerce, Washington, D.C.

Wander, W., and Dunne, P. J., 1981, A preliminary assessment of the spring shorebird migration along the Delaware Bayshore of New Jersey, Unpublished report to U.S. Fish and Wildlife Service, Washington, D.C.

Widrig, R. S., 1979, The shorebirds of Leadbetter Point, Privately published report, R. S. Widrig, Ocean Park, Wash.

Widrig, R. S., 1980, The shorebirds of Leadbetter Point, *Wader Study Group Bull.* **29**:30–36.

Williams, A. J., and Pringle, S., 1982, Holarctic waders observed at Tierra del Fuego, November 1977, *Wader Study Group Bull.* **35**:34.

Williams, T. C., and Williams, J. M., 1978, An oceanic mass migration of land birds, *Sci. Am.* **239**: 166–176.

Williams, T. C., Williams, J. M., Ireland, L. C., and Teal, J. M., 1977, Autumnal bird migration over the western North Atlantic Ocean, *Am. Birds* **31**:251–267.

Williams, T. C., Williams, J. M., Ireland, L. C., and Teal, J. M., 1978, Estimated flight time for transatlantic autumnal migrants, *Am. Birds* **32**:275–280.

Wilson, J. R., 1981, The migration of High Arctic shorebirds through Iceland, *Bird Study* **28**:21–32.

Witts, B. F., 1982, Waders, Pp. 4.D. 1–27 in: *The Report of the Joint Services Expedition to Princess Marie Bay, Ellesmere Island, 1980* (S. R. Williams, ed.), AUWE Report Production Department, Portland, England.

Witts, B. F., and Morrison, R. I. G., 1980, Joint Services Expedition to Princess Marie Bay, Ellesmere Island, 1980: Preliminary report, *Wader Study Group Bull.* **30**:34–35.

Chapter 4

FORAGING AND ACTIVITY PATTERNS IN WINTERING SHOREBIRDS

Gillian M. Puttick*

Percy Fitzpatrick Institute of African Ornithology
University of Cape Town
Rondebosch 7700, South Africa

I. INTRODUCTION

Work on shorebird activity in the past 20 years or so has largely comprised descriptive studies of general feeding ecology (Holmes, 1966; Recher, 1966; Holmes and Pitelka, 1968; Thomas and Dartnall, 1971; Smith and Evans, 1973; Puttick, 1978; Hartwick and Blaylock, 1979), foraging behavior (Goss-Custard, 1969; Baker and Baker, 1973; Baker, 1974; Evans, 1975; Hartwick, 1976; Puttick, 1979), and foraging related to prey availability (Bengtson and Svensson, 1968; Goss-Custard, 1970; O'Connor and Brown, 1977; Evans, 1979; Sutherland, 1982a,b). As a result, some general patterns of activity are beginning to emerge, although few detailed studies of activity patterns have been published. In addition, some information is available on the effects of abiotic and biotic influences on activity. In this chapter, I plan to describe general activity patterns among wintering shorebirds and then to concentrate mostly on foraging activity. I will also describe some of the abiotic and biotic influences that affect shorebird foraging, and discuss these effects. Finally, I will briefly discuss optimal foraging theory with respect to studies done on shorebird foraging. I have chosen to discuss foraging and activity patterns in wintering, rather than breeding, shorebirds for two reasons. First, in the past two decades, research on the basic ecology and breeding behavior of breeding birds

* Present address: Museum of Comparative Zoology, Harvard University, Cambridge, Massachusetts 02138.

has advanced much faster than on wintering birds (Pitelka, 1979) and, second, other chapters in this volume will cover aspects of foraging in breeding shorebirds.

Shorebirds require energy, in the form of food, for three main purposes while on their wintering grounds: self-maintenance; flying to and from their feeding grounds; and foraging (Evans, 1976). Procuring energy for these needs is subject to several major abiotic and biotic influences. Of the former, I intend to concentrate on tides, daylength, and temperature (see also Burger and Myers, both in this volume). In addition, I will discuss the following biotic influences on foraging: prey population fluctuations; bird density; and bird age and sex. Shorebird activities are affected by an interacting array of factors including several in addition to those mentioned above, e.g., wind (Evans, 1976), substrate (Myers *et al.*, 1980), territoriality (Myers *et al.*, 1981), and predation (Page and Whitacre, 1975). Very little has been published on the effects of the first three of these on foraging activity in shorebirds. The fourth, territoriality, has been well studied in Sanderling *Calidris alba* and is discussed elsewhere in this volume.

II. ACTIVITY PATTERNS

In general, most wintering shorebirds show similar daily activity patterns. They usually arrive on feeding areas as these are uncovered by the tide (Heppleston, 1971; Burger *et al.*, 1977; Kelly and Cogswell, 1979; Rands and Barkham, 1981), and the number of birds foraging usually peaks around low water (Heppleston, 1971; Burger *et al.*, 1977; Hockey, 1982). In some species there may be a bimodal distribution of foraging activity with one peak occurring on ebb and one on flow tides (Burger *et al.*, 1977; Hartwick and Blaylock, 1979). In the Northern Hemisphere, most birds spend almost all daylight hours foraging and many species forage at night also (Goss-Custard, 1969; Evans, 1976). In the Southern Hemisphere, two studies so far have shown that birds spend fewer hours foraging than birds in the Northern Hemisphere (Puttick, 1979; Hockey, 1982). Birds roost for most of the rest of their daily activity periods.

Activity patterns of the Curlew Sandpiper *Calidris ferruginea* in the Southern Hemisphere follow this general pattern. Curlew Sandpipers are Palearctic migrants that winter in the Southern Hemisphere from September to April. I studied the feeding ecology of Curlew Sandpipers at Langebaan Lagoon (33°S, 18°E), South Africa, where a population of 37,000–55,000 shorebirds winters during the austral summer (Pringle and

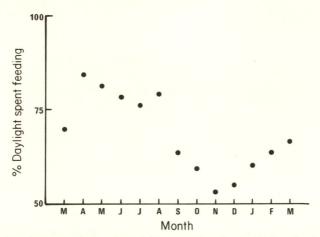

Fig. 1. Monthly changes in the amount of time spent foraging by Curlew Sandpipers during daylight hours. [From Puttick (1979). Reproduced with permission from the Netherlands Ornithological Society.]

Cooper, 1975; Summers, 1977). Curlew Sandpipers account for nearly two-thirds of this population. In the austral winter (May–August), the Curlew Sandpiper population is as high as 12,000 individuals (Pringle and Cooper, 1975), and most of these are first-year birds spending their first boreal summer in the wintering grounds (Elliott *et al.*, 1976). Other common species at Langebaan are Red Knot *Calidris canutus*, Black-bellied Plover *Pluvialis squatarola*, Sanderling *C. alba*, and Ruddy Turnstone *Arenaria interpres*. Curlew Sandpipers showed marked changes in activity seasonally, which appeared to be responses to seasonal variations in energy requirements (Puttick, 1980). Hence, they foraged for 55–65% of daylight hours in spring (October–December) and summer (January–March) but up to 80% in autumn (April–June) and winter (July–September) (Fig. 1). The birds foraged on the sandflats as soon as these were exposed, regardless of the time of day. The birds' foraging and success rates reflected the density and distribution of potentially available prey. Hence, rates were highest at the top half of the shore where prey density was higher (Puttick, 1977) and decreased down the shore (Puttick, 1979). During autumn and winter, the birds foraged in marsh areas during high tides instead of roosting there (Fig. 2). Birds roosted on marsh areas during high water of springs; during high water of neaps they remained roosting on sand spits at the high-water mark. About a quarter of the population foraged in the marshes at low tide in all seasons. Birds roosted for up to 25% of daylight hours during the longer days of spring and summer. Another South African study showed that the proportion of the population

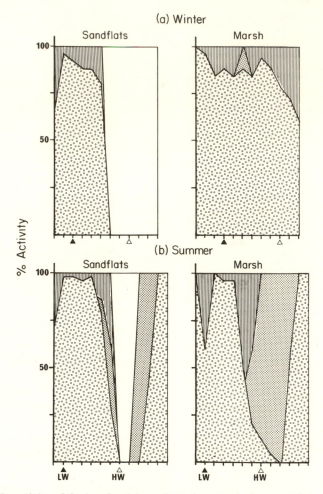

Fig. 2. Daily activity of Curlew Sandpipers from dawn to dusk on sandflats and marshes during (a) winter (first-year birds) and (b) summer (total population). Open circles indicate foraging, vertical lines roosting, and dots standing and/or preening. Blank spaces indicate that no birds were present. [Adapted from Puttick (1979).]

of Black Oystercatchers *Haematopus moquini* foraging on a rocky shore was highest at low tide levels, and a progressively larger proportion roosted as the tide rose (Fig. 3). The birds foraged for 37% of daylight hours in winter (Hockey, 1982). In contrast to these two studies in the Southern Hemisphere, shorebirds in the Northern Hemisphere fed for more than 90% of daylight hours, except for oystercatchers in one study (Table I). What may explain this difference? Daylength in winter at the

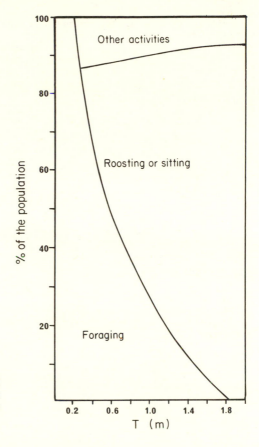

Fig. 3. Activity partitioning of African Black Oystercatchers related to the predicted tidal height. [From Hockey (1982).]

Northern Hemisphere sites is shorter than that at the Southern Hemisphere sites, as latitudes are fairly different (53°N at the Wash and 57°N at Ythan vs. 33°S at Langebaan). In midwinter, the period sunrise to sunset at the former latitudes is 7–7½ hr while it is 10 hr at the South African sites. This difference may account for apparent differences in the percentage of total daylight hours spent foraging in Northern and Southern Hemisphere sites, respectively. Other possible factors could be differences in ambient temperatures or differences in prey density and/or availability. Lower ambient temperatures in the Northern Hemisphere would increase the birds' energy requirements relative to those in the Southern Hemisphere. Lower temperatures would also affect prey availability (Goss-Custard, 1969). Lower prey availability would increase the length of time birds would need to forage to satisfy any given energy requirement.

Table I. Foraging Activity (Percent of Daylight Hours) of Some Shorebird Species

Species	Season	Foraging (% daylight)	Place	Reference
Curlew Sandpiper (*Calidris ferruginea*)	S	60	Langebaan Lagoon, South Africa	Puttick (1979)
	W	80	Langebaan Lagoon, South Africa	Puttick (1979)
African Black Oystercatcher (*Haematopus moquini*)	W	37	Marcus Island, South Africa	Hockey (1982)
Common Redshank (*Tringa totanus*)	W	100	Ythan Estuary, Scotland	Goss-Custard (1969)
	W	100	East Lothian coast, Scotland	Baker (1981)
Eurasian Oystercatcher (*H. ostralegus*)	W	70–80	The Wash, east England	Goss-Custard (1979)
	W	93	Ythan Estuary, Scotland	Heppleston (1971)
Ruddy Turnstone (*Arenaria interpres*)	W	100	East Lothian coast, Scotland	Baker (1981)
Red Knot (*C. canutus*)	W	100	The Wash, east England	Goss-Custard (1979)
Mixed assemblage	S	93	New Jersey coast, USA	Burger *et al.* (1977)

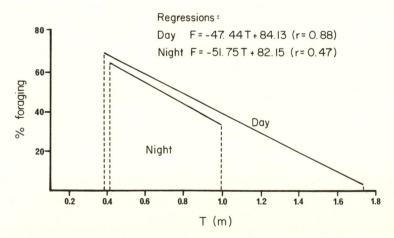

Regressions:
Day   F = -47.44 T + 84.13 (r = 0.88)
Night F = -51.75 T + 82.15 (r = 0.47)

Fig. 4. Diurnal and nocturnal foraging activity of territorial African Black Oystercatchers related to predicted tidal height. [From Hockey (1982).]

III. FACTORS AFFECTING FORAGING ACTIVITY

Foraging activity may be affected by several environmental factors. These factors act directly by affecting the birds' foraging behavior and/or physiology, or indirectly by affecting prey availability. In the following section (A), I will discuss three important abiotic factors that affect shorebird foraging activity—tide, daylength, and temperature (see also Burger, this volume). In the subsequent section (B), I will discuss the influence of four biotic factors on foraging—prey population fluctuations, bird density, and bird age and sex.

A. Abiotic Factors

1. Tides

The habitat used by shorebirds wintering in marine coastal areas is dominated by tidal cycles, and therefore tides are obviously important in affecting shorebird foraging and other activities (Heppleston, 1971; Prater, 1972; Burger *et al.*, 1977; Puttick, 1979; Morrell *et al.*, 1979; Hartwick and Blaylock, 1979; Connors *et al.*, 1981; Hockey, 1982). So, for example, the foraging activity of African Black Oystercatchers on a rocky shore showed a clear linear relationship to tidal height (Fig. 4).

Tides influence shorebird foraging directly by their effect on the length of time, and on the amount of space, available for foraging each day. The interaction of tides and daylight in affecting time available for foraging is also important, depending on the relative success of foraging by day and by night. This, in turn, depends on whether birds forage by touch or by sight (Evans, 1976). Tides dictate patterns of habitat use and hence the daily movements of birds because of the above direct effects. For example, Curlew Sandpipers switched foraging and roosting areas subject to tidal changes (Puttick, 1979). They foraged on marsh areas when sandflats were inundated and also moved to the upper reaches of the lagoon, where the tides were delayed, from already covered flats nearer the mouth (personal observation). Sanderlings at Langebaan Lagoon foraged on ocean sandy beaches as well as on the mudflats of the lagoon itself (Summers, 1977). Tides on the two feeding areas were not synchronous and the birds regularly flew 12–15 miles each day in order to forage in both areas. Similarly, Sanderlings wintering at Bodega Bay, California, also switched foraging areas on a regular tidal schedule by foraging on outer beaches at high and mid-tide levels and then moving to protected sandflats as the tide receded (Connors *et al.*, 1981). Also, Burger *et al.* (1977) found that a mixed assemblage of shorebirds feeding in three intertidal habitats in New Jersey showed a temporal pattern of foraging characteristic of each of the three habitats related to tide time.

The indirect effects of tides on foraging have been fairly well documented. Tides affect the activity rhythms of invertebrates and thus prey availability for shorebirds. Lugworms *Arenicola marina* (Smith, 1975, in Evans, 1979), snails *Hydrobia ulvae* (Newell, 1962; Little and Nix, 1976), amphipods *Bathyporeia pilosa* (Preece, 1971), bivalves *Macoma balthica* (Brafield and Newell, 1961), crabs *Carcinus maenas* (Naylor, 1958), and ragworms *Nereis diversicolor* (Vader, 1964) all show activity rhythms mediated by tides.

Curlew Sandpipers fed faster on incoming than on outgoing tides (Puttick, 1979), presumably in response to increased prey activity after quiescence during the preceding exposure period (Vader, 1964). Similarly, Bar-tailed Godwit *Limosa lapponica* followed the tide edge closely on both ebb and flow, for their prey, the lugworm *A. marina*, formed casts at the tide edge and this provided a detectable cue for the birds (Smith, 1975, in Evans, 1979).

The effect of tidal range interacting with prey distribution affects prey availability for shorebirds also. Larger prey organisms tend to occur lower down on the shore and are only available for a short time each tidal cycle. *Cerastoderma edule*, the main prey of the Eurasian Oystercatcher *Haematopus ostralegus* at Traeth Melynog bay, Anglesey, move downshore as they grow (Sutherland, 1982a). Also, prey density may be higher at lower shore levels. Connors *et al.* (1981) found that the "energy density" (kcal/m^2) of Sanderling prey increased down a sandflat shore at Bodega Bay and this was due to increased prey density. The zonation of intertidal invertebrates is not only affected by tides but also by slope, exposure, and substratum as well, which in turn affect their availability as prey items (Puttick, 1979; Myers *et al.*, 1980).

Several responses by shorebirds to some of the indirect effects described above might be predicted. Birds may switch prey items as these are uncovered by the tide or as the activity rhythms of preferred prey change. Birds may forage faster or use alternative foraging methods in response to different prey at various shore levels. Finally, they may be expected to concentrate on one shore level if potential prey availability does not alter much there during a tidal cycle. Curlew Sandpipers and Sanderlings both switched prey as changing tide levels exposed different species as potential prey (Evans, 1979; Puttick, MS). The intensity of foraging activity of African Black Oystercatchers increased during neap tides when the tidal range was decreased (Hockey, 1982) and Eurasian

Oystercatchers foraged for a greater proportion of available foraging time when feeding areas were exposed for a shorter time at neaps (Heppleston, 1971). Finally, Black-bellied Plovers stayed around a defended area at midtide level through the low-water period (Evans, 1979). However, the availability of ragworms, their main prey, diminished with time after exposure of the site, and the birds searched progressively larger territories as they foraged. Some birds were unable to maintain a uniform rate of prey intake, in spite of this adaptation (Evans, 1979).

In summary, shorebird activity does appear to change markedly in response both to direct and to indirect effects of tidal patterns. These changes are reflected in the birds' choice of different habitats in which to forage, time spent foraging, foraging rate and method, and shore level at which the birds feed.

The varied array of responses observed in shorbirds to the direct effects of tides and to the common problem of temporal variations in food supply caused by tidal cycles is not surprising considering the complex interaction of factors that give each intertidal habitat a unique set of local tidal conditions. Tidal regimes vary among geographical areas in amplitude, periodicity, and regularity. These are complicated by local differences in tidal drainage, which may cause large time lags within one lagoon or estuary, and local wind, wave, and current conditions (Connors *et al.*, 1981).

2. Daylength

Daylength affects the activity of wintering shorebirds by determining the time available for foraging, particularly in the higher latitudes where daylength is markedly reduced. Wintering shorebirds appear to have several "options" for meeting their energy needs during the shorter days of winter. They may forage for longer periods during the day (assuming they are not already using all available daylight hours to forage), they may forage at night, or they may increase their ingestion rate by foraging faster or by selecting energetically more rewarding prey. In fact, different shorebirds employ all these tactics under different circumstances.

Many shorebirds forage during all, or almost all, daylight hours (Goss-Custard, 1969, 1979; Baker, 1981) and forage sequentially in habitats over which tidal exposures are not in phase in order to do so (Heppleston, 1971; Baker and Baker, 1973; Hartwick and Blaylock, 1979). Also, birds can increase their ingestion rates by foraging sequentially in habitats with the highest foraging profitability (Connors *et al.*, 1981; Baker, 1981). Some birds forage at night (Goss-Custard, 1969; Heppleston, 1971; Baker and Baker, 1973; Evans, 1976; Hockey, 1982) but their foraging behavior may

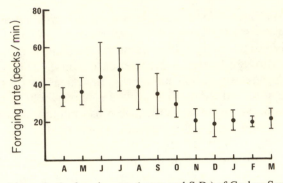

Fig. 5. Monthly changes in the foraging rate (mean and S.D.) of Curlew Sandpipers. Means from May to August represent first-year birds only.

change in order to do so, depending on alterations in prey behavior at night (Vader, 1964), or on whether they feed by touch or by sight (Evans, 1976).

Nonbreeding individuals in a number of northern species remain at mid- or southern latitudes during the austral winter (Pitelka, 1979). Curlew Sandpipers are among these, and during the austral winter they foraged for longer periods during the day and expanded their use of different feeding areas. Numbers and individual sizes of prey items in marsh feeding areas increased during winter, and this increased foraging profitability (Puttick, 1977, 1978). The birds also foraged faster in winter (Fig. 5), but their rate of prey intake did not appear to increase in spite of this.

Predators hunting at night are less efficient, especially those that hunt by sight (Evans, 1976). Black-bellied Plovers, for example, foraging at Teesmouth Estuary, England, fed at only two-thirds the daytime rate when foraging at night. African Black Oystercatchers only foraged for half the available foraging time at night, though Hockey (1982) suggests that reduced nocturnal feeding may be due to exposure to predators on the upper half of the shore (for strandline vegetation may harbor mammalian predators) rather than to less efficient feeding at night.

3. Temperature

Low temperatures affect foraging directly by increasing the birds' energy demands, resulting in the need to increase energy intake. In addition, migratory species appear not to have well-developed powers of thermoregulatory acclimatization (King, 1974). As far as I am aware, very little information dealing with the thermal biology of shorebirds is available, but heat loss due to lowered temperatures during winter may be considerable (Evans, 1976).

Low temperatures affect shorebird foraging indirectly by altering the activity levels of prey, which in turn alters prey availability (temperature extremes at the surface of the substratum may be avoided by invertebrates to some extent by burrowing), or by reducing prey availability entirely when the exposed substratum freezes. However, foraging will still be possible along the water's edge when the exposed substratum freezes, if sea temperatures remain above freezing (Evans, 1976).

Several studies have shown how foraging has been affected by temperature-mediated changes in prey behavior and/or activity. Mud temperature in winter was an important cause of changes in the foraging and ingestion rates of Common Redshank *Tringa totanus* on the Ythan Estuary because it changed the behavior of the main prey *C. volutator* (Goss-Custard, 1969). The frequency with which *Corophium* protruded from their burrow entrances decreased as mud temperatures decreased, and this presumably changed their relative availability. Below 6°C, redshank took increasing numbers of the bivalve *Macoma balthica* or the worm *Nereis diversicolor* instead of *Corophium*. The numbers of bivalves taken by Red Knots on the Wash, Norfolk, also declined with decreasing mud temperature (Goss-Custard, 1979). Similarly, Bar-tailed Godwits at Lindisfarne, Northumberland, also switched diets at low temperatures and again the switch was caused by changes in the activity of lugworms, their main prey. Also, capture rates of the polychaete *Scoloplos armiger* by Ringed and Black-bellied Plovers were negatively correlated with decreasing temperature (Evans, 1979).

There are indications that wintering shorebirds have difficulty in meeting their energy requirements during winter. First-year Curlew Sandpipers had substantial energy deficits during the austral winter (Puttick, 1980), though this may have been due to factors other than the effects of reduced prey availability coupled with increased energy demands (see Section III). Goss-Custard (1980) concludes that food shortage, in part due to lower prey availability at lower temperatures, may be at least a contributory factor to winter mortality at British estuaries but the evidence is circumstantial. Most of the birds present at British estuaries in midwinter appear to endure adverse foraging conditions rather than try to find better foraging areas elsewhere (Evans, 1976). It seems probable that young individuals are the most likely to be at risk, and hence that winter mortality induced by an inability to obtain food may be potentially important in affecting wader population dynamics.

B. Biotic Factors

Many studies of the effects of prey population fluctuations, bird density, and the age and sex of shorebirds on foraging patterns or foraging behavior have as their underlying assumption that birds were optimizing their food or energy intake. This did not necessarily imply that food was limiting, and therefore that competition was the proximate cause of the observed foraging differences. Conflicting hypotheses exist in the literature concerning the importance to shorebirds of competition on their wintering grounds, and very few of these have been tested (cf. Baker and Baker, 1973; Duffy *et al.*, 1981). Circumstantial evidence shows that food shortage may be implicated in the winter mortality of shorebirds in Britain (Goss-Custard, 1979), but there is no direct evidence as yet that food is generally limiting for wintering shorebirds and this should be borne in mind while reading the rest of this chapter.

1. Prey Population Fluctuations

Marked seasonal fluctuations in the density and biomass of intertidal invertebrates, especially those on sandy or muddy shores, have been recorded (George, 1964; Brafield and Chapman, 1967; Hughes, 1970; Beukema, 1974; Chambers and Milne, 1975; Rees, 1975; Puttick, 1977). These changes can be expected to influence foraging patterns in shorebirds. Smaller sizes of newly-recruited or settled prey individuals are energetically less rewarding than larger and older individuals, and fluctuations in numbers and biomass of prey items due to recruitment, immigration, and growth will change the profitability of different areas to foraging shorebirds as the seasons change. This in turn may have major effects on shorebird foraging and activity patterns. Apart from fluctuations in prey populations due to their own life cycle patterns, shorebird depredation itself can be expected to influence prey populations and hence food availability by depleting prey numbers or by altering the size distribution of prey available as the season progresses. Seasonal fluctuations in the populations of prey species potentially available to Curlew Sandpipers at Langebaan Lagoon appeared to have an impact on the birds' foraging patterns.

The biomass and numbers of invertebrates on the intertidal sandflats at Langebaan Lagoon increased in autumn (Puttick, 1977) at the time that Curlew Sandpipers were accumulating fat prior to migration (Elliott *et al.*, 1976). The biomass of individual invertebrates occurring in the salt marshes increased during the winter, at the time that wintering first-year birds had fewer daylight hours in which to forage and when their energy needs were possibly elevated because of the lower temperatures. An increase in prey density will increase the probability that foraging will be successful and larger individual prey will enhance the energy reward per prey item (up to a point). Both these factors appeared to enhance Curlew Sandpiper foraging success at times in the annual cycle when energy needs were elevated (Puttick, 1979). *Assiminea globulus* were recruited into the intertidal population in spring; however, the proportion of these snails in the birds' diet dropped in spring (Puttick, 1978). Presumably, at this time the smaller size of individual snails made them energetically less rewarding, and the birds instead selected more nereid worms and crabs then.

Some intertidal invertebrates migrate downshore in winter or as they grow larger. For example, the polychaete *Nerine cirratulus* was the most important prey of Sanderlings in northeastern England. It moved from just below midtide in summer to a lower position on the shore in winter, and Sanderlings altered their food spectrum as a result (Evans, 1979). Cockles *Cerastoderma edule* at Traeth Melynog bay moved downshore as they matured. As the profitability of different sites to Eurasian Oystercatchers depended on cockle size, birds adjusted their foraging patterns with respect to this shift (Sutherland, 1982a,b).

Evans (1979) suggests that some invertebrates may show cyclical behavior patterns not in response to environmental changes but to avoid avian predators. For example, *M. balthica* in the Wash moved downward in the substratum during winter. In summer, 90% were at a depth that was accessible to the Red Knot, their major shorebird predator. However, in winter only 4% were found at depths that were accessible to Red Knots. The most marked downward movement coincided with the presence of Red Knots, and so this seasonal movement did not appear to occur at the Ythan Estuary, Scotland, where Red Knots did not occur.

Prey populations may be substantially depleted by shorebird depredation. Curlew Sandpipers, for example, took an estimated 34–51% of the standing crop of intertidal invertebrates at Langebaan Lagoon (Puttick, 1981). Red Knots took about 14–34% of the *C. edule* and *M. balthica* standing crop at the Wash in winter (Goss-Custard, 1977b), and the whole wader population of the Wash took 25–45% of the total amount of food available in winter, which reduced their average intake by 5–30% during this period (Goss-Custard, 1980). The depletion of prey populations has been shown to affect subsequent foraging patterns. So, for example, Eurasian Oystercatchers severely depleted the stocks of *C. edule* in Strangford Lough, Northern Ireland, each winter. They initially concentrated their feeding in areas containing second-year and larger cockles, but when numbers of these were sufficiently depleted they began to accept smaller cockles and to forage again in areas that had previously been depleted of the originally preferred sizes of prey (O'Connor and Brown, 1977).

2. Bird Density

There are broad variations both within and among shorebird species in dispersion patterns while foraging. These may range from a loose scattering of birds foraging over a sandflat to the territoriality of sandy beach Sanderlings (see Myers, this volume). In general, aggregation appears to be related to enhanced foraging efficiency in particular localities (Murton, 1971; Krebs et al., 1972), and may be expected to affect foraging patterns. The densities of foraging shorebirds often correlate positively with the densities of their prey (Prater, 1972; Bryant, 1979; Goss-Custard, 1979; Rands and Barkham, 1981; Burger, 1982; Puttick, MS), thus confirming the suggestion of Royama (1971) and Hassell and May (1974) that predators should collect in areas of high prey density. Studies have shown that foraging efficiency was enhanced at higher prey densities (Goss-Gustard, 1970, 1977a; Puttick, MS), or at least that shorebirds foraging in flocks were more successful than those that foraged singly (Smith and Evans, 1973; Silliman et al., 1977).

However, prey density alone was insufficient to account for the distribution of foraging shorebirds in at least two studies. The density of Curlew Sandpipers foraging at Langebaan Lagoon increased initially as prey density increased but declined at high prey density (Fig. 6). Similarly, the density of Eurasian Oystercatchers feeding on cockles at Traeth Melynog bag dropped sharply at cockle densities above about 60/m^2 (Sutherland, 1982b). In both these studies, individual prey at higher than intermediate densities were found to be smaller than those at the low and intermediate densities. Thus, prey size was the important factor affecting bird densities, for fewer birds chose to feed on the smaller, energetically less rewarding prey items. Most birds instead gathered together at those prey densities where prey items were larger and profitability, in terms of energy intake per unit time, was highest.

In the studies mentioned till now, depression of food intake, i.e., reduced foraging efficiency, affected shorebird foraging at higher bird densities. What other factors may affect foraging at higher bird densities by depressing food intake? One may be that of prey depletion. Shorebirds may have considerable impact on invertebrate populations over the course of a season and this does have an impact on foraging success (see Section III.B.1). However, depletion is unlikely to provide cues to birds in the short term that bird densities are too high. Second, prey themselves might affect foraging patterns by reacting to the presence of larger numbers of

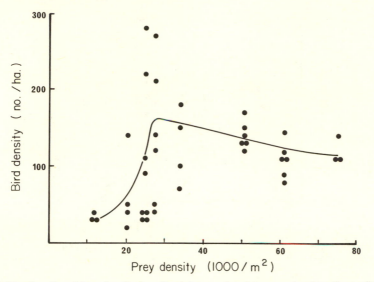

Fig. 6. The density of foraging Curlew Sandpipers related to prey density. Curve fitted by eye.

birds on the surface of the sand or mud. For example, *Corophium* activity was arrested by Common Redshanks walking over the mud surface (Goss-Custard, 1980). Third, aggressive encounters over food have been shown to increase as bird density increased and birds foraged closer together (Hamilton, 1959; Goss-Custard, 1977a; Silliman *et al.*, 1977; Burger *et al.*, 1979). Aggressive interactions among Curlew Sandpipers increased as bird density increased (Fig. 7). Encounters occurred mainly over food and this probably decreased the average foraging rate. Recher and Recher (1969) suggested that aggression among dense aggregations of shorebirds would be suppressed if high concentrations of food were present. In contrast to the situation at Langebaan (Fig. 7), Curlew Sandpipers feeding on an extremely dense aggregation of dipteran larvae in the wrack on an ocean beach appeared to suppress their aggression, so enhancing foraging efficiency in this situation and capitalizing on the availability of a rich food supply (Puttick, 1981).

Goss-Custard *et al.* (1982) have described in detail the relationship between aggression and density among Eurasian Oystercatchers foraging on mussel beds in the Exe Estuary, southern Devon. A population of immature oystercatchers fed on two preferred mussel beds during spring and summer, but the population of oystercatchers on the estuary increased sixfold as adults returned there from breeding. Birds spread to less suitable mussel beds presumably as interference, or a decline in the rate of

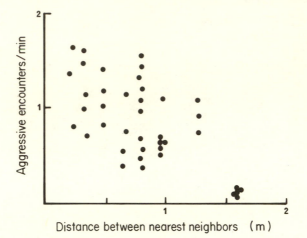

Fig. 7. The frequency of aggressive encounters between Curlew Sandpipers while foraging on sandflats, related to the distance separating individuals.

prey taken per predator, and competition took place. Spacing was achieved because immature birds were displaced from preferred mussel beds by adults. Adults were more aggressive than immatures and avoided other birds much less often. Adults were also much more successful in defending their food against attacks. Consequently, the immature birds spread out to low-ranking beds and only returned to preferred mussel beds when adults left the estuary at the end of winter. Immatures changed from least preferred to more preferred beds as they got older, until their status increased sufficiently to enable them to stay on preferred beds.

Birds may achieve a density at which interference among themselves is minimized by some form of avoidance behavior. Spacing by mutual avoidance has been described for Eurasian Oystercatchers (Vines, 1980). The spacing pattern observed among individuals feeding on mussels *Mytilus edulis* at the Ythan Estuary in winter was significantly nonrandom, but the birds maintained a preferred nearest-neighbor distance. This pattern, described as "spaced-out gregariousness," was not maintained by means of agonistic encounters but rather by means of mutual avoidance. Eighty-six percent of agonistic encounters that did occur involved food: the goal of attack appeared to be to steal the attacked bird's mussel and/ or its position on the feeding area. Mutual avoidance, then, appeared to decrease the possibility that individuals that found food would be attacked. Spaced-out gregariousness as an associative pattern predicts that a preference for specific nearest-neighbor distance would tend to establish an upper limit to density, but Vines' study did not conclusively show this. In any event, the adaptive significance of avoiding neighbors appeared to be that attack success declined as the initial distance between attacker and victim increased. For each individual there was a compromise between the advantages of avoiding being robbed and of getting close enough to launch a successful theft. The degree of spacing in different areas differed, and it was not clear whether the "food stealing" model could account for these. Vines (1980) suggests that pressure from individuals attempting to settle in areas where feeding conditions were particularly favorable might lead to smaller distances between individuals. Higher feeding rates may then compensate for higher attack rates. It would, in fact, be interesting to know what effect a particular oystercatcher density had on the foraging rate of birds at that density.

Whatever the mechanisms of interference among foraging shorebirds, it seems clear that the tendency for birds to aggregate in response to the enhanced feeding efficiency presumably associated with aggregating would be counteracted by the interference that appears to follow when birds get too close. What too close means varies, depending on many different factors attending the individual birds so far studied. These factors were the age or social status of the individual birds, the behavior of prey, prey density or size, the senses needed to detect prey, and probably others.

An important factor I have not mentioned yet with respect to bird density is the birds' response to the risk of predation. Aggregation may occur in response to this risk, despite the potential disadvantage of foraging interference occurring among close neighbors. The likelihood of shorebirds foraging in a Californian lagoon being taken by raptors differed by the ratio 3.2:1 depending on whether the bird was foraging singly or in a flock (Page and Whitacre, 1975). On the other hand, Goss-Custard (1979) and Prater (1979) regarded the chances of a wintering shorebird in Britain being taken by a raptor as negligible. Stinson (1980) presents a model of flocking that suggests that even infrequent attacks on foraging birds can select for flocking to enhance predator avoidance. He found that the spatial dispersion of foraging Black-bellied Plovers on St. Catherines Island, Georgia, was consistent with the hypothesis that the threat of predators was a major selective pressure influencing flocking in foraging shorebirds, although the evidence he presents that the birds' spatial dispersion was not due to other factors, such as prey distribution, is not entirely convincing. Without quantitative studies of how flocking affects survival rates, it will be difficult to demonstrate the effect that predator avoidance has on shorebird aggregation and/or flocking.

Table II. Mean Foraging Rates (Pecks or Probes
per min) of Adult and Immature Curlew
Sandpipers[a,b]

	Mean	S.D.	N
Adult	43.13	18.67	15
Immature	28.44	8.31	15

[a] Modified from Puttick (1979).
[b] Each value is the mean foraging rate of 15 separate 2-hr
periods of observation. Mann–Whitney U-test, U = 58,
$p \leq 0.001$.

3. Age and Sex

Age-related differences in foraging success have been studied in many
seabirds (Orians, 1969; Recher and Recher, 1969; Buckley and Buckley,
1974; Verbeek, 1977; Morrison *et al.*, 1978; Searcy, 1978), and in all these
cases juveniles were less successful at foraging than adults. Similarly,
juvenile shorebirds were relatively less successful in the following species: Ruddy Turnstone (Groves, 1978), Black-necked Stilt *Himantopus
americanus* (Burger, 1980), and Curlew Sandpiper (Puttick, 1979). Immature Curlew Sandpipers foraged significantly slower than adults did
(Table II), which I attribute to an inability to locate prey. They had significantly less foraging success (Table III), and also took smaller prey
items than adults (Puttick, 1978). Finally, they spent significantly more
time foraging each day than adults did. Although data on foraging success
were lacking in a study of juvenile Eurasian Oystercatchers foraging on
mussel beds on the Exe Estuary, dominant adults ousted juveniles from
preferred beds during winter and their attacks on juveniles over food items
were more successful than vice versa (Goss-Custard *et al.*, 1982). Both
these factors must have reduced juvenile foraging efficiency.

Table III. Success Rates (No. Prey Objects per
min) of Adult and Immature Curlew
Sandpipers[a,b]

	Mean	S.D.	N
Adult	10.08	3.20	24
Immature	4.33	1.44	24

[a] Modified from Puttick (1979).
[b] Each value is the mean success rate of 24 separate 2-hr
periods of observation. z [standard normal variate calculated from the Mann–Whitney U-test (Siegel, 1956)]
= 5.71, $p \leq 0.00003$.

Some of the above authors suggested that inexperience was the main reason why immature birds foraged relatively inefficiently. If this is so, then it should follow that disparities between juvenile and adult foraging success should diminish with time as juveniles gain more experience. So far, this has not been shown, but as Groves (1978) notes, it is probable that the most inefficient juveniles do not reach adulthood. Those that did survive their first few months would improve their foraging efficiency as they matured. Evidence of age-related mortality among shorebirds showed that first-year birds accounted for 77% of winter mortality among Eurasian Oystercatchers at the Ythan Estuary during one winter (Heppleston, 1971).

Few studies of sex-related differences in foraging among shorebirds have been published. Selander and Giller (1963) suggested that evolutionary divergence of the sexes in foraging behavior accompanied by increased sexual dimorphism in size would permit a wider and more thorough exploitation of available food resources by individuals within the species. Males and females may employ similar foraging techniques but take food of different sizes, employ different foraging techniques, or forage in different microhabitats. There is evidence for all three occurring among shorebirds. Curlew Sandpiper females are larger than males (Thomas and Dartnall, 1971), and the female's bill is 10% longer on average than that of the male. Females exploit a slightly different prey spectrum than males by taking a wider size range of prey items and fewer prey species (Puttick, 1978). Furthermore, males and females occurred more often in segregated than in mixed groups, though it was not possible to determine whether they were foraging in different microhabitats. Agonistic behavior was a function of between-bird distance within foraging groups, rather than a function of encounters between males and females (Puttick, 1981), which I interpret to mean that competition was in fact minimal between males and females. Finally, females appeared to forage faster and certainly more successfully than males did.

Similarly, Smith and Evans (1973) found spatial segregation among foraging male and female Bar-tailed Godwit at Lindisfarne, and suggested that this was important in reducing competition in areas where bird density was high in relation to available foraging area. Finally, Pitelka (1950) showed that sexual dimorphism and different microhabitat utilization in two dowitchers, *Limnodromus griseus* and *L. scolopaceus*, increased the potential area that the two species could utilize for foraging.

Age and sex differences in foraging behavior, foraging success, habitat use, social dominance, and mortality seem to imply that competition may be an important factor affecting shorebird populations in wintering areas. However, the evidence in the above studies suggesting that competition, both among age classes and between sexes either for food or for foraging areas, may be important, is circumstantial. Before invoking competition as one of the reasons why age and sex differences in foraging behavior may occur, it should be established that food is, in fact, limiting.

IV. OPTIMAL FORAGING

So far, I have considered the effects of abiotic and biotic factors on foraging efficiency, that is, factors over which a bird has no control (with the exceptions of prey density and bird density). In this section, I will discuss the choices an animal must make once it has "decided" to forage under a given set of environmental circumstances.

The assumption that energy intake is being maximized or that maximum profitability is the goal of the feeding choices an animal makes has been implicit in the preceding sections. For example, a bird choosing to feed at a particular shore level where prey density is high is presumably doing so because it can maintain a high rate of prey intake there. At a more fine-grained level, optimal foraging theory explores the choices an animal makes once it is foraging. These choices affect foraging behavior and hence also foraging and activity patterns, and it is in this context that I will discuss two aspects of optimal foraging theory.

Animals face particular problems while foraging. These include what prey types to choose, as these will differ in net energy yield, choice of search strategy appropriate to prey type and foraging area, where to forage, as different patches of their environments will probably have different levels of profitability, and what time allocation to make there (Krebs and Cowie, 1976). In most studies, it has been assumed that energy intake is being maximized (but see Glander, 1981). Schoener (1971) has described as energy maximizers those animals that, limited by a fixed amount of time in which to forage and whose fitness is increased with increasing amounts of energy obtained, strive to maximize the amount of energy they obtain in that time. Because for most wintering shorebirds foraging time is limited both by tidal cycles and by short hours of daylight, they may be described as energy maximizers. Optimal foraging theory has provided models of how animals should forage so as to maximize energy intake. The approach has been to identify compromises and constraints operating when an animal is foraging and to construct cost/benefit models to incorporate these. Cost/benefit analysis provides a standard against which the behavior of an animal can be compared, and has stimulated work in most areas of predator behavior especially with regard to prey

Table IV. Seasonal Changes in (a) the Profitability $(P)^a$ of the Main Prey Items of Curlew Sandpipers and (b) the Percent Occurrence of These (by Mass) in the Diet

	Worms		Crabs		Snails	
(a)	P	Rank	P	Rank	P	Rank
Spring	0.026	1	0.023	2	0.012	3
Summer	0.014	2	0.020	1	0.012	3
Autumn	0.021	1.5	0.021	1.5	0.016	3
Winter	0.028	1	0.017	2	0.012	3
(b)	%	Rank	%	Rank	%	Rank
Spring	50	1	25	2	15	3
Summer	80	1	10	2	5	3
Autumn	75	1	15	2	5	3
Winter	70	1	20	2	10	3

$^a P = E/H$, where E is energy content of prey item and H is handling time.

choice and patch choice. I will discuss these briefly below. The subject has been reviewed by Pyke *et al.* (1977). Optimal foraging theory has so far comprised relatively simple models that have not yet enabled us to consider more than one aspect of foraging at the same time. However, recently some authors have begun to do this (Pyke *et al.*, 1977; Krebs *et al.*, 1981). In fact, shorebirds are potentially promising subjects for attempting a synthesis of this sort, for their activities or choices are usually easy to monitor, the distribution and density of prey are relatively easy to determine, and the basic feeding ecology of several species is already fairly well known.

A. Prey Choice

Simple models predict that animals should choose prey items that maximize their rate of energy intake (Pyke *et al.*, 1977). For each predator, each prey type has a dietary value or profitability determined by its energy yield (E) and the handling time (H) spent by the predator on procuring it. I calculated the profitability ($P = E/H$) of three main prey items in the diet of Curlew Sandpipers during four seasons and found a good relationship between P and the rank of these items as they appeared in stomach contents (Table IV). These data were not obtained in a direct test of the optimal prey choice prediction but they seem to indicate that Curlew Sandpipers mainly chose prey that provided the highest energy

Table V. Seasonal Changes in the Foraging Diversity of
Curlew Sandpipers on Sandflats and Marshes[a]

	Diversity $(\bar{H})^b$	
	Flats	Marsh
Spring	1.54	1.98
Summer	1.39	1.36
Autumn	1.48	1.41
Winter	1.50	1.62

[a] From Puttick (1979). Reproduced with permission from the Netherlands Ornithological Society.
[b] Calculated from the formula $\bar{H} = -\Sigma_1 p_1 \log_e p_1$, where p_1 = the proportion of observations in each category of foraging (MacArthur and MacArthur, 1961).

intake. The highest ranking prey should always be eaten when encountered and lower ranking prey should only be taken when higher ranking prey are too scarce to meet the energy requirements of the predator, irrespective of their own relative abundance. The proportions of worms, crabs, and snails at Langebaan Lagoon were 6:1:12 in summer and in autumn. The inclusion of worms in the diet in summer and in autumn when their profitability was less than that of crabs occurred possibly because crabs were too scarce to meet the birds' energy requirements.

Goss-Custard (1977c) provides good field evidence for optimal prey choice. Common Redshanks preferred those size classes of worms that provided the highest reward per unit handling time (P), i.e., large worms. Foraging rate on large worms depended on the density of large worms in the mud. Foraging rate on small worms depended more on the biomass of large worms ingested than on the density of small worms. Small prey were least preferred, for they were only taken when large prey were scarce.

If prey density decreases within a preferred feeding area, predators have two possible choices while foraging there: either to take a new prey item not previously taken, because of unpalatability for example; or to exploit the regular prey item(s) more thoroughly, for example by taking previously overlooked smaller individuals. In other words, dietary breadth should expand, i.e., selectivity should decrease, as food density decreases (MacArthur and Pianka, 1966; Schoener, 1971). Selectivity may be indicated by the degree of diversity of foraging behavior displayed by a predator. Curlew Sandpipers displayed high foraging diversity on the marshes at Langebaan Lagoon (Table V), when prey density was low in spring (Puttick, 1977), and low foraging diversity in summer, when prey density was high. Baker and Baker (1973) also showed that foraging diversity among six shorebird species showed seasonal changes in response to food density in accordance with this model.

B. Patch Use

The density of shorebird prey shows both spatial and temporal variations in the short term and the long term. If prey density decreases in an area, predators have the choice of moving elsewhere to find preferred prey. In order to do this efficiently, predators need to decide which areas or patches to visit and when to move from one patch to the next. Prey capture rate within an area is likely to be a decreasing function of time spent there (Cowie and Krebs, 1979), so if a bird is attempting to maximize the rate of food intake, its capture rate may eventually fall below the rate it could achieve by finding another area to feed in. In fact, shorebirds show a strong tendency to concentrate in areas where their intake is highest (Goss-Custard, 1977b,c, 1979; Sutherland, 1982b; Puttick, MS), but these are not always where prey density is highest. Profitabilities in different patches may change regularly, due to many different factors such as prey depletion (O'Connor and Brown, 1977), temperature- or tide-induced changes in prey behavior (Connors et al., 1981), interference (Goss-Custard, 1980), etc. It is implicit in optimal foraging models that predators can keep track of differing intake rates in different patches, or in other words, that they have perfect knowledge about the relative profitability of all potential foraging areas. Whatever the cause of changes in profitability, predators can generally only keep track of changes in patches by actually feeding there. The density of foraging Curlew Sandpipers was highest in those areas where profitability was highest, i.e., at intermediate prey densities (Puttick, MS). At higher prey densities, as I noted earlier, prey size was inversely correlated with density and this possibly decreased profitability in these patches. However, some birds foraged in the latter areas possibly to sample profitability. Similarly, Common Redshanks did not only concentrate their foraging efforts in highly profitable areas at all times but they also spent time feeding in poorer areas (Goss-Custard, 1977b). Goss-Custard concluded that this was because birds have to sample alternative feeding areas to keep track of changes in relative profitability. Finally, in a study of Eurasian Oystercatchers feeding on cockles, O'Connor and Brown (1977) showed that the birds sampled cockle patches apparently by assessing the local densities of cockles around the anvils where cockles were smashed open and abandoning patches when they became unprofitable. Birds would return to these previously unprofitable patches later in the winter to feed on smaller cockles when the most profitable cockle sizes had been taken in the preferred feeding areas.

In summary, the foraging behavior of the shorebirds mentioned in the above studies matches the qualitative predictions of optimal foraging theory with regard to choice of prey and choice of areas in which to forage. However, much of the information available so far provides only circumstantial evidence in support of optimal foraging predictions. In addition, very little information is available on shorebird foraging in the light of predictions made by optimal foraging theory in two other respects, namely what search strategy birds should use that would be most appropriate to prey type and foraging area, and what time allocation birds should make in a particular foraging area. In general, optimal foraging models have been reasonably well supported by available data (Pyke *et al.*, 1977). Shorebird studies, in particular, need to progress from descriptive accounts to quantitative tests of optimal foraging predictions, by means of field, and possibly laboratory, experiments.

V. CONCLUSION

Studies on the activity of wintering shorebirds so far have involved investigation of general feeding ecology, foraging behavior, and the relationship of foraging behavior to the distribution and availability of prey.

Studies have focused on spatial and temporal variability in shorebird environments and on how shorebirds have responded to these. A good deal is now known about the effects of biotic and abiotic influences on foraging. It is not yet clear whether, in fact, shorebirds are limited by food availability on their wintering areas. The assumption that food is limiting has influenced much of the work to date on foraging behavior and activity, but evidence has been circumstantial. Enough descriptive accounts and models of shorebird activity and foraging now exist. Experiments need to be conducted to test predictions based on this large body of information. These should concern: the predictions of optimal foraging theory with respect to the selection of prey items, foraging periods, foraging habitat, and foraging methods, especially comprising models that take into account more of the complexities of the bird's environment; the degree of competition occurring both among wintering shorebirds, and among wintering and resident shorebirds; the degree to which juvenile foraging improves with age, i.e., the extent to which "inexperience" renders juvenile birds less successful at foraging; and the importance of predator avoidance in determining the spatial patterns observed in foraging shorebirds.

ACKNOWLEDGMENTS

A. Burger, P. Hockey, R. Holmes, and W. R. Siegfried made valuable suggestions on an earlier version of the manuscript. P. Hockey kindly sent me his manuscript before it was published. D. Wheye, B. Leon, and L. Harwood helped with typing, illustrations, and word-processing. H. Mooney tolerated me in his laboratory while I was writing this chapter. I wish to thank all these people. I especially thank J. Glyphis for sharpening my pencils.

REFERENCES

Baker, J. M., 1981, Winter feeding rates of redshank *Tringa totanus* and turnstone *Arenaria interpres* on a rocky shore, *Ibis* **123**:85.

Baker, M. C., 1974, Foraging behavior of Black-bellied Plovers (*Pluvialis squatarola*), *Ecology* **55**:162.

Baker, M. C., and Baker, A. E. M., 1973, Niche relationships among six species of shorebirds on their wintering and breeding ranges, *Ecol. Monogr.* **43**:193.

Bengtson, S.-A., and Svensson, B., 1968, Feeding habits of *Calidris alpina* L. and *C. minuta* Leisl. (Aves.) in relation to the distribution of marine shore invertebrates, *Oikos* **19**:152.

Beukema, J. J., 1974, Seasonal changes in the biomass of the macro-benthos of a tidal flat area in the Dutch Wadden Sea, *Neth. J. Sea Res.* **8**:94.

Brafield, A. E., and Chapman, G., 1967, Gametogenesis and breeding in a natural population of *Nereis virens*, *J. Mar. Biol. Assoc. U.K.* **47**:619.

Brafield, A. E., and Newell, G. E., 1961, The behavior of *Macoma balthica* (L.), *J. Mar. Biol. Assoc. U.K.* **41**:81.

Bryant, D. M., 1979, Effects of prey density and site character on estuary usage by over-wintering waders (Charadrii), *Estuarine Coastal Mar. Sci.* **9**:369.

Buckley, F. G., and Buckley, P. A., 1974, Comparative feeding ecology of wintering adult and juvenile Royal Terns (Aves: Laridae, Sterninae), *Ecology* **55**:1053.

Burger, A. E., 1982, Foraging behavior of Lesser Sheathbills *Chionis minor* exploiting invertebrates on a sub-Antarctic island, *Oecologia (Berlin)* **52**:236.

Burger, J., 1980, Age differences in foraging Black-necked Stilts in Texas, *Auk* **97**:633.

Burger, J., Howe, M. A., Hahn, D. C., and Chase, J., 1977, Effects of tide cycles on habitat selection and habitat partitioning by migrating shorebirds, *Auk* **94**:743.

Burger, J., Hahn, D. C., and Chase, J., 1979, Aggressive interactions in mixed-species flocks of migrating shorebirds, *Anim. Behav.* **27**:459.

Chambers, M. R., and Milne, H., 1975, Life cycle and production of *Nereis diversicolor* D. F. Müller in the Ythan Estuary, Scotland, *Estuarine Coastal Mar. Sci.* **3**:133.

Connors, P. G., Myers, J. P., Connors, S. W., and Pitelka, F. A., 1981, Interhabitat movements by Sanderlings in relation to foraging profitability and the tidal cycle, *Auk* **98**:49.

Cowie, R. J., and Krebs, J. R., 1979, Optimal foraging in patchy environments, in: *Population Dynamics* (R. M. Anderson, B. T. Turner, and L. R. Taylor, eds.), pp. 183–205, Blackwell, Oxford.

Duffy, D. C., Atkins, N., and Schneider, D. C., 1981, Do shorebirds compete on their wintering grounds?, *Auk* **98**:215.

Elliott, C. C. H., Waltner, M., Underhill, L. G., Pringle, J. S., and Dick, W. J. A., 1976, The migration system of the Curlew Sandpiper *Calidris ferruginea* in Africa, *Ostrich* **47**:191.

Evans, P. R., 1975, Notes on the feeding of waders on Heron Island, *Sunbird* **6**:25.

Evans, P. R., 1976, Energy balance and optimal foraging strategies in shorebirds: Some implications for their distributions and movements in the non-breeding season, *Ardea* **64**:117.

Evans, P. R., 1979, Adaptations shown by foraging shorebirds to cyclical variations in the activity and availability of their intertidal invertebrate prey, in: *Cyclic Phenomena in Marine Plants and Animals* (E. Naylor, and R. G. Hartnoll, eds.), pp. 357–366, Pergamon Press, Elmsford, N.Y.

George, J. D., 1964, The life history of the cirratulid worm, *Cirriformia tentaculata*, on an intertidal mudflat, *J. Mar. Biol. Assoc. U.K.* **44**:47.

Glander, K. E., 1981, Feeding patterns in mantled howler monkeys, in: *Foraging Behavior: Ecological, Ethological and Psychological Approaches* (A. C. Kamil, and T. D. Sargent, eds.), pp. 231–258, Garland STPM Press, New York.

Goss-Custard, J. D., 1969, The winter feeding ecology of the redshank *Tringa totanus, Ibis* **111**:338.

Goss-Custard, J. D., 1970, The responses of redshank (*Tringa totanus* [L.]) to spatial variations in the density of their prey, *J. Anim. Ecol.* **39**:91.

Goss-Custard, J. D., 1977a, The ecology of the Wash. III. Density-related behavior and the possible effects of a loss of feeding grounds on wading birds (Charadrii), *J. Appl. Ecol.* **14**:721.

Goss-Custard, J. D., 1977b, Predator responses and prey mortality in redshank, *Tringa totanus* (L.), and a preferred prey, *Corophium volutator* (Pallas), *J. Anim. Ecol.* **46**:21–35.

Goss-Custard, J. D., 1977c, Optimal foraging and the size selection of worms by redshank (*Tringa totanus*), in the field, *Anim. Behav.* **25**:10.

Goss-Custard, J. D., 1979, Effect of habitat loss on the numbers of overwintering shorebirds, in: *Studies in Avian Biology No. 2* (F. A. Pitelka, ed.), p. 167, Cooper Ornithological Society, Allen Press, Lawrence, Kans.

Goss-Custard, J. D., 1980, Competition for food and interference among waders, *Ardea* **68**:31.

Goss-Custard, J. D., Durell, S. E. A., McGrorty, S., and Reading, C. J., 1982, Use of mussel *Mytilus edulis* beds by oystercatchers *Haematopus ostralegus* according to age and population size, *J. Anim. Ecol.* **51**:543.

Groves, S., 1978, Age-related differences in Ruddy Turnstone foraging and aggressive behavior, *Auk* **95**:95.

Hamilton, W. J., 1959, Aggressive behavior in migrant Pectoral Sandpipers, *Condor* **61**:161.

Hartwick, E. B., 1976, Foraging strategy of the Black Oystercatcher (*Haematopus bachmani* Audubon), *Can. J. Zool.* **54**:142.

Hartwick, E. B., and Blaylock, W., 1979, Winter ecology of a Black Oystercatcher population, in: *Studies in Avian Biology No. 2* (F. A. Pitelka, ed.), pp. 207–215, Cooper Ornithological Society, Allen Press, Lawrence, Kans.

Hassell, M. P., and May, R. M., 1974, Aggregation in predators and insect parasites and its effect on stability, *J. Anim. Ecol.* **43**:567.

Heppleston, P. B., 1971, The feeding ecology of oystercatchers (*Haematopus ostralegus*) in winter in northern Scotland, *J. Anim. Ecol.* **40**:651.

Hockey, P., 1982, Behaviour patterns of non-breeding African Black Oystercatchers *Haematopus moquini* at offshore islands, *Proc. Vth Pan-Afr. Ornithol. Congr.* (J. A. Ledger, ed.).

Holmes, R. T., 1966, Feeding ecology of Red-backed Sandpiper (*Calidris alpina*) in Arctic Alaska, *Ecology* **47**:32.

Holmes, R. T., and Pitelka, F. A., 1968, Food overlap among coexisting sandpipers on northern Alaskan tundra, *Syst. Zool.* **17**:305.

Hughes, R. N., 1970, Population dynamics of the bivalve *Scrobicularia plana* (da Costa) on an intertidal mudflat in north Wales, *J. Anim. Ecol.* **39**:333.

Kelly, P. R., and Cogswell, H. L., 1979, Movements and habitat use by wintering populations of Willets and Marbled Godwits, in: *Studies in Avian Biology No. 2* (F. A. Pitelka, ed.), pp. 69–82, Cooper Ornithological Society, Allen Press, Lawrence, Kans.

King, J. R., 1974, Seasonal allocation of time and energy resources in birds, in: *Avian Energetics* (R. A. Paynter, Jr., ed.), pp. 6–70, Publ. Nuttall Ornithol. Club 15.

Krebs, J. R., and Cowie, R. J., 1976, Foraging strategies in birds, *Ardea* **64**:98.

Krebs, J. R., MacRoberts, M. H., and Cullen, J. M., 1972, Flocking and feeding in the Great Tit *Paris major*—An experimental study, *Ibis* **114**:507.

Krebs, J. R., Houston, A. I., and Charnov, E. L., 1981, Some recent developments in optimal foraging, in: *Foraging Behavior: Ecological, Ethological and Psychological Approaches* (A. C. Kamil and T. D. Sargent, eds.), pp. 3–18, Garland STPM Press, New York.

Little, C., and Nix, W., 1976, The burrowing and floating behaviour of the gastropod *Hydrobia ulvae*, *Estuarine Coastal Mar. Sci.* **4**:537.

MacArthur, R., and MacArthur, J., 1961, On bird species diversity, *Ecology* **42**:594.

MacArthur, R. H., and Pianka, E. R., 1966, On optimal use of a patchy environment, *Am. Nat.* **100**:603.

Morrell, S. H., Huber, H. R., Lewis, T. J., and Ainley, D. G., 1979, Feeding ecology of Black Oystercatchers on South Farallon Island, California, in: *Studies in Avian Biology No. 2* (F. A. Pitelka, ed.), pp. 185–186, Cooper Ornithological Society, Allen Press, Lawrence, Kans.

Morrison, M. L., Slack, R. D., and Shanley, E., Jr., 1978, Age and foraging ability relationships of olivaceous cormorants, *Wilson Bull.* **90**:414.

Murton, R. H., 1971, The significance of a specific search image in the feeding behavior of the Wood-pigeon, *Behaviour* **40**:10.

Myers, J. P., Williams, S. L., and Pitelka, F. A., 1980, An experimental analysis of prey availability for Sanderling (Aves: Scolopacidae) feeding on sandy beach crustaceans, *Can. J. Zool.* **58**:1564.

Myers, J. P., Connors, P. G., and Pitelka, F. A., 1981, Optimal territory size and the Sanderling: Compromises in a variable environment, in: *Foraging Behavior: Ecological, Ethological and Psychological Approaches* (A. C. Kamil and T. D. Sargent, eds.), pp. 135–158, Garland STPM Press, New York.

Naylor, E., 1958, Tidal and diurnal rhythms of locomotory activity in *Carcinus maenas*, *J. Exp. Biol.* **35**:602.

Newell, R. C., 1962, Behavioral aspects of the ecology of *Peringia* (= *Hydrobia*) *ulvae* (Pennant), *Proc. Zool. Soc. London* **138**:49.

O'Connor, R. J., and Brown, R. A., 1977, Prey depletion and foraging strategy in the oystercatcher *Haematopus ostralegus, Oecologia (Berlin)* **27**:75.

Orians, G. H., 1969, Age and hunting success in the Brown Pelican, *Anim. Behav.* **17**:316.

Page, G., and Whitacre, D. F., 1975, Raptor predation on wintering shorebirds, *Condor* **77**:73.

Pitelka, F. A., 1950, Geographic variation and the species problem in the shorebird genus *Limnodromus, Unif. Calif. Berkeley Publ. Zool.* **50**:1.

Pitelka, F. A. (ed.), 1979, *Studies in Avian Biology No. 2*, Cooper Ornithological Society, Allen Press, Lawrence, Kans.

Prater, A. J., 1972, The ecology of Morecambe Bay. III. The food and feeding habits of Knot (*Calidris canutus* L.) in Morecambe Bay, *J. Appl. Ecol.* **9**:179.

Prater, A. J., 1979, Shorebird census studies in Britain, in: *Studies in Avian Biology No. 2* (F. A. Pitelka, ed.), pp. 157–166, Cooper Ornithological Society, Allen Press, Lawrence, Kans.

Preece, G. S., 1971, The swimming rhythm of *Bathyporeia pilosa* (Crustacea: Amphipoda), *J. Mar. Biol. Assoc. U.K.* **51**:777.

Pringle, J. S., and Cooper, J., 1975, The Palaearctic wader population of Langebaan Lagoon, *Ostrich* **46**:213.

Puttick, G. M., 1977, Spatial and temporal variations in intertidal animal distribution at Langebaan Lagoon, South Africa, *Trans. R. Soc. S. Afr.* **42**:403.

Puttick, G. M., 1978, The diet of the Curlew Sandpiper at Langebaan Lagoon, South Africa, *Ostrich* **49**:158.

Puttick, G. M., 1979, Foraging behaviour and activity budgets of Curlew Sandpipers, *Ardea* **67**:111.

Puttick, G. M., 1980, Energy budgets of Curlew Sandpipers at Langebaan Lagoon, South Africa, *Estuarine Coastal Mar. Sci.* **11**:207.

Puttick, G. M., 1981, Sex-related differences in foraging behaviour of Curlew Sandpipers, *Ornis Scand.* **12**:13.

Pyke, G. H., Pulliam, H. R., and Charnov, G. L., 1977, Optimal foraging: A selective review of theory and tests, *Q. Rev. Biol.* **52**:137.

Rands, M. R. W., and Barkham, J. P., 1981, Factors controlling within-flock feeding densities in three species of wading birds, *Ornis Scand.* **12**:28.

Recher, H. F., 1966, Some aspects of the ecology of migrant shorebirds, *Ecology* **47**:393.

Recher, H. F., and Recher, J. A., 1969, Comparative foraging efficiency of adult and immature Little Blue Herons (*Florida caerulea*), *Anim. Behav.* **17**:320.

Rees, C. P., 1975, Life cycle of the amphipod *Gammarus palustris* Bousfield, *Estuarine Coastal Mar. Sci.* **3**:413.

Royama, T., 1971, Evolutionary significance of predators' response to local differences in prey density: A theoretical study. Proc. Adv. Study Inst. Dynamics Numbers Popul. (Oosterbeek, 1970).

Schoener, T. W., 1971, Theory of feeding strategies, *Annu. Rev. Ecol. Syst.* **2**:369.

Searcy, W. A., 1978, Foraging success in three age classes of Glaucous-winged Gulls, *Auk* **95**:587.

Selander, R. K., and Giller, D. R., 1963, Species limits in the woodpecker genus *Centurus* (Aves), *Bull. Am. Mus. Nat. Hist.* **124**:213.

Siegel, S., 1956, *Nonparametric Statistics for the Behavioural Sciences*, McGraw–Hill, New York.

Silliman, J., Mills, G. S., and Alden, S., 1977, Effect of flock size on foraging activity in wintering Sanderlings, *Wilson Bull.* **89**:434.

Smith, P. C., and Evans, P. R., 1973, Studies of shorebirds at Lindisfarne, Northumberland. I. Feeding ecology and behaviour of the Bar-tailed Godwit, *Wildfowl* **24**:135.

Stinson, C. H., 1980, Flocking and predator avoidance: Models of flocking and observations on the spatial dispersion of foraging winter shorebirds (Charadrii), *Oikos* **34**:35.

Summers, R. W., 1977, Distribution, abundance and energy relationships of waders (Aves: Charadrii) at Langebaan Lagoon, *Trans. R. Soc. S. Afr.* **42**:483.

Sutherland, W. J., 1982a, Spatial variation in the predation of cockles by oystercatchers at Traeth Melynog, Anglesey. I. The cockle population, *J. Anim. Ecol.* **51**:481.

Sutherland, W. J., 1982b, Spatial variation in the predation of cockles by oystercatchers at Traeth Melynog, Anglesey, II. The pattern of mortality, *J. Anim. Ecol.* **51**:491.

Thomas, D. G., and Dartnall, A. J., 1971, Ecological aspects of the feeding behavior of two calidridine sandpipers wintering in southeastern Tasmania, *Emu* **71**:20.

Vader, W. J. M., 1964, A preliminary investigation into the reaction of the infauna of tidal flats to tidal fluctuations in water level, *Neth. J. Sea Res.* **2**:189.

Verbeek, N. A. M., 1977, Comparative feeding behaviour of immature and adult Herring Gulls, *Wilson Bull.* **87**:415.

Vines, G., 1980, Spatial consequences of aggressive behavior in flocks of oystercatchers *Haematopus ostralegus* L., *Anim. Behav.* **28**:1175.

Chapter 5

INTAKE RATES AND FOOD SUPPLY IN MIGRATING AND WINTERING SHOREBIRDS

J. D. Goss-Custard

The Institute of Terrestrial Ecology
Furzebrook Research Station
Wareham, Dorset BH20 5AS, England

I. INTRODUCTION

Much of the work done on shorebirds on migration and in winter has concentrated on their feeding activities. Research on individual adaptation has focused on the varied morphological and behavioral characteristics of shorebirds, and studies at the population level have selected food supply as the factor outside the breeding season most likely to have a significant effect on numbers. It became clear very early in these studies that distinguishing between the invertebrate food organisms that were present in the substrate and those that were actually available to the birds was of enormous importance (Goss-Custard, 1969; Smith, 1975). A bird may be unable to find a prey because its sense organs are unable to detect it or be unable to eat one that has been located because, for example, it is beyond the reach of the bill. The number of prey that can be both detected and obtained determines the proportion available to the bird (Goss-Custard *et al.*, 1977c). The absolute abundance of the prey and the proportion that is vulnerable to the birds, hereafter referred to as their availability, together determine the density of prey that the birds actually encounter as they search, and how the birds respond to this in turn determines the observed intake rates. This chapter reviews the many factors that, by affecting the density of available prey and the birds' response to it, seem to determine intake rate. Some conclusions on the relevance of these findings to studies of individual adaptation and population size are then drawn.

A variety of terms have been used to refer to the rates at which various feeding activities are carried out. The terms used in this article are as follows. *Pecking* and *probing* rates refer to the numbers of discrete feeding attempts made per unit time, *feeding success* is the proportion successful, and *feeding* or *capture rate* the numbers of prey ingested per unit time. *Intake rate* and *ingestion rate* refer to the biomass, usually expressed as dry weight or ash-free dry weight, consumed per unit time. It is sometimes useful to define rates as number per unit distance searched, but this is always specified when it is done.

II. FOOD AVAILABILITY

The ability of shorebirds to detect and obtain prey is affected by two processes. One includes environmental factors that have a direct effect by simply preventing the birds from either detecting or reaching their food. The other includes factors that have an indirect effect through changing the behavior of the prey themselves.

A. Factors Having a Direct Effect

The obvious example is the tide, which covers the mudflats and prevents shorebirds from feeding there. Ice-flows on the mud surface, or the freezing of the mud surface itself, have a similarly decisive effect on the birds' access to prey.

There are more subtle effects, however. Myers *et al.* (1980) studied the availability of isopods (*Excirolana* spp.) and sand crabs (*Emerita analoga*) to captive Sanderlings (*Calidris alba*). Dead prey were placed at various depths in a tray of sand. The compactness of the sand was varied by banging the tray and the penetrability determined by measuring the depth to which a steel rod dropped from a constant height sunk into the sand. Figure 1 shows that the depth to which Sanderling probed at the first attempt was related to substrate penetrability. Subsequent probes deepened the hole but probe depth was still deeper in the softer substrates. As prey near to the surface were more at risk than those buried deeper and as a bird must probe more times to reach a given depth in harder substrates, the authors expected intake rate to decrease in the firmer substrates. By partialling out the effects of prey density and depth, the authors showed that this was indeed the case: intake rate was significantly

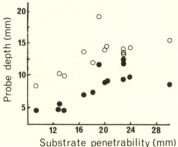

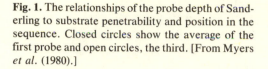

Fig. 1. The relationships of the probe depth of Sanderling to substrate penetrability and position in the sequence. Closed circles show the average of the first probe and open circles, the third. [From Myers *et al.* (1980).]

higher in soft substrates. Further observations showed that variation in substrate penetrability in the field was similar to that obtained in the laboratory. The level of the beach, moisture content, and particle size were considered likely to affect penetrability. The sand is particularly easy to penetrate immediately after it has been flooded by a wave, and this may partly explain why Sanderlings forage so closely to the water's edge.

In the study of Myers *et al.*, variations in penetrability affected the proportion of the prey the bird could reach. Smith's (1975) study on Bartailed Godwits (*Limosa lapponica*) eating lugworms (*Arenicola marina*) suggested that the background formed by the substrate can also affect the ability of a visually hunting shorebird to detect its prey. Lugworms are usually too deep in the sand for godwits to reach and they only become vulnerable when they move back up in their burrows to defecate—the well-known worm casts of sandflats. However, lugworms are so dense on the sandflats at Lindisfarne, northeastern England, where Smith did his work, that casts being formed may easily be missed among the masses that accumulate on the flats during the low-water period. To test this, Smith removed all the old casts from an area of sand and then compared the birds' behavior, when they returned to feed in this plot, with their behavior in the same place on the next day when all the casts formed during low tide were allowed to remain. The results indicate that the godwits were more successful at catching worms when the area had been smoothed so that new casts were not hidden against a background of old ones, although Smith himself points out that further tests of this are desirable.

These studies illustrate how the substrate may influence the ability of shorebirds to both detect (godwits) and obtain (Sanderlings) their prey. Other studies have suggested that the substrate affects intake rate, without showing specifically how this is achieved. Tjallingi (1969) showed that Pied Avocets (*Recurvirostra avosetta*) catch ragworms (*Nereis diversi-*

color) faster in soft substrates even though the worms may be relatively scarce. The feeding rate of common Redshank (*Tringa totanus*) on the Ythan decreased down the shore independently of changes in prey density, and it is presumed that this was related to variations in some feature of the substrate (Goss-Custard, 1970c).

Both rainfall and wind may affect intake rate in visually hunting shorebirds. Redshank (Goss-Custard, 1970b) and plovers (Charadriidae) (Pienkowski, 1981) feed more slowly in rain, perhaps in part because turbidity of the water on the surface makes prey less easily seen. Smith (1975) suggests Godwits wading at the tide edge may feed less well in strong winds because they cannot see their prey in the turbid water. A. Baker (1974) also found that strong winds and heavy rain affect the feeding of the South Island Pied Oystercatcher (*Haematopus ostralegus finschi*) and Variable Oystercatcher *H. unicolor* in New Zealand. The birds responded to increased surface turbidity by changing from detecting prey by sight to touch, though any effect on intake rate was not measured.

As well as affecting the visibility of surface cues, wind may have a direct effect on the birds. Goss-Custard (1966) found that, by seeking shelter in gales, Redshank fed in places where intake rate was unusually low. Feare (1966) compared the behavior of Purple Sandpipers (*Calidris maritima*) feeding in northern England at the tide edge when the sea was calm and rough. The birds had to fly or run out of the way of waves almost five times more frequently when the sea was rough, resulting in 8 min of foraging time being lost per hour. Furthermore, in December when the tide normally left the shore uncovered for only six daylight hours, storms and onshore winds held up the tide for 30 min or more. In total, then, 25% of feeding time was lost in stormy weather due to the direct effect of waves and tide on the access of birds to the food.

Another factor likely to affect the ability of shorebirds to find prey is light intensity. Shorebirds that detect prey by sight may feed more slowly at night, unless the behavior of the prey changes so that they actually become more accessible (Dugan, 1981a; McClennan, 1979). The peck rate of Grey Plovers (*Pluvialis squatarola*) and Ringed Plovers (*Charadrius hiaticula*) at Lindisfarne during the night was about half that recorded in the same place during the day (Pienkowski, 1982), but unfortunately feeding success and prey size could not be measured. In fact, larger prey may become available at night and Grey Plover may achieve higher intake rates in darkness than during the day (Dugan, 1981a). Although several studies of night feeding in free-ranging shorebirds that normally detect their prey by sight have been made using a variety of techniques, none until recently have provided really convincing evidence of how intake rate may be affected by darkness, let alone separated out the effect of low visibility in any reduction that may occur.

The recent availability of night-viewing devices should change this, however. Field studies by Zwarts and Drent (1981) in Holland and preliminary observations with an image-intensifier on the Exe Estuary in southwestern England suggest that at night Eurasian Oystercatchers find *Mytilus* at less than half the daytime rate. Sutherland (1982c) reports a similar reduction in Eurasian Oystercatchers feeding on cockles. The oystercatchers switched from detecting prey by sight to touch feeding and, in doing so, found cockles of a smaller size at a lower rate than they did during the day. However, Hulscher (1976) obtained different results from a captive bird. He studied a Eurasian Oystercatcher feeding in a small pen placed on tidal flats. He was thus able to watch a bird in controlled conditions and, by using infrared binoculars, could also study its behavior at night. In daylight the bird usually detected cockles by sight by pecking at surface cues. At night it switched to touch feeding by pushing its beak through the mud and continuously moving it up and down, the so-called "sewing" action. This change actually resulted in a substantial increase in gross intake rate when prey density was high, although the effect on the net rate of intake of using the probably more energetically expensive sewing technique has not been evaluated. At low prey densities, the birds often used the sewing technique even in daylight and gross intake rates were similar to those achieved during the night.

In a similar study on Eurasian Oystercatchers eating *Macoma balthica*, Hulscher (1982) obtained a different result. The birds found this prey by sewing both day and night, but were less successful in the dark. By smoothing the surface of the sand so that all cues as to the presence of *Macoma* beneath were removed, Hulscher found that daytime intake rate dropped to that recorded at night. The implication is that the birds use surface cues to direct them to spots where *Macoma* might be found, but then located them by touch with the sewing method: a similar suggestion was made for oystercatchers in New Zealand by A. Baker (1974).

More studies of the effect of light intensity are badly needed. Though several authors have noticed that visually hunting shorebirds may change to touch feeding at night, little has been done to compare intake rates. One difficulty is that the diet may change. For example, Common Redshank in the Ythan normally pecked at the mud surface for *Corophium* in daylight but "swished" their bills from side to side in darkness and ate small gastropods (Goss-Custard, 1969). The possible effect of this on intake rate has yet to be evaluated.

To conclude, several factors may have a direct effect on the ability of waders either to detect or obtain prey that is otherwise accessible.

Tide, ice, wind, rain, and substrate penetrability have been identified, but the possible effect of darkness needs clarification.

B. Factors Having an Indirect Effect through Prey Behavior

A number of environmental factors are thought to influence intake rate in shorebirds indirectly through their effect on prey behavior. For example, the behavior of Common Redshank feeding by sight on the amphipod *Corophium volutator* on the Ythan was much affected by temperature and rainfall (Goss-Custard, 1969, 1970b). In one winter, Common Redshank fed almost entirely on *Corophium* at all mud temperatures but made more pecks at very low temperatures to find the same amount (Fig. 2). This suggests that the birds were pecking at less reliable cues when the temperature dropped, perhaps because the prey were less active and so more difficult to detect or were buried more deeply in the mud. An indication that this is indeed so was obtained by watching the activity of *Corophium* at the mud surface. *Corophium* are visible to the human eye when their appendages emerge briefly from their burrows and, as expected, this happened less often at low temperatures (Fig. 2). In another winter when the bivalve mollusc *M. balthica* was abundant, redshank switched to these at low temperatures perhaps because *Corophium* were less available. Redshank took fewer *Corophium* during rain, when again few of the prey were visible at the surface. Part of this disappearance may have been due to the rain stirring up the mud surface, but the absence of any sign of *Corophium* at the surface suggests their behavior had also changed.

Intake rate may have dropped in this study either because prey were less detectable, due to being less active, or because they were less obtainable, due to being buried deeper. These possibilities were not distinguished in Common Redshank, but studies on Grey Plover suggest both factors may be involved in some situations. Pienkowski (1980a) showed that the proportion of the polychaetes *Scoloplos armiger* and *Notomastus latericeus* in the upper 5 cm and the proportion appearing at the surface (as outflows of water from holes) decreased as mud temperatures decreased. At the same time, the numbers caught by Grey Plovers decreased (Fig. 3). C. Baker (1974) also showed that feeding success in Grey Plovers eating *Nereis succinea* in New Haven Harbor, Connecticut, was also related to temperature, but prey activity was not measured. The rate at which lugworms form casts decreased when mud temperatures fell below 3°C, and the number taken by Bar-tailed Godwits was also reduced

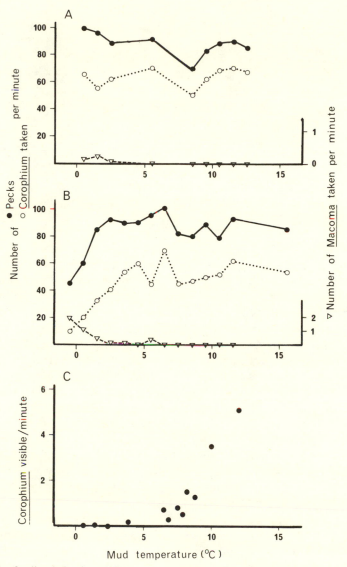

Fig. 2. The feeding behavior of Common Redshank and the behavior of their main prey, *Corophium*, in relation to mud temperature. (A) and (B) show the rates at which redshank pecked at the mud and took *Corophium* and *Macoma* in the winter of 1965–66 (A) and 1964–65 (B); (C) shows the number of *Corophium* appearing at the mud surface. [From Goss-Custard (1969).]

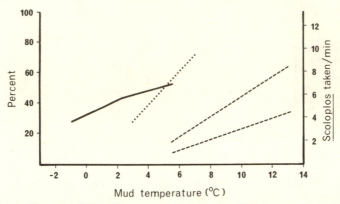

Fig. 3. The feeding rate of Grey Plovers and the behavior of their prey. *Scoloplos armiger*, in relation to mud temperature. The proportion of *Scoloplos* appearing at the mud surface (-----) was measured in two areas, the proportion present in the upper 5 cm of substrate (———) in one. The dotted line shows the number of *Scoloplos* eaten by the plovers. [From Pienkowski (1980a).]

(Smith, 1975). In this case also, prey were less detectable and obtainable at low temperatures.

Bar-tailed Godwits in Smith's study switched to *Scoloplos* and *Notomastus* at low temperatures, which they found by the sewing technique and, thus, presumably, by touch. Interestingly, the rate of feeding on this prey also decreased eventually, but at much lower temperatures than the reduction that occurred in the visually hunting Grey Plovers feeding on the same prey (Pienkowski, 1980a). By being able to feed by touch, the godwits were able to catch *Scoloplos* and *Notomastus* even though they became less active and buried deeper in the substrate.

The intake rate of Red Knot (*Calidris canutus*) detecting the bivalves *Macoma* and *Cerastoderma edule* by touch was also affected by mud temperature (Fig. 4). The response to low temperature was immediate: as the mud warmed up following a cold night, intake rate rose sharply (Goss-Custard *et al.*, 1977c). As the burying depth of *Macoma*, the main prey, was not related to short-term changes in temperature (Reading and McGrorty, 1978), it may have been a change in detectability alone that affected the birds' intake rate. For example, *Macoma* may respond more in warm mud when a probing beak touches them, so giving their position away.

The study of Sanderling by Myers *et al.* (1980) isolated prey depth as a factor that can alone influence intake rate. The crustaceans used in this study had been killed and so could not be detected through activity. Similarly, a very high proportion of the *Macoma* eaten by Red Knot on

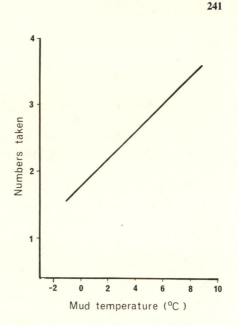

Fig. 4. The number of molluscs eaten per minute by Red Knots in relation to mud temperature. [From Goss-Custard *et al.* (1977c).]

the Wash are buried too deeply in the substrate for the birds to reach in winter (Reading and McGrorty, 1978), so their obtainability was clearly reduced.

A number of other studies have linked changes in intake rate with changes in prey behavior associated with environmental shifts. Rainfall reduces the number of polychaetes appearing at the mud surface and plovers feed more slowly (Pienkowski, 1980b). Both Bar-tailed Godwits (Smith 1975) and plovers (Pienkowski, 1981) feed more slowly in strong wind and the reduced activity of the prey at the surface in these conditions suggests that this is at least partly due to prey detectability being reduced.

Several authors have pointed out the possible importance of substrate wetness in influencing prey behavior and thus their availability to waders. *Corophium* do not appear at the surface in dry areas of sand or mud, and it may be because of this that Common Redshank avoid such areas (Goss-Custard, 1970a), a tendency noted by others (Burger *et al.*, 1977; Evans, 1976; Prater, 1972; Recher, 1966). Smith (1975) showed that Bar-tailed Godwits found more lugworms in wet areas, where the worms form casts more frequently, and avoided sandy areas that dry out at low tide. The wetness of substrates recently exposed by the receding tide may partly explain why many waders follow the tide, though the higher penetrability of the substrate in this area may also be involved. The numbers of *Nereis* (Dugan, 1981a) and lugworms (Smith, 1975) visible at the surface declines steadily after the tide has passed, in line with the drying out of the substrate. *Nereis* may also burrow deeper in the substrate after the tide has gone out (Vader, 1964), and so be less obtainable. The feeding rate of Eurasian Oystercatchers, which obtain mussels (*Mytilus edulis*) by stabbing (Norton-Griffiths, 1967) into gaping individuals, may be highest as the tide recedes and advances over the mussel bed (Zwarts and Drent, 1981) when most mussels are gaping (Hartwick, 1976). The wetness of the substrate at the tide edge may have a particularly significant effect on prey availability in sandy areas that are most likely to dry out at low tide.

The tide edge may have an effect on prey availability over and above that due to the wetness of the substrate. Boates (1980) studied Semipalmated Sandpipers (*Calidris pusilla*) eating *Corophium* in the Bay of Fundy. On the receding tide, the birds bunch along the water's edge and follow it out closely. Their intake rate is very high because feeding success is almost 100% and they catch particularly large adult *Corophium*. This is because on the receding tide-line, larger-sized adults, especially the males, crawl on the surface in pursuit of mates. This is a short-lived phenomenon and within half an hour of the tide passing a spot there are only 5% of the initial number of *Corophium* on the surface, even in the wetter areas. The receding tide edge is not only associated with wet substrate but may also provide a stimulus to amphipod behavior that results in them becoming very accessible to the birds.

These studies suggest that several factors may alter the behavior of prey and thus their detectability and/or obtainability to waders. Temperature, wind, rain, substrate wetness, and the tide-edge itself have all been suggested. As discussed earlier, several of these may also have a direct effect on shorebirds, but so far few studies have investigated the precise way in which an environmental factor exerts its influence on intake rate.

Not all studies, however, have shown correlations between intake rate and these factors. Goss-Custard *et al.* (1977c) found no association on the Wash between intake rate and both mud temperature and wind strength in Eurasian Oystercatchers eating *Cerastoderma*, Bar-tailed Godwits eating *Lanice conchilega*, and Eurasian Curlews (*Numenius arquata*) eating *Nereis*. Similarly, Dugan (1981b) found intake rate and temperature to be unrelated in Grey Plovers eating *Nereis diversicolor* on the Tees Estuary. Sutherland (1982a) found that intake rate in Eurasian Oystercatchers eating *Cerastoderma* was not related to how recently the sand had been uncovered by the receding tide. Clearly, the circumstances in which variations in environmental conditions and prey behavior are correlated with intake rate are not yet fully understood.

III. PREY DENSITY AND SIZE

One of the difficulties of interpreting these results is that any change in prey availability may not be revealed by a change in bird behavior. Even though the proportion available may decrease, the density of available prey may still be sufficient for the birds to maintain their intake rate at the same level. It is striking that in some studies of prey behavior and mud temperature (e.g., Goss-Custard, 1969), both feeding and ingestion rates level off even though prey activity, and presumably the proportion available, and thus the number of prey encountered per unit distance searched, continue to increase. This raises the question as to what might be the relationship between feeding rate or intake rate and the density of available prey.

A. Feeding Rate and Prey Density

The results of several field studies on Eurasian Oystercatchers (Goss-Custard, 1977a; Sutherland, 1982a) and Common Redshanks (Goss-Custard, 1970c, 1977b) feeding on small, homogeneous areas have produced a similar curve relating feeding rate to prey density. After an initially rapid rise in feeding rate as prey density increases, it levels off or may even decline slightly (Fig. 5). Though never tested, both authors have assumed that the proportion of prey available is constant at all prey densities. In other words, it is assumed that variations in the total density of prey in the substrate do reflect variations in the density of those actually available to the birds. These curves are therefore thought to show how feeding rate responds to changes in the density of available prey.

The general shape of these curves is similar to that termed the type II functional response by Holling (1959). However, there is an important difference. Holling's equation for this curve assumes that the maximum feeding rate occurs where prey are so dense and prey found so quickly that most of the available time is taken up handling prey, so that handling time limits feeding rate. This is not so in these studies. In the study of Common Redshank eating *Corophium* (Fig. 5A), for example, less than half of the foraging time was spent in pecking at and handling prey (Goss-Custard, 1977c). The rest was spent in searching. Nor was feeding rate limited by the maximum rate at which Common Redshank could peck, because birds on the Ythan often pecked at twice the highest rate recorded in southwestern England.

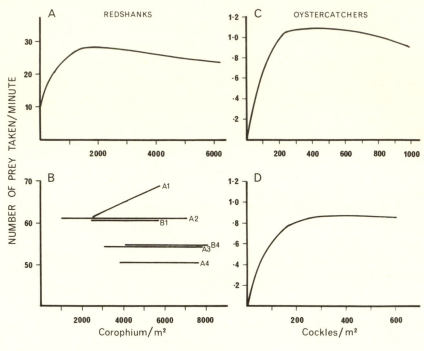

Fig. 5. The feeding rate of Common Redshanks on *Corophium* (A, B) and Eurasian Oystercatchers on *Cerastoderma* (C, D) in relation to prey density. In (B), data from six areas
are shown. [(A) from Goss-Custard (1977c); (B) from Goss-Custard (1970c); (C) from Goss-
Custard (1977a); (D) from Sutherland (1982a) (*Journal of Animal Ecology*).]

B. Intake Rate and Prey Density

How prey density affects intake rate depends on the relationship
between the density and size of the prey. In Goss-Custard's (1970a,c)
study of Common Redshank, *Corophium* were larger where they were
numerous. So even though the numbers of prey consumed were more or
less constant over a wide range of prey densities, intake rate increased
as prey density increased because the birds ate larger *Corophium* (Fig.
6). This was also the case when prey density was expressed as biomass
rather than numerical density. In contrast, Sutherland (1982a,b) found
that intake rate was highest in Eurasian Oystercatchers at low numerical
densities of prey because cockles were older, and therefore larger where
they were sparse. Animals of a given length also had a higher dry weight,
and therefore food value. Intake rate was highest at densities of 50/m²,

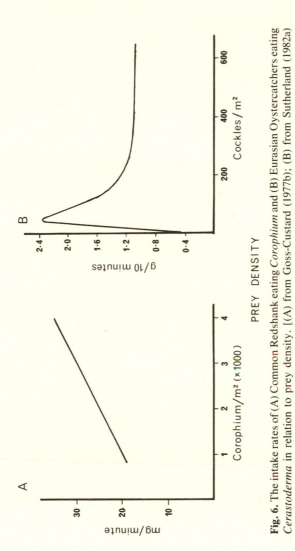

Fig. 6. The intake rates of (A) Common Redshank eating *Corophium* and (B) Eurasian Oystercatchers eating *Cerastoderma* in relation to prey density. [(A) from Goss-Custard (1977b); (B) from Sutherland (1982a) (*Journal of Animal Ecology*).]

in a population where density often reached 600/m². Clearly, changes in prey size with prey density can have an important effect on the relationship between intake rate and prey density.

C. Size Selection and Prey Density

In both these studies, the birds were easily able to eat the largest prey they encountered so the size consumed increased as the size present increased. Taking the size distribution of the prey popultion as a whole, both species took prey much larger than the average. However, this is by no means the general case in either of these species or in shorebirds as a whole. Few studies indicate that the size taken reflects exactly the size present and it seems that most shorebirds select prey from a part of the full range of sizes present. (That is why only prey within a restricted size range were considered in Figs. 5 and 6—see original papers for details.)

In general, the larger shorebirds eat the larger prey, as studies on the Wash in eastern England illustrate (Goss-Custard *et al.*, 1977a). The upper limit is no doubt often set by gape size: without the behavioral adaptation needed to open bivalves, Red Knot simply cannot swallow the large *Macoma* regularly consumed by Eurasian Oystercatchers. But this is not the whole story. Common Redshank, for example, can swallow very large polychaete worms and yet seldom do so. Clearly, some selection process is involved.

Selection may be either passive or active (Hulscher, 1982). In passive selection, the proportion of grey encountered by a bird varies between size classes as a direct result of the method of location. As a result, the bird feeds disproportionately on particular size classes. Hulscher showed that the selection of large *Macoma* taken by touch-feeding Eurasian Oystercatchers could be explained this way. He measured the effective surface area of each size class, and the depth at which they occur in the substrate and the depth to which his captive bird probed for prey. By also measuring the width of the bill tip and assuming that this measured the size of the area within which a *Macoma* could be detected, he was able to compare the sizes taken with the sizes of *Macoma* the birds would by chance encounter while feeding by touch. Compared with the population of *Macoma* as a whole, the captive bird took larger than average individuals, the difference between mean sizes present and taken being highly statistically significant (Fig. 7). However, large *Macoma* occur deeper in the substrate than small ones, and when this was taken into account the difference between size taken and size present was less, though still sig-

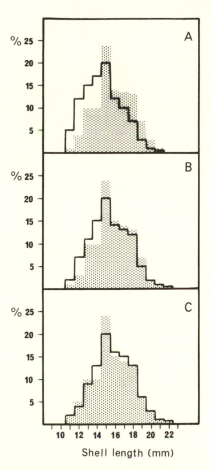

Fig. 7. Comparison between the sizes of *Macoma* taken by a captive Eurasian Oystercatcher (shaded histograms) and the sizes present in the sand (not shaded) defined in three ways: (A) all the *Macoma* present, (B) only those within reach of the bill, and (C) those within reach of the bill but corrected for effective touch area. [From Hulscher (1982).]

nificant. However, when the number of *Macoma* of each size class within reach of the probing bill was multiplied by the effective touch area of that size class, the difference between size present and size taken disappeared almost completely. Though the birds may actively reject the very small prey as uneconomic, above a certain minimum prey size they are taken in proportion to the frequency in which they are encountered. Because large *Macoma* at a given depth have a larger surface area, they are more likely to be encountered and so figure disproportionately in the diet. Myers *et al.* (1980) also argued that passive selection of this kind accounted for the tendency of captive Sanderling to eat larger than average prey.

Passive selection may explain why Common Redshank on the Ythan ate larger *Corophium* as prey density and prey size increased. The dif-

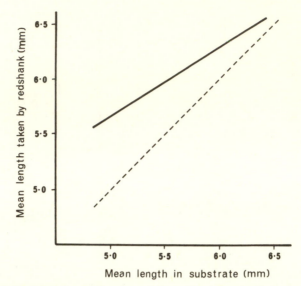

Fig. 8. The relationship between the sizes of *Corophium* taken by Common Redshank and the size present in the mud. The dashed line shows the relationship that would occur were the birds not to select for large prey. [From Goss-Custard (1969).]

ference between the mean length of those eaten by the birds and those in the substrate was larger where *Corophium* were small (Fig. 8), and this could reflect a greater chance that large individuals would be detected: large *Corophium* presumably provide larger visual cues at the surface for Common Redshank to see. This interpretation is supported by the finding that the probability that Common Redshank would take small *Corophium* was unaffected by the abundance of large *Corophium* or by the rate at which Common Redshank ingested them (Goss-Custard, 1977c). There is thus no evidence of active selection taking place in this case.

This is not so when Common Redshank eat *Nereis*, which vary in size from 0.5 to over 100 mg dry weight in the areas studied by Goss-Custard (1977d). This study showed that, as the abundance of the larger worms in the habitat increases, Common Redshank eat more of them while the risk of the small worms being taken decreases (Fig. 9). Because there is no reason to believe that the availability of the small worms decreases as the abundance of the large ones increases, this result indicates that small prey were actively rejected when large prey were abundant. This argument is supported by experiments on captive Common Redshank feeding on dead prey whose availability to the birds was held constant (Goss-Custard, 1978). As in the field, the birds did not take many

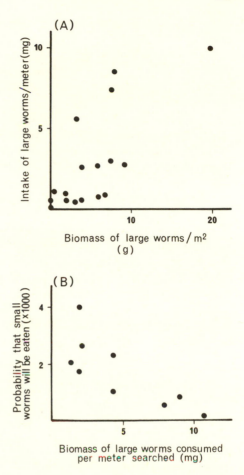

Fig. 9. The biomass of large *Nereis* eaten per meter searched by Common Redshank increases as the abundance of large worms in the environment increases (A), while the probability that they will eat small worms decreases (B). [From Goss-Custard (1977d).]

small prey when large ones were abundant but did so when they were scarce.

The argument that active selection was involved is again supported by the finding that the behavior of Common Redshank eating worms was similar to that of predators making optimal decisions in theoretical studies. The numbers of small worms eaten by Common Redshank, both in the field and in captivity, depended on the abundance of the large prey and very little on their own density. Similarly, the preferred size classes of worms had the highest ratio of energy content to handling time, as is also predicted by optimal foraging theory (Pyke *et al.*, 1977). Though there were discrepancies between actual Common Redshank behavior and that of the theoretically optimal predator, the weight of evidence suggested that Common Redshank were making active choices (Goss-Custard, 1981a).

The mechanism of active choice is not known, however. One possibility is that the birds saw small worms and rejected them when it was more profitable to move on quickly to find the next profitable, large prey. This would be active selection in the simplest sense. But another possibility is that Common Redshank searched faster when large worms were abundant and, as a result, were less likely to detect the smaller worms, while the large cues associated with the large worms would still be visible. Under this hypothesis, the birds may have taken every prey they detected but they were less able to detect small ones when large ones were abundant. In one sense, therefore, this is passive selection, but as walking rate is presumably under the bird's control, the choice of search speed would be made actively.

Sutherland's (1982c) studies of size selection in Eurasian Oystercatchers eating cockles produced similar findings, though pace rate was unrelated to cockle size. These birds also took more of the large, relatively profitable cockles as their abundance increased, and ate fewer of the smaller ones. Active selection of the larger cockles therefore seemed to be involved, so that the peak in intake rate recorded at cockle densities of $50/m^2$ (Fig. 6) was in part due to the selection for the more profitable prey.

D. Determinants of Feeding Rate and Intake Rate

Studies on prey size selection suggest an explanation that accounts for maximum feeding rates being much less than those predicted by Holling's equation for the type II functional response. This equation assumes that all prey are equally likely to be taken because they all have a similar value and handling time, whereas in nature this is not true. In both Eurasian Oystercatcher and Common Redshank studies, the measures of prey density included a wide range of prey sizes. In the case of Eurasian Oystercatchers, which may actively select the larger, more profitable individuals, taking every prey encountered may have reduced average intake rate substantially. It was therefore a better strategy for the bird to spend time searching for the most profitable prey, and as a result the maximum numbers of prey taken were not determined solely by handling time. A tendency to become more selective as density increased may explain why the feeding rate of Eurasian Oystercatchers on the Wash may have decreased slightly at very high prey densities (Goss-Custard, 1977a). Because the food value of cockles increases disproportionately with their length, it may have been better for the birds to select a smaller number of the especially large cockles.

Under this explanation, the feeding rate levels off as prey density increases because the birds select a smaller proportion of the prey as their choice increases. But this does not seem to be the whole story because further, unestablished internal factors seem also to be involved. In Common Redshank, for example, there is no evidence to suggest that the birds select actively for the larger *Corophium* so it is difficult to understand why even more *Corophium* were not taken at very high densities of prey when over half the foraging time was still devoted to searching. Furthermore, Common Redshank on the Ythan collected *Corophium* at almost twice the rate of birds in southwestern England (Fig. 5). Even though *Corophium* of a given length from the Ythan weighed approximately half as much as those from the southern study areas, the feeding rate was not limited by the ability of the gut to digest food: Common Redshank eating worms ingested prey biomass at a much higher rate than birds eating *Corophium* and, within the Ythan, intake rates varied a great deal and were clearly not maximal in most circumstances (Fig. 6).

This is confirmed by some data from Eurasian Oystercatchers. An individual bird feeding in the same place increased its intake rate threefold during a period of 16 days as its breeding effort increased, though possible parallel changes in the food were not measured (Hulscher, 1982). Similarly, parent birds feeding their young collected mussels 1.7 times faster than adults without young (Zwarts and Drent, 1981). These observations do suggest a bird may vary its intake rate, though there was no evidence that Common Redshank eating worms increased their feeding rate as daylength decreased during the winter (Goss-Custard, 1981a). But the data from oystercatchers do suggest that shorebirds may be able to adjust their intake rates according to their needs and so not always feed at the maximum rate, as is predicted by simple models of optimal foraging (Goss-Custard, 1981a).

This may partly be understood in terms of the risks, other than starvation, which also influence the birds' behavior. For example, birds may frequently feed more slowly than they might because of the need to devote some attention to the possibility that predators may attack (Krebs, 1980). The maximum intake rate may not always be optimal for fitness given the variety of selection pressures operating in nature. Similarly, the observed rate may therefore reflect internal as well as external factors, and experiments on captive birds may well be needed to disentangle their separate effects. Given the possible dependence of intake rate on factors other than the abundance of available prey, it will take more research to fully understand the relationships between intake rate on the one hand and prey abundance and factors that affect its availability on the other.

IV. OTHER PREY CHARACTERISTICS

Studies of Eurasian Oystercatchers have shown that several factors apart from prey size may influence intake rate. Hulscher (1982) argues that Eurasian Oystercatchers check the *Macoma* they find before swallowing them because both free-living oystercatchers and a captive bird frequently rejected *Macoma* after they had opened them. These *Macoma* were invariably parasitized by the trematode parasite *Parvatrema affinis*. The sporocysts occupy the gonads and digestive gland and birds are the final host (Swennen and Ching, 1974). Infected *Macoma* are particularly likely to emerge onto the mud surface and crawl about (Hulscher, 1973; Swennen, 1969), a response to infection that might increase the chances that infected prey will be eaten and the life cycle of the parasite continued (Swennen, 1969).

Hulscher fed a captive Eurasian Oystercatcher infected and uninfected *Macoma*, in which the shells had been broken. The birds ate significantly more of the uninfected *Macoma* and rejected significantly more of the infected ones. Further observations on the same bird showed that it consumed about 50% of the small infected *Macoma* but only 20% of the large ones, which have on average far more sporocysts (Swennen and Ching, 1974). Though the bird did eat parasitized *Macoma*, it is clear that it selected strongly against them.

Hulscher argues that Eurasian Oystercatchers must open the shells before they can detect the infestation because so many are brought to surface and opened before they are rejected. Bad taste or touching the hard, granular sporocysts may be the stimulus to which the birds respond. The benefit from rejecting parasitized prey is unknown as the effect of *Parvatrema* on its avian hosts has not been studied, but Hulscher suggests that a heavy infestation may reduce the chances of oystercatchers surviving severe weather.

Common Redshank also reject a small proportion of the worms that they pull to the surface (Goss-Custard, 1977d), and it is possible that these might also be parasitized. The dog whelks (*Nucella lapittus*) eaten also by Eurasian Oystercatchers on rocky shores in England and are also infected with a trematode, *Parorchis aranthus*. Infected animals join the winter aggregations in crevices, where they are safe from Eurasian Oystercatchers, later than uninfected ones (Feare, 1971), and may be vulnerable to the birds for longer. However, it is not known if this results in more or fewer of them being eaten. Further work is needed to see how widely infestation by parasites influences shorebirds' choice of prey. Apart from the direct effect of parasites on the condition of the bird, the frequency of infection may affect intake rate because time is wasted in rejecting unacceptable prey.

Some Eurasian Oystercatchers attack mussels by pulling them off the substrate, carrying them to a firm spot, and hammering through the shell on the ventral surface. If the bissus attachment of the mussel to the substrate is too strong, the birds cannot tear them free and so select the more weakly attached individuals (Norton-Griffiths, 1967). Clearly, intake rate could be affected by the average strength of bissus threads, which changes seasonally (Price, 1980) and perhaps from place to place.

Norton-Griffiths also showed that Eurasian Oystercatchers hammer into the thinner parts of the shell and that the thickness of the shell has a considerable effect on how many blows are required to break into the mussel. Data obtained on the Exe Estuary in 1979 show that Eurasian Oystercatchers that hammer into mussels ventrally open mussles that have particularly thin shells on the ventral side (Fig. 10). As the birds cannot presumably detect the thickness of the shell on the ventral surface before the mussel has been pulled up and turned over, the birds make many mistakes. Ventral hammerers carry and reject far more mussels than do those birds that hammer mussels *in situ* on the dorsal side or stab into gaping mussels (Goss-Custard, *et al.*, 1982a). As expected, mussels that were carried and rejected by ventral hammerers had much thicker shells than those of similar size that were opened successfully (Fig. 10).

The average thickness of mussel shells varies considerably from place to place, and could affect intake rate quite considerably. Collections of shells in different parts of one bed show that the birds take smaller mussels where the shells are thick (Fig. 11A). In contrast, size taken was not correlated with the size present. Clearly, the birds select the thinner-shelled mussels and, where shells are on average thick, they eat the smaller mussels, which have thinner shells. The effect of this is that intake rate was less where shells were thick, both in different parts of this mussel bed (Fig. 11B) and when different mussel beds are compared (Goss Custard *et al.*, 1981).

V. OTHER BIRDS

The presence of other birds, usually of the same species, may have a significant effect on intake rate and influence the way in which indi-

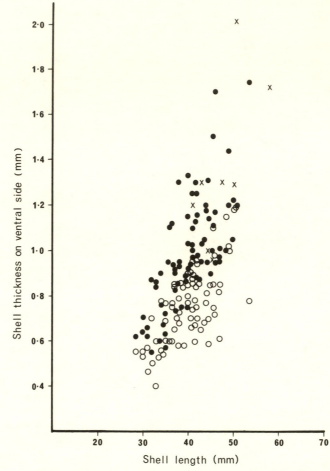

Fig. 10 The thickness of the shells of all mussels (●) and of those opened by Eurasian Oystercatchers (○) of different sizes from bed 4 (Durell and Goss-Custard, 1984) of the Exe Estuary. Crosses show mussels that oystercatchers gave up trying to break open.

viduals respond to their food supply. This section reviews the evidence and discusses some of the mechanisms that may be involved.

A. Field Evidence

The feeding rate of Common Redshank detecting their prey by sight decreased as birds fed closer together (Goss-Custard, 1976). This response

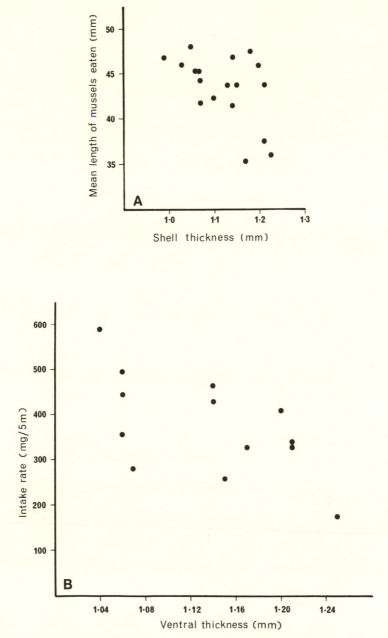

Fig. 11. (A) Mean length of mussels opened by Eurasian Oystercatchers in different parts of bed 4 of the Exe Estuary in relation to the mean shell thickness of a mussel standard length. (B) Intake rate in relation to shell thickness on the same mussel bed.

to the proximity of other birds was immediate and reversible: as soon as bird density decreased, feeding rate recovered. Therefore, the presence of birds did not have its effect by depleting the standing stock of prey but some form of interference, associated with the close proximity of other birds, was involved. Interference has also been measured in Eurasian Oystercatchers feeding on mussels (Ens and Goss-Custard, 1984; Koene, 1978; Zwarts and Drent, 1981) and in Eurasian Curlew eating *Nereis* (Zwarts, 1978), and it has been postulated in Grey Plovers by Stinson (1977). However, both Goss-Custard (1977a) and Sutherland (1982a) found no interference among Eurasian Oystercatchers eating cockles. Similarly, the intake rates of Common Redshank (Goss-Custard, 1976) and Red Knot (Goss-Custard, 1980) feeding by touch were unrelated to bird density. It seems that interference occurs in some situations but not in others, and studying the mechanism by which it is brought about may be necessary before it is possible to understand the circumstances under which it occurs.

Two broad categories have been postulated. In one, interference occurs because the presence of other birds reduces the density of available prey. In the other, the birds have a direct effect on each other's intake rate.

Within the first category, several mechanisms have been proposed. Goss-Custard (1970b) showed experimentally that *Corophium* disappear from the mud surface when a Common Redshank walks over them, and this might reduce their availability to the birds. Alternatively, the number of prey available at any one instance may be so low that the birds actually deplete the available fraction to a significant extent, the more so when bird density is high. It was estimated that less than 1% of *Corophium* were available to Common Redshank at any one time and that it was possible for Common Redshank to have a significant impact on this fraction. A similar possibility exists with Bar-tailed Godwits eating lugworms (Goss-Custard, 1980). These suggestions are consistent with the finding that interference occurred in visually hunting Common Redshank, but not touch-feeding Common Redshank or Red Knot. Visually hunting birds would be affected by the loss of prey from the mud surface but those detecting buried prey by touch would not.

Neither of these possibilities provides a convincing explanation for the interference recorded in Eurasian Oystercatchers eating mussels. Although most of the birds studied by Koene (1978) and Zwarts and Drent (1981) obtained food by stabbing into gaping mussels which might close up and become unavailable when bird density is high, most Eurasian Oystercatchers on the Exe hammered into mussels and yet interference still occurred. As Eurasian Oystercatchers are most unlikely in the short-term to affect shell thickness in the mussels, interference is probably a direct response to the presence of other birds.

Zwarts and Drent (1981) were able to show that part of the reduction in intake rate was due to increased kleptoparasitism by Herring Gulls (*Larus argentatus*) or Common Gulls (*Larus canus*) at high Eurasian Oystercatcher densities, but there was apparently an additional effect due to the oystercatchers themselves. Oystercatchers on the Exe frequently interact aggressively with each other, at a rate that increases with bird density—a finding also reported in several other shorebird studies (Burger *et al.*, 1979; Goss-Custard, 1977a; Hamilton, 1959; Recher and Recher, 1969; Silliman *et al.*, 1977; Vines, 1980). Goss-Custard *et al.* (1982a) showed that many individuals restrict their foraging to small areas of the mussel bed for long periods, whereas others range more freely. Those that feed regularly in one place are the most aggressive, and are more often successful in encounters—they are also more likely to steal more mussels than they have stolen from them by other birds. In a further study, Ens and Goss-Custard (1984) showed that, in terms of success in encounters, a linear dominance hierarchy exists among birds whose feeding areas overlap and that the most dominant birds may not suffer from interference. This further suggests that direct social factors are involved, rather than a general reduction in the density of available prey.

Although Eurasian Oystercatchers were more often aggressive as density increased, mussels lost in encounters accounted for only a part of the interference that occurred. The greatest change was in the number captured as bird density increased. How this happened is not clear, but perhaps the subdominants were distracted from looking for potential prey when there were large numbers of other, potentially aggressive birds in the vicinity. Alternatively, they may have made turns to avoid other birds rather than to locate prey, and so fed less effectively (Goss-Custard, 1970b). Indeed, because Eurasian Oystercatchers seem to space themselves over mussel beds by avoidance (Vines, 1980), and the food supply might vary from place to place on a very small scale, it is also possible that the subdominants fed more often in poorer places at high bird densities in an attempt to keep away from the more aggressive birds. It might be because of this that subdominant birds ranged over a larger area than did the dominant individuals.

Though not tested at this very small scale, there is evidence on a larger scale that high densities of birds in an area may deter other birds from feeding there. Studies on several species have shown that an increasing proportion of birds feed in the less preferred parts of the estuary as bird numbers rise (Goss-Custard, 1977a,b; Zwarts and Drent, 1981), and a study on Eurasian Oystercatchers on the Exe (Goss-Custard *et al.*,

1981, 1982b) provides an example. Only a few hundred immature birds feed on the mussel beds in summer when most of the birds are far away on the breeding grounds. Over half of these young birds feed on two mussel beds that seem to be preferred because they are close to the roost used at high tide, have a firm substrate, and contain many small mussels, which young birds eat at that time of the year. The adults return to the Exe during August and September and the population feeding on mussels increases sixfold. Though the density of birds on the two most preferred beds does increase during this period, it does not do so in proportion to the increase in the population as a whole. The proportion of the whole population feeding on these preferred beds declines from over 50% to almost 20%. At the same time, there is a changeover in the individuals using them. Gradually, most of the young birds that had been there during the summer move to other, less preferred mussel beds or onto the mud-flats. By October, most of the birds on these beds are adults. As young Eurasian Oystercatchers are in general subdominant to adults and actively avoid them, it does seem that the young birds leave the preferred beds in autumn because of competition from the returning adults. As would be expected on this hypothesis, in winter there are more young birds on the least preferred mussel beds and, in spring, they gradually return to the most preferred beds as the adults leave for the breeding grounds.

The result of the change in distribution as the population size rises is that an increasing proportion of the birds feed in less preferred habitats. As the gross intake rate (Goss-Custard, 1970a, 1977a,b; Goss-Custard *et al.*, 1977b; Sutherland, 1982a; Zwarts and Drent, 1981), and perhaps net intake rate (Goss-Custard *et al.*, 1981) of shorebirds may be relatively low in such areas, average intake rate in the population will be reduced. Whether the intake rate of all the birds, or only the subdominant ones, is affected is not yet clear (Goss-Custard, 1980). If dominant birds can maintain their high intake rate at all bird densities, then only the sub-dominants may suffer a reduced intake rate: in this case, the "ideal-despotic" model of habitat distribution proposed by Fretwell and Lucas (1970) may be appropriate. If, on the other hand, there is little that individuals can do to protect themselves from interference because, for instance, it is brought about by changes in prey behavior, all birds may suffer equally. In this case, the "ideal-free" model of Fretwell and Lucas may be relevant.

In both models, the balance between high-quality feeding areas attracting many birds on the one hand and interference reducing their attractiveness on the other would result in the density of birds being correlated with habitat quality. Insofar as prey density defines habitat quality, this has been shown in several studies: Common Redshank eating *Cor-*

ophium (Goss-Custard, 1970a, 1977b); Eurasian Oystercatchers eating *Cerastoderma* (Goss-Custard, 1977a; Rands and Barkham, 1981; Sutherland, 1982a), *Mytilus* (Goss-Custard *et al.*, 1981; Zwarts and Drent, 1981), and *Macoma* (Hulscher, 1982); Dunlin (Rands and Barkham, 1981) and Sanderling (Connors *et al.*, 1981) eating several prey; and several shorebird species eating several invertebrates (Bryant, 1979). In view of the variety of factors in addition to prey density that may affect gross or net intake rates, prey density is only one of the factors that may determine habitat quality, and a multivariate approach in defining the quality of feeding areas would be desirable (Goss-Custard *et al.*, 1981; Myers *et al.*, 1980). However, some data are inconsistent with the ideal-free model. First, some dominant Eurasian Oystercatchers on the Exe seem to be little affected by interference. Second, several studies have shown large differences in intake rate between areas of an estuary, and this would not be expected in the ideal-free model (Sutherland, 1982a). Further tests of the applicability of these models to feeding shorebirds seem to be worthwhile (Zwarts and Drent, 1981).

B. Age, Dominance, and Intake Rate

The intake rate of individual Eurasian Oystercatchers on the Exe varied considerably, even at low bird densities when interference is least (Goss-Custard *et al.*, 1982a). Intake rate is related to dominance (Ens and Goss-Custard, 1984; Goss-Custard *et al.*, 1984), but it is not clear why this should be. One possibility is that dominance secures the best feeding sites within the mussel bed, and this is reflected in increased rates of intake. Another possibility is that dominant birds tend to be older, and therefore more experienced in feeding, and so are able to feed at a faster rate. As the dominant birds occur more frequently in a particular part of a mussel bed (Goss-Custard *et al.*, 1982a), it is possible that their greater experience in that place may be the critical factor. But dominance and experience may be correlated in a more general sense. In this study, young birds were usually subdominant to adults. Though frequently as aggressive, they lost more encounters and avoided adults more frequently than did adults (Goss-Custard *et al.*, 1982b). A general inexperience in feeding, rather than subdominance per se, may have caused these birds to have lower intake rates at low bird densities. At first sight, this might appear unlikely as most of the birds were at least several years old. However, the feeding behavior of oystercatchers on the Exe changes during the first few years of life (Goss-Custard and Durell, 1983) so it may be several years before they achieve full feeding proficiency. Indeed, birds in their second winter still have relatively low weights and a high risk of dying (Goss-Custard *et al.*, 1982c), and this may reflect inexperience at feeding (Dare, 1977). Groves (1978) found that juvenile Ruddy Turnstones (*Arenaria interpres*) forage more slowly than adults and suggests that inexperience in foraging may have been responsible, though being subdominant to adults was not ruled out. Harrington and Groves (1977), however, found that juvenile Semipalmated Sandpipers (*Calidris pusilla*) were not necessarily subdominant to adults, though this seems likely to be an exception to the general case. Taken together, studies on age-related differences in foraging and social behavior underline the difficulty in disentangling the various factors that probably affect intake rate in shorebirds. Not only do the density, availability, and suitability of prey vary, but the birds feeding on them may differ in experience and status and, perhaps, condition (Smith and Evans, 1973).

VI. PREY DEPLETION

Shorebirds may affect their own intake rate in the longer term through prey depletion as well as in the short term through interferences. Several studies have shown that shorebirds remove a substantial proportion of the standing crop of prey during winter (Goss-Custard, 1980). Only the work of Duffy *et al.* (1981) differs from this general result, perhaps because their comparisons were made over only a few weeks rather than the several months typical of the other studies. Similarly, Schneider (1978) showed by enclosure experiments that shorebirds may also remove a substantial proportion of their prey on stopover points during fall migration. Mortality from July to September varied between 36 and 84%, according to the initial abundance of the prey.

Figure 12 summarizes the data from winter studies. The results are separated into those from known preferred feeding areas and those from all feeding areas combined. This is done because the impact of shorebirds on their prey seems to be greatest in the preferred areas where they concentrate their feeding (Goss-Custard, 1977b; Horwood and Goss-Custard, 1977; Schneider, 1978; Sutherland, 1982a,b) and where prey depletion will affect the greatest number of birds. The highest value is the 90% recorded by Evans *et al.* (1979) in the Tees Estuary where much mudflat has been removed by reclamation during the last few years. In addition, Zwarts and Drent (1981) calculated that Eurasian Oystercatchers removed 40% of the mussels during the winter that were present at the start, and Sutherland (1982a) calculated that Eurasian Oystercatchers ate up to 28%

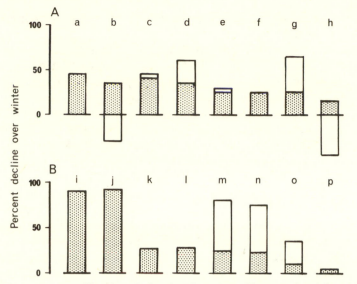

Fig. 12. The percentage of food loss during the winter (open histograms) and that part caused by wader predation (stippled histograms), in (A) main feeding areas, and (B) all areas. Histograms below the line indicate food increases due to immigration. All data from northwestern Europe. The studies cited species involved and are as follows: a, *Lanice conchilega* eaten by *P. squatarola, N. arquata,* and *L. lapponica;* b, several invertebrates eaten by all waders present; c, *C. volutator* eaten by *T. totanus.;* d, *C. edule* eaten by *C. canutus;* e, several invertebrates eaten by all shorebirds present; f, *A marina* eaten by *L. lapponica;* g, *M. balthica* eaten by *C. canutus;* h, *C. edule* eaten by *H. ostralegus;* i, *H. ulvae* eaten by several shorebird species; j, *N. diversicolor* eaten by several shorebird species; k, *M. edulis* eaten by *H. ostralegus;* l—o, *c. edule* eaten by *H. ostralegus;* p, *M. balthica* eaten by *C. canutus.* References: a, b, d, e, g, h, Goss-Custard (1977a); c, Goss-Custard (1969); f, Smith (1975); i, j, Evans *et al.* (1979); k, Goss-Custard *et al.* (1980); l–o, Horwood and Goss-Custard (1977); p, Prater (1972).

of the cockles in his study area. In contrast, Hartwick and Blaylock (1979) calculate that American Black Oystercatchers (*H. bachmani*) removed only a tiny fraction of the mussels in their study area in Vancouver Island. But in general, prey are reduced by 25–45%, mainly through predation by the shorebirds themselves.

The effect of such decreases on the average intake rate of shorebirds has been estimated in a few cases. Goss-Custard (1980) calculated that a prey depletion of 25–45% would reduce intake rate by 8–23% in Common Redshank eating *Corophium*, by 11–33% in redshank eating *Nereis*, and by 5–18% in Eurasian Oystercatchers eating *Cerastoderma*, depending on the initial abundance of prey and the proportion removed during the winter. J. P. Myers (personal communication) calculated that prey depletion by Sanderling during winter would reduce their caloric intake rate in two areas by 41%.

However, these studies assume, of course, that the birds would continue feeding on the same prey species as its abundance decreased. This may often be true, but the birds would be expected to turn to alternative prey if the density of the preferred one declined sufficiently. In many cases, this would be likely also to result in a reduction in intake rate because the alternative prey is likely to be less profitable. Thus, Hartwick (1976) showed that the preferred prey of American Black Oystercatchers, the mussel *M. californianus*, was a much more profitable prey than the main alternative, the limpets *Acmaea digitalis* and *A. persona*. However, this is not always so. Common Redshank eat more *Nereis* as their preferred prey *Corophium* becomes scarce, and can collect worm biomass at a greater rate (Goss-Custard, 1977c). Unless *Nereis* fail to provide a critical nutrient, which on present evidence appears unlikely, the intake rate of Common Redshank may actually increase as they deplete *Corophium*, assuming *Nereis* is present in adequate abundance.

It can be assumed in the majority of these studies that most, if not all, of the individuals in the prey populations would be available during the winter, even if only a small fraction is vulnerable at any one time. For instance, it is likely that most of the *Arenicola* defecate during the winter and so become available to Bar-tailed Godwits. However, this is not so in some cases. For example, a high proportion of *Macoma* on the Wash bury themselves too deeply in the sand for Red Knot to reach, and many individuals may remain inaccessible for long periods. In cases like this, measuring the proportion of the whole prey population eaten by shore birds may seriously underestimate the birds' actual impact on that portion that actually comprises their own food supplies.

VII. CONCLUSIONS

Research during the last decade or so has identified many factors that affect the intake rates of shorebirds. Most progress has been made in studying external factors. These may be subdivided into (1) characteristics of the food supply and (2) other environmental factors that influence the birds' ability to exploit their food. The main sources of variation within the food supply itself are its species composition, density, and prey size though other factors, such as infection by parasites or shell-thickness in the prey of oystercatchers, may prove to be of critical importance. The second category, the other environmental factors, includes a wide range of factors that can be subdivided into three groups, depending on the way in which they affect intake rate. One group physically prevents the birds from either detecting or reaching their food: turbid water or hard substrate are examples. The second group affects the behavior of the prey and thus the ability of shorebirds to detect and reach them: mud temperature is a well-worked example. Between them, these two groups of factors influence the proportion of the prey available and, in combination with prey density, determines the abundance of available prey that the bird will encounter. The third group of environmental factors acts directly on the bird itself. For instance, kleptoparasites steal prey found by others, as Black-headed Gulls (*Larus ridibundus*) and Common Gulls (*L. canus*) from Common Lapwings (*Vanellus vanellus*) (Kallander, 1977); dominant Eurasian Oystercatchers steal from subdominants and may exclude them from the best feeding places; high winds buffet shorebirds too much to allow them to forage. In some cases, a single factor can influence intake rate in two or three of these ways. Thus, rain may reduce prey availability by making the surface turbid and by driving prey below the surface. Interference from conspecifics could arise in all three ways: by keeping birds off the best feeding area, by driving prey beneath the surface, or by stealing prey already captured. This further complicates the difficult task of disentangling the separate effects of a number of factors that are frequently confounded anyway. So far, most studies have been based on simple correlations between intake rate and variations in one or more independent variables. With the notable exception of Myers *et al.* (1980), most studies have not tried to separate out the individual effects of particular factors, either statistically or experimentally. This is certainly an area of research to which more attention should be given in the future.

It is likely that the internal state of the bird will receive increasing attention now that the considerable variation that exists between and within individuals is becoming apparent. In Eurasian Oystercatchers on the Exe, for example, adults may specialize on *Mytilus* or on a mixture of *Nereis* and *Scrobicularia*, or on *Littorina* spp. or on *Cerastoderma*, and these specializations may change with age (Goss-Custard and Durell, 1983). Among those eating mussels, three feeding techniques have been identified that individuals use consistently (Goss-Custard *et al.*, 1982a). As these specializations may be acquired (Norton-Griffiths, 1968), interpreting field behavior requires a knowledge of birds' individual histories. Experience may also affect the intake rate in species that apply particular skills in feeding, such as the Eurasian Oystercatcher. It may take a long time for young birds to gain the experience to feed as efficiently as adults, and this may critically affect their survival. There is also some evidence to suggest that intake rate may be controlled by the bird in accordance with the time available for feeding or parental commitments, though the evidence on this is sketchy. Though much of the variation in intake rate may be explained by variations in external factors, a full understanding depends on the study of internal factors as well. In particular, it may be necessary to revise the idea that the birds always feed at the maximum rate, as would be predicted by simple models of optimal foraging (Goss-Custard, 1981a).

Because one of the main activities of shorebirds in winter is foraging, studying variations in intake rate is vital to understanding their adaptations at that time of the year. Such studies have focused on sex differences in morphology and feeding behavior (Smith and Evans, 1973), seasonal patterns of fat deposition in relation to seasonal factors that affect prey availability (Davidson, 1981; Evans and Smith, 1975), prey selection (Evans, 1976; Goss-Custard, 1977c), and dispersion (Goss-Custard, 1970b; Myers *et al.*, 1981). Intake rates are assumed in all these studies to be of paramount importance to understanding morphology and behavior, with optimal foraging theory (Pyke *et al.*, 1977) providing the theoretical base. In most studies, the authors are forced to deal only with gross rates of intake though it is usually acknowledged that net rate of intake is the critical measure. Some authors have used pacing or pecking rates to provide some measure of expenditure (Goss-Custard, 1970a, 1977c), but further progress is unlikely to be made until techniques such as respirometry, heart telemetry (Ferns *et al.*, 1979) or the doubly labeled water (D_2O^{18}) technique (Bryant and Westerterp, 1980) are more widely applied. Caloric content is usually taken to be the important unit of prey value, though some studies throw some doubt on this (Goss-Custard, 1977c, 1981a; Hartwick, 1976).

Many of the studies on intake rates are designed to understand the role winter food supplies play in determining the numbers of shorebirds. Broad comparisons between the foraging behavior of species in winter and at other times (Baker and Baker, 1973; Duffy *et al.*, 1981) and crude estimates of the impact of shorebirds on their food supplies (Hale, 1981) do not provide critical tests of this possibility because neither approach may detect the subtle processes that could be involved. Measuring the gross impact of the birds on their food may greatly underestimate the real effect depletion has on their intake rate, because only a proportion of the prey may be vulnerable. Intake rates may vary considerably between individuals so that only a small proportion may be in difficulties at any one time.

Studies on seasonal changes in intake rates, the time available in which to feed, and food requirements suggest that in many areas energy balance may be difficult to achieve in certain conditions in winter (Evans,

1976; Feare, 1966; Goss-Custard, 1969; Heppleston, 1971). Mortality can be quite severe in bad weather (e.g., Heppleston, 1971), and the loss of body condition of individuals during periods of weather in which intake rates are known to be depressed further implicates food shortage in some winter mortality (Davidson, 1981). But the main question is what effect this has on population size. Where it has been studied, winter mortality usually seems to be low and may only be substantial during prolonged periods of bad weather, but then only in some species in some areas (Davidson, 1981; Goss-Custard, 1981b). On the Exe, for example, less than 2% of adult Eurasian Oystercatchers die in winter and most of the annual mortality occurs elsewhere (Goss-Custard et al., 1982c). On the other hand, up to 15% of the young birds may die and the mortality among this relatively small section of the population may have an important effect on population size (Goss-Custard and Durell, in press). This possibility further underlines the need for more studies on individual variations in intake rate in shorebirds.

ACKNOWLEDGMENTS

I am very grateful to S. Boates, S. Durell, Dr. M. G. Morris, and Dr. W. Sutherland for criticizing the manuscript, to the British Ecological Society for permission to reproduce figures from the *Journal of Animal Ecology*, and to Peter Smith for permission to quote from his Ph.D. thesis.

REFERENCES

Baker, A. J., 1974, Prey-specific feeding methods of New Zealand Oystercatchers, *Notornis* 21:219–233.

Baker, C. B., 1974, Foraging behavior of Black-billed Plovers (*Pluvialis squatarola*), *Ecology* 55:162–167.

Baker, M. C., and Baker, A. E. M., 1973, Niche relationships among six species of shorebirds on their wintering and breeding ranges, *Ecol. Monogr.* 43:193–212.

Boates, J. S., 1980, Foraging semipalmated sandpipers *Calidris posilla*. L. and their major prey *Corophium volutator* (Pallas) on the Starrs Point mudflat Minas Basin, Unpublished M.S. thesis, Acadia University, Canada.

Bryant, D. M., 1979, Effects of prey density and site character on estuary usage by over-wintering waders (Charadrii), *Estuarine Coastal Mar. Sci.* 9:369–384.

Bryant, D. M., and Westerterp, K. R., 1980, The energy budget of the House Martin (*Delichon urbica*), *Ardea* 68:91–102.

Burger, J., Howe, M. A., Hahn, D. C., and Chase, J., 1977, Effects of tide cycles on habitat selection and habitat partitioning by migrating shorebirds, *Auk* 94:743–758.

Burger, J., Hahn, D. C., and Chase, J., 1979, Aggressive interactions in mixed-species flocks of migrating shorebirds, *Anim. Behav.* **27**:459–469.

Connors, P. G., Myers, J. P., Connors, C. S. W., and Pitelka, F. A., 1981, Interhabitat movements by Sanderlings in relation to foraging profitability and the tidal cycle, *Auk* **98**:49–64.

Dare, P. J., 1977, Seasonal changes in body weight of oystercatchers *Haematopus ostralegus*, *Ibis* **119**:494–506.

Davidson, N. C., 1981, Survival of shorebirds (Charadrii) during severe weather: The role of nutritional reserves, in: *Feeding and Survival Strategies of Estuarine Organisms* (N. V. Jones and W. J. Wolff, eds.), pp. 231–249, Plenum Press, New York.

Duffy, D. C., Atkins, N., and Schneider, D. C., 1981, Do shorebirds compete on their wintering grounds? *Auk* **98**:215–229.

Dugan, P. J., 1981a, The importance of nocturnal foraging in shorebirds—A consequence of increased invertebrate prey activity, in: *Feeding and Survival Strategies of Estuarine Organisms* (N. V. Jones and W. J. Wolff, eds.), pp. 251–260, Plenum Press, New York.

Dugan, P. J., 1981b, Seasonal movements of shorebirds in relation to spacing behavior and prey availability, Unpublished Ph.D. thesis, University of Durham, U.K.

Durell, S. E. A., Goss-Custard, J. D., 1984, Prey selection within a size-class of mussels, *Mytilus edulis*, by oystercatchers, *Haematopus ostralogus, Anim. Behav.* **32**.

Ens, B. J., and Goss-Custard, J. D. 1984, Interference among oystercatchers, *Haematopus ostralegus* L., feeding on mussels, *Mytilus edulis*, L., on the Exe Estuary, *J. Anim. Ecol.* **53**:217–232.

Evans, P. R., 1976, Energy balance and optimal foraging strategies in shorebirds: Some implications for their distributions and movements in the non-breeding season, *Ardea* **64**:117–139.

Evans, P. R., and Smith, P. C., 1975, Studies of shorebirds at Lindisfarne, Northumberland. II. Fat and pectoral muscles as indicators of body condition in the Bar-tailed Godwit, *Wildfowl* **26**:64–76.

Evans, P. R., Herdson, D. M., Knights, P. J., and Pienkowski, M. W., 1979, Short-term effects of reclamation of part of Seal Sands, Teesmouth, on wintering waders and Shelduck, *Oecologia (Berlin)* **41**:183–206.

Feare, C. J., 1966, The winter feeding of the Purple Sandpiper, *Br. Birds* **59**:165–179.

Feare, C. J., 1971, Predation of limpets and dog whelks by oystercatchers, *Bird Study* **18**:121–129.

Ferns, P. N., MacAlpine-Leny, I. H., and Goss-Custard, J. D., 1979, Telemetry of heart rate as a possible method of estimating energy expenditure in the redshank, *Tringa totanus* (L.), in: *A Handbook of Biotelemetry and Radio Tracking* (C. J. Amlaner and D. W. MacDonald, eds.), pp. 595–601, Pergamon Press, Elmsford, N.Y.

Fretwell, S. D., and Lucas, H. L., 1970, On territorial behavior and other factors influencing habitat distribution in birds. I. Theoretical development, *Acta Biotheor.* **19**:16–36.

Goss-Custard, J. D., 1966, The feeding ecology of redshank, *Tringa totanus* L, in winter, on the Ythan Estuary Aberdeenshore, Unpublished Ph.D. Thesis, Aberdeen University.

Goss-Custard, J. D., 1969, The winter feeding ecology of the redshank *Tringa totanus*, *Ibis* **111**:338–356.

Goss-Custard, J. D., 1970a, The responses of redshank (*Tringa totanus* (L.)) to spatial variations in the density of their prey, *J. Anim. Ecol.* **39**:91–113.

Goss-Custard, J. D., 1970b, Feeding dispersion in some overwintering wading birds, in: *Social Behaviour in Birds and Mammals* (J. H. Crook, ed.), pp. 3–35, Academic Press, New York.

Goss-Custard, J. D., 1970c, Factors affecting the diet and feeding rate of the redshank (*Tringa totanus*), in: *Animal Populations in Relation to Their Food Resources* (A. Watson, ed.), pp. 107–110, Blackwell, Oxford.

Goss-Custard, J. D., 1976, Variation in dispersion of redshank *Tringa totanus* on their winter feeding grounds, *Ibis* **118:**257–263.

Goss-Custard, J. D., 1977a, The ecology of the Wash. III. Density-related behaviour and the possible effects of a loss of feeding grounds on wading birds (Charadrii), *J. Appl. Ecol.* **14:**721–739.

Goss-Custard, J. D., 1977b, Predator responses and prey mortality in redshank, *Tringa totanus* (L.), and preferred prey, *Corophium volutator* (Pallas), *J. Anim. Ecol.* **46:**21–35.

Goss-Custard, J. D., 1977c, The energetics of prey selection by redshank, *Tringa totanus* (L.), in relation to prey density, *J. Anim. Ecol.* **46:**1–19.

Goss-Custard, J. D., 1977d, Optimal foraging and the size selection of worms by redshank, *Tringa totanus*, in the field, *Anim. Behav.* **25:**10–29.

Goss-Custard, J. D., 1978, Sequential choice for prey size by captive redshank. *Ibis* **120:**230–232.

Goss-Custard, J. D., 1980, Competition for food and interference among waders, *Ardea* **68:**31–52.

Goss-Custard, J. D., 1981a, Feeding behavior of redshank, *Tringa totanus*, and optimal foraging theory, in: *Foraging Behavior* (A. C. Kamil and T. D. Sargent, eds.), pp. 115–133, Garland, STPM Press, New York.

Goss-Custard, J. D., 1981b, Role of winter food supplies in the population ecolgy of common British wading birds, *Verh. Ornithol. Ges. Bayern* **23:**125–146.

Goss-Custard, J. D., and Durell, S. E. A., in press, Feeding ecology, winter mortality and the population dynamics of oystercatchers, *Haematopus ostralegus*, on the Exe Estuary, in: *Coastal Waders and Wildfowl in Winter* (P. R. Evans, J. D. Goss-Custard, and W. G. Hale, eds.), Cambridge University Press, Cambridge.

Goss-Custard, J. D., and Durell, S. E. A., 1983, Individual and age differences in the feeding ecology of oystercatchers, *Haematopus ostralegus*, wintering on the Exe Estuary, S. Devon, *Ibis* **125:**155–171.

Goss-Custard, J. D., Jones, R. E., and Newbury, P. E., 1977a, The ecology of the Wash. I. Distribution and diet of wading birds (Charadrii), *J. Appl. Ecol.* **14:**681–700.

Goss-Custard, J. D., Kay, D. G., and Blindell, R. M., 1977b, The density of migratory and overwintering redshank, *Tringa totanus* (L.), and curlew, *Numenius arquata* (L.), in relation to the density of their prey in south-east England, *Estuarine Coastal Mar. Sci.* **15:**497–510.

Goss-Custard, J. D., Jenyon, R. A., Jones, R. E., Newbery, P. E., and Williams, R. B., 1977c, The ecology of the Wash. II. Seasonal variation in the feeding conditions of wading birds (Charadrii), *J. Appl. Ecol.* **14:**701–719.

Goss-Custard, J. D., McGrorty, S., Reading, C. J., and Durell, S. E. A., 1980, Oystercatchers and mussels on the Exe Estuary, in: *Essays on the Exe Estuary*, Devon Assoc. Special Vol. 2, pp. 161–185.

Goss-Custard, J. D., Durell, S. E. A., McGrorty, S., Reading, C. J., and Clarke, R. T., 1981, Factors affecting the occupation of mussel (*Mytilus edulis*) beds by oystercatchers (*Haematopus ostralegus*) on the Exe Estuary, Devon, in: *Feeding and Survival Strategies of Estuarine Organisms* (N. V. Jones and W. J. Wolff, eds.), pp. 217–229, Plenum Press, New York.

Goss-Custard, J. D., Durell, S. E. A., and Ens, B. J., 1982a, Individual differences in aggressiveness and food stealing among wintering oystercatchers, *Haematopus ostralegus* L., *Anim Behav.* **30**:917–928.

Goss-Custard, J. D., Durell, S. E. A., McGrorty, S., and Reading, C. J., 1982b, Use of mussel *Mytilus edulis* beds by oystercatchers *Haematopus ostralegus* according to age and population size, *J. Anim Ecol.* **51**:543–554.

Goss-Custard, J. D., Durell, S. E. A., Sitters, H. P., and Swinfen, R., 1982c, Age-structure and survival of a wintering population of oystercatchers, *Haematopus ostralegus*, *Bird Study* **29**:83–98.

Goss-Custard, J. D., Clarke, R. T., and Durell, S. E. A., 1984, Rates of food intake and aggression of oystercatchers *Haematopus ostralogus* on the most and least preferred mussel *Mytilus edulis* beds of the Exe Estuary, *J. Anim. Ecol.* **53**:233–246.

Groves, S., 1978, Age-related differences in Ruddy Turnstone foraging and aggressive behaviour, *Auk* **95**:95–103.

Hale, W. A., 1981, *Waders*, Collins, Glasgow.

Hamilton, W. J., 1959, Aggressive behaviour in migrant Pectoral Sandpipers, *Condor* **61**:161–179.

Harrington, B. A., and Groves, S., 1977, Aggression in foraging migrant semipalmated Sandpipers, *Wilson Bull.* **89**:336–338.

Hartwick, E. B., 1976, Foraging strategy of the Black Oystercatcher (*Haematopus bachmani* Audubon), *Can. J. Zool.* **54**:147–155.

Hartwick, E. B., and Blaylock, W., 1979, Winter ecology of a Black Oystercatcher population, in: *Studies in Avian Biology No. 2* (F. A. Pitelka, ed.), pp. 207–216, Cooper Ornithological Society, Allen Press, Lawrence, Kans.

Heppleston, P. B., 1971, The feeding ecology of oystercatchers (*Haematopus ostralegus* L.) in winter in northern Scotland, *J. Anim Ecol.* **40**:651–672.

Holling, C. S., 1959, Some characteristics of simple types of predation, *Can. Entomol.* **91**:385–398.

Horwood, J., and Goss-Custard, J. D., 1977, Predation by the oystercatcher, *Haematopus ostralegus* (L.), in relation to the cockle, *Cerastoderma edule* (L.) , fishing in the Burry Inlet, south Wales, *J. Appl. Ecol.* **14**:139–158.

Hulscher, J. B., 1973, Burying depth and trematode infection in *Macoma balthica, Neth. J. Sea Res.* **6**:141–156.

Hulscher, J. B., 1976, Localisation of cockles (*Cardium edule* L.) by the oystercatcher (*Haematopus ostralegus* L.) in darkness and daylight, *Ardea* **64**:292–310.

Hulscher, J. B., 1982, The oystercatcher as a predator of *Macoma, Ardea* **70**:89–152.

Kallander, H., 1977, Piracy by Black-headed Gulls on lapwings, *Bird Study* **24**:186–194.

Koene, P., 1978, De scholekster: auntalseffecten op de voedsclopname, Doktorandlonderzoek, Zool. Lab. R, University of Groningen.

Krebs, J. R., 1980, Optimal foraging, predator risk and territory defence, *Ardea* **68**:83–90.

McClennan, J. A., 1979, The formation and function of mixed-species wader flocks in fields, Unpublished Ph.D. thesis, University of Aberdeen, U.K.

Myers, J. P., Williams, S. L., and Pitelka, F. A., 1980, An experimental analysis of prey availability for Sanderlings (Aves: Scolopacidae) feeding on sandy beach crustaceans, *Can. J. Zool.* **58**:1564–1574.

Myers, J. P., Connors, P. G., and Pitelka, F. A., 1981, Optimal territory size and the Sanderling: Compromises in a variable environment, in: *Foraging Behaviour* (A. C. Kamil and T. D. Sargent, eds.), pp. 135–158, Garland STPM Press, New York.

Norton-Griffiths, M., 1967, Some ecological aspects of the feeding behaviour of the oystercatcher, *Haematopus ostralegus*, on the edible mussel, *Mytilus edulis*, *Ibis* **109**:412–424.

Norton-Griffiths, M., 1968, The feeding behaviour of the oystercatcher, *Haematopus ostralegus*, Unpublished Ph.D. thesis, University of Oxford, U.K.

Pienkowski, M. W., 1980a, Differences in habitat requirements and distribution patterns of plovers and sandpipers as investigated by studies of feeding behaviour, *Verh. Ornithol. Ges. Bayern* **23**:105–124.

Pienkowski, M. W., 1980b, Aspects of the ecology and behavior of Ringed and Grey Plovers *Charadrius hiaticula* and *Pluvialis squatarola*, Ph.D. thesis, University of Durham, U.K.

Pienkowski, M. W., 1981, How foraging plovers cope with environmental effects on invertebrate behaviour and availability, in: *Feeding and Survival Strategies of Estuarine Organisms* (N. V. Jones and W. J. Wolff, eds.), pp. 179–192, Plenum Press, New York.

Pienkowski, M. W., 1982, Diet and energy intake of Grey and Ringed Plovers, *Pluvialis squatarola* and *Charadrius hiaticula*, in the non-breeding season, *J. Zool.* **197**:511–549.

Prater, A. J., 1972, The ecology of Morecambe Bay. III. The food and feeding habits of knot (*Calidris canutus* L.) in Morecambe Bay, *J. Appl. Ecol.* **9**:179–194.

Price, H. A., 1980, Seasonal variation in the strength of byssal attachment of the common mussel *Mytilus edulis* L., *J. Mar. Biol. Assoc. U.K.* **60**:1035–1037.

Pyke, G. H., Pulliam, H. R., and Charnov, E. L., 1977, Optimal foraging: A selective review of theory and tests, *Q. Rev. Biol.* **52**:137–154.

Rands, M. R. W., and Barkham, J. P., 1981, Factors controlling within-flock feeding densities in three species of wading bird, *Ornis Scand.* **12**:28–36.

Reading, C. J., and McGrorty, S., 1978, Seasonal variations in the burying depth of *Macoma balthica* (L.) and its accessibility to wading birds, *Estuarine Coastal Mar. Sci.* **6**:135–144.

Recher, H. F., 1966, Some espects of the ecology of migrant shorebirds, *Ecology* **47**:393–407.

Recher, H. F., and Recher, J. A., 1969, Some aspects of the ecology of migrant shorebirds. II. Aggression, *Wilson Bull.* **81**:140–154.

Schneider, D., 1978, Equalisation of prey numbers by migratory shorebirds, *Nature (London)* **271**:353–354.

Silliman, J., Mills, G. S., and Alden, S., 1977, Effect of flock size on foraging activity in wintering Sanderlings, *Wilson Bull.* **89**:434–455.

Smith, P. C., 1975, A study of the winter feeding ecology and behavior of the Bar-tailed Godwit (*Limosa lapponica*), Unpublished Ph.D. thesis, University of Durham, U.K.

Smith, P. C., and Evans, P. R., 1973, Studies of shorebirds at Lindisfarne, Northumberland. I. Feeding ecology and behaviour of the Bar-tailed Godwit, *Wildfowl* **124**:135–139.

Stinson, C. H., 1977, The spatial distribution of wintering Black-bellied Plovers, *Wilson Bull.* **89**:470–472.

Sutherland, W. J., 1982a, Spatial variation in the predation of cockles by oystercatchers at Traeth Melynog, Anglesey. II. The pattern of mortality, *J. Anim. Ecol.* **51**:491–500.

Sutherland, W. J., 1982b, Spatial variation in the predation of cockles by oystercatchers at Traeth Melynog, Anglesey. I. The cockle population, *J. Anim. Ecol.* **51**:481–490.

Sutherland, W. J., 1982c, Do oystercatchers select the most profitable cockles? *Anim. Behav.* **30**:857–861.

Swennen, C., 1969, Crawling-tracks of trematode infected *Macoma balthica* (L.), *Neth. J. Sea Res.* **4**:376–379.

Swennen, C., and Ching, H. L., 1974, Observations on the trematode *Parvatrema affinis*, causative agent of crawling tracks of *Marcoma balthica, Neth. J. Sea Res.* **8**:108–115.

Tjallingii, S. T., 1969, Habitat-kuize en Gebruik van de Kluut, Doktoranlonderzoek, Zool. Lab. R, University of Groningen.

Vader, W. J. M., 1964, A preliminary investigation into the reactions of the infauna of the tidal flats to tidal fluctuations in water level, *Neth. J. Sea Res.* **2**:189–222.

Vines, G., 1980, Spatial consequences of aggressive behaviour on flocks of oystercatchers, *Haematopus ostralegus* L., *Anim. Behav.* **28**:1175–1183.

Zwarts, L., 1978, Intra- and interspecific competition for space in estuarine bird species in a one-prey situation, *Int. Ornithol. Congr.* **17**:1045–1050.

Zwarts, L., and Drent, R. H., 1981, Prey depletion and the regulation of predator density: Oystercatchers (*Haematopus ostralegus*) feeding on mussels (*Mytilus edulis*, in: *Feeding and Survival Strategies of Estuarine Organisms* (N. V. Jones and W. J. Wolff, eds.), pp. 193–216, Plenum Press, New York.

Chapter 6

SPACING BEHAVIOR OF NONBREEDING SHOREBIRDS

J. P. Myers

Academy of Natural Sciences
Philadelphia, Pennsylvania 19103

Bodega Marine Laboratory
University of California
Bodega Bay, California 94923

and Biology Department
University of Pennsylvania
Philadelphia, Pennsylvania 19104

I. INTRODUCTION

Through spacing behavior, individuals alter their spatial relationships with conspecifics and with the environment. It is a broad class of behavior, spanning spatial scales as small as individual distances within a flock or as global as annual migration. At every position along this continuum in scale, shorebirds show remarkable behavioral variability. The question is whether such variability is simply random chaos, or whether it represents a series of predictable responses by individual shorebirds to ecological and evolutionary processes.

Table I defines four distinct levels of organization in spacing behavior in terms of their spatial and temporal scales. Table I is a reductionist's attempt to partition variability in shorebird spacing behavior into a series of manageable chunks. The distinctions among levels are not always precise because it is, indeed, a continuum of scales. Moreover, the table expands the scope of spacing behavior beyond its traditional treatment, usually levels 1 and 2. This emphasizes the hierarchical nature of spacing behavior, first, to bring out common ecological and evolutionary threads that connect all levels, and second, to emphasize that the requirements

Table I. Levels of Organization in Shorebird Nonbreeding Spacing Behavior[a]

Level	Behavioral continuum	Suggested ecological correlates	Possible selective bases	Sources
Individual	Solitary-flocking (including flock size, tightness, and cohesion)	Food patchiness and density	Increased foraging efficiency	Blick (1980), Rands and Barkham (1981), Zwarts and Orent (1981)
		Feeding style	Sensitivity to feeding interference	Recher and Recher (1969), Goss-Custard (1970, 1976), Pienkowski (1980a)
		Predator occurrence	Minimizing predation risk	Smith and Evans (1973), Page and Whitacre (1975), Silliman et al. (1977), Kus (1980), Myers (1980a,b), Stinson (1980)
		Nonmigratory population and proximity of relatives (for cohesion)	Increase inclusive fitness; cohesive group defense of potential breeding site	Myers (1983a)
	Aggressiveness	Food density, dispersion, and renewal rate	Increased foraging rate	Recher and Recher (1969), Harrington and Groves (1977), Myers and Myers (1979), Gochfeld and Burger (1980), Vines (1980), Goss-Custard et al. (1982)

Territoriality	Shorebird density	Reduced feeding interference	Recher and Recher (1969), Goss-Custard (1976, 1980), Burger et al. (1979), Vines (1980), Zwarts (1981)
	Food and competitor density, resource stability, and patchiness	Energetic costs and benefits of defense	Recher and Recher (1969), Burger et al. (1979), Myers et al. (1979a,b, 1980a), Myers (1980a,b), Dugan (1981a), Johnson et al. (1981), Mallory (1981, 1982), Townshend et al. (1983)
	Predator occurrence	Alter predation risk	Myers et al. (1979a, 1984), Myers (1980a,b)
Local Roost dispersion	Spatiotemporal pattern of food distribution	Increased feeding efficiency	Blick (1980)
	Limited roosting sites	Information exchange	Blick (1980), Myers (1980b)
	Thermal conditions	Enhance survivorship	[Lack (1968)]
		Increase thermoregulatory ability	[Kelty and Lustic (1977)]
Single vs. multiple	Predator occurrence	Reduced predation risk	Blick (1980)
	Tidal pattern of good availability	Increased foraging success	Heppleton (1971), Prater (1972), Burger et al. (1977), Connors et al. (1981), Townshend (1981)

(Continued)

Table I. (*Continued*)

Level	Behavioral continuum	Suggested ecological correlates	Possible selective bases	Sources
Regional	Nomadism vs. site fidelity	Spatiotemporal variability of resources	Maximize survivorship in nonbreeding season	Pienkowski and Clarke (1979), Tree (1979), Evans (1981), Myers (1980a)
Global	Migratory degree	Spatial segregation between optimal breeding and wintering grounds, including resource and climatic factors	Maximizing reproductive output	Evans (1976, 1981), Myers et al. (1984)
	Amount of latitudinal segregation of age/sex classes in winter	Constricted breeding season Length of breeding season competition for breeding sites	Minimizing winter mortality Premium on first arrival at breeding ground	[Greenberg (1980)] Myers (1981)
	Degree of philopatry	Spatiotemporal resources	Adaptation to local environment Ontogenetic familiarity with area Tracking food	Harrington and Morrison (1979) [Yasukawa (1979)] Pitelka et al. (1974)

[a] The sources discuss the continuum, the ecological correlates, or the selective bases, but their conclusions may not support the proposed relationships. Fuller detail is provided in the text. Where no shorebird sources are available, I provide a reference from elsewhere, in brackets.

for spacing behavior include compatibility with other spacing behaviors both within and among levels.

1. Individual: Moment-to-moment adjustments in behavior and position in the immediate foraging area, flock size or tightness, etc., at a spatial scale of centimeters to hundreds of meters.
2. Local: Processes involving switches between foraging sites within the home range, roosting patterns, etc., at a spatial scale of hundreds to thousands of meters over a day or tidal cycle.
3. Regional: Site faithfulness vs. nomadism among different wintering areas, involving tens to hundreds of kilometers and periods of days to months.
4. Global: Migration, philopatry, with distances exceeding thousands of kilometers over periods of a year plus.

Within each level, the scheme in Table I posits a series of behavioral axes that, to varying degrees, are independent of one another. In many cases they share ecological correlates or selective bases, and no doubt are empirically correlated. It is useful to consider them independently, nevertheless, because they represent distinct behavioral processes and because each raises unique issues that otherwise would be, and have been, muddled.

In this chapter I review what we know, assume, and hypothesize about the first three levels of spacing in nonbreeding shorebirds. Other chapters examine migration (Pienkowski and Evans, Morrison), philopatry (Oring and Lank, Vol. 5 of this series), and foraging (Goss-Custard, Puttick). I concentrate on evidence from shorebirds because it is useful to ask within a single, well-studied taxonomic group, where we stand in linking data and theory. My approach is to dissect each behavioral axis and its explanatory hypothesis into their basic components and underlying logic. At times this reductionism may appear overdone. Certainly it would be more pleasing to weave adaptive stories. But the reductionism has an important goal: to lay bare what can and cannot be shown empirically about shorebird winter spacing behavior and its ecological control. A Chantilly lace with dismaying patchiness emerges—some notable successes, some marked failures, and a fabric whose basic pattern is defined as much by its holes as by its threads.

II. THE INDIVIDUAL LEVEL

At the individual level of spacing behavior, individuals adjust their position and behavior with respect to others in their immediate foraging area. Processes at this level have been reduced to a single continuum, flocking–territoriality (e.g., Brown and Orians, 1970), but this obscures important differences among three distinct behavioral processes (Table I): flocking, aggressiveness, and territoriality. Shorebird examples can be found throughout most of the three-dimensional space defined by these axes. Solitary, unaggressive, nonterritorial Sanderlings *Calidris alba*, for example, can be observed along the same kilometer of beach with highly aggressive territorial individuals, as well as unaggressive and aggressive flock members (personal observation with color-marked birds). Individual Pectoral Sandpipers *C. melanotos* defend territories in a clustered array around a central area used by nonterritorial birds, and effectively are members of the local flock (Hamilton, 1959). This is also true for Buff-breasted Sandpipers *Tryngites subruficollis* (Myers, 1980a). Some combinations of behavioral attributes seem unlikely, nevertheless. For example, it would be impossible to have an unaggressive territorial bird.

A. Flocking

Shorebirds exhibit a considerable range in flocking behavior during the nonbreeding season (Table II). A few species, such as Spotted Sandpiper *Actitis macularia*, Wandering Tattler *Heteroscelus incanus*, and Solitary Sandpiper *Tringa solitaria*, almost never appear in winter flocks, while others, notably Stilt Sandpiper *Calidris himantopus* and the Short and Long-billed Dowitchers *Limnodromus griseus* and *L. scolopaceus*, almost invariably move in tight groups. The majority of species, however, shows great flexibility in winter dispersion, so much so that deducing any general ecological pattern from interspecific comparisons resists a simple solution. The frustration this presents is amply conveyed by Goss-Custard (1970).

Shorebird flocks differ in size, spatial tightness, and spatiotemporal cohesiveness. Most work has concentrated on the effects that size and tightness have on predator avoidance and foraging efficiency (references in Table I and below). The overall implication of this work is that flocking improves predator avoidance while it has more complicated, possibly contradictory effects on foraging efficiency. Goss-Custard (1970) has interpreted variability in flock size and tightness as due to both predation and foraging effects simultaneously: all species of similar sizes share similar predation risks in the same site, but they are differentially sensitive to feeding interference. Thus, species for which feeding interference is minimal can move in tighter, larger flocks. I will review the evidence for these interpretations below.

Table II. Flocking Behavior of Nonbreeding Shorebirds while Foraging in California, Argentina, Chile, Ecuador, Paraguay, and Peru[a]

	Flocking type		
	I (tight)	II	III (solitary)
Wattled Jacana	U	C	C
South American Painted Snipe	U	C	C
Black-necked Stilt	U	C	C
American Avocet	C	C	U
Black Oystercatcher	U	C	C
Blackish Oystercatcher	U	C	C
American Oystercatcher	U	C	C
Southern Lapwing		C	C
Black-bellied Plover		C	C
Lesser Golden Plover		C	C
Semipalmated Plover		C	C
Snowy Plover		C	C
Wilson's Plover			C
Killdeer		R	C
Two-banded Plover		C	C
Rufous-chested Dotterel		C	C
Tawny-throated Dotterel		C	U
Marbled Godwit	C	C	C
Hudsonian Godwit	C	C	C
Whimbrel	U	C	C
Spotted Sandpiper			C[b]
Lesser Yellowlegs	U	C	C
Greater Yellowlegs		R	C
Solitary Sandpiper			C[b]
Wandering Tattler			C[b]
Willet	U	C	C
Ruddy Turnstone	U	C	C
Black Turnstone	U	C	C
Red Knot	C	U	
Pectoral Sandpiper	U	C	C
White-rumped Sandpiper	U	C	C
Baird's Sandpiper	U	C	C
Western Sandpiper	C	C	R
Least Sandpiper	C	C	C
Dunlin	C	C	R
Sanderling	C	C	C
Stilt Sandpiper	C		
Buff-breasted Sandpiper		C	C
Surfbird	C	R	
Short-billed Dowitcher	C	U	
Long-billed Dowitcher	C	U	
Wilson's Phalarope	C	C	C

(*Continued*)

Table II. (*Continued*)

	Flocking type		
	I (tight)	II	III (solitary)
Red Phalarope	U	C	
Common Snipe		R	C
Least Seedsnipe	R	C	

[a] From unpublished observations and Myers and Myers (1979). Categories indicate the frequency of observations of different flocking types, from most common to least: C, common; R, regular; U, unusual; blank, rarely if ever. Flock types are a modification of Goss-Custard (1970), as in Myers and Myers (1979). Type I: most individuals in flock within 5 body lengths of one another, tight and cohesive movements. Type II: looser but still cohesive flock structure. Type III: scattered birds without cohesive movements, including solitary individuals.
[b] Individuals almost invariably territorial.

1. Flocking in Relation to Predation

In theory, individuals can reduce predation risk by joining a flock for at least two reasons:

1. By joining a flock, an individual spreads its risks among others (Hamilton, 1971; Vine, 1971). This may be because others provide cover from, or alternative targets to, the predator.
2. Predators may not be as effective in hunting individuals within flocks, because flocks are more difficult to locate (e.g., Vine, 1973; Triesman, 1975), because flocks may detect predators more readily than do individuals (e.g., Pulliam, 1973; Kenward, 1978), or because predators are confused (e.g., Milinski, 1977) or risk physical damage (Tinbergen, 1951; Kruuk, 1964) by attacking individuals within flocks.

Evidence is available for a wide variety of animals on these points (e.g., Powell, 1974; Milinski, 1977; Kenward, 1978; Caraco *et al.*, 1980). For shorebirds, the available data address two critical points. First, individual risk within flocks is lower than risk to solitary birds. Second, when predators appear, shorebirds join flocks and alter flock characteristics in ways that suggest facultative responses to predation risk. The behavior of individuals in vs. out of flocks is consistent with some of the hypotheses in (1) and (2) above. These two points are developed more fully below.

Risk in Flocks vs. Risk for Solitary Birds. Page and Whitacre (1975) showed that hunting success, defined as the number of successes relative to the number of attempts, by a female Merlin *Falco columbarius* wintering on a California estuary was high when it attacked solitary small shorebirds and when it attacked large flocks. Hunting success was low for intermediate flock sizes. By taking into account the relative abundance of different flock sizes as well as the number of individuals within each flock, Page and Whitacre demonstrated that solitary individuals ran roughly three times the risk of being eaten as did birds within flocks.

Kus (1980) subsequently studied the same raptor–shorebird system in more detail and corroborated Page and Whitacre's basic result: individual risk for shorebirds declines with flock size. Kus' work provides further insight into the processes involved in risk reduction. She found that while its hunting success decreased with flock size, the Merlin attacked large flocks at a higher frequency than small flocks relative to their respective occurrences on the lagoon. These opposing trends balanced, such that the probability of capture of one bird was roughly equal across all flock sizes for any particular Merlin attack. The protection for individuals in larger flocks thus came about through dilution: in large flocks the probability is lowered that a particular individual will be the one captured.

Kus *et al.* (1984) show that juveniles are more likely to be taken than adults, given their relative abundances. Kus' data (personal communication) suggest that within a flock, juveniles are located more peripherally, that peripheral and relatively isolated birds within flocks incur the highest predation rate, and that juveniles tend to occur in smaller flocks than do adults.

The Response of Shorebirds to Predators. Shorebirds respond to the appearance of predators by flocking. It may be an immediate, temporary response. For example, territorial Buff-breasted Sandpipers temporarily abandon their territories and flock when a predator flies over their feeding area (Myers, 1980a). An important observation here was that flocking was initiated even after the predator appeared. This indicates that benefits for flocking include more than simply enhancing the likelihood of predator detection.

Alternatively, rather than causing short-term responses, predation may also effect long-term adjustments in dispersion. In fact, Stinson (1980) argues that the overall nonrandom, clumped dispersion of shorebirds is a behavioral response to predators (but see Myers, 1980a). At Bodega Bay in central coastal California, Sanderlings often defend feeding territories during winters when no Merlin is resident locally. During some winters, however, one or more Merlin defend their own feeding territories along the ocean beaches. When this occurs, Sanderling cease defending territories and instead feed in flocks (Myers, 1980b; Table III). This behavior contrasts with the Buff-breasted Sandpipers, above, because the

Table III. Occurrence of Merlin and Territorial Sanderling during
the Fall at Bodega Bay, California[a]

Fall	Resident Merlin	Territorial Sanderling
1974	Absent	Regular
1975	Absent	Regular
1976	Absent	Abundant
1977	Absent	Abundant
1978	Present	Absent
1979	Present	Absent
1980	Absent then present	Abundant then absent
1981	Absent then present	Regular then absent
1982	Absent then present	Irregular then absent

[a] These categories refer to the frequency with which we observed Sanderling defending territories during high tide periods along Bodega Bay beaches. *Absent,* defense almost never observed, and if observed, did not persist for more than a few hours. No more than 1–4 individuals of a local population of 600+ involved. *Irregular,* defense observed intermittently, with territories scattered through area and not persisting long in time. *Regular,* defense invariably encountered but not necessarily widespread. Ten to seventy-five individuals usually involved. *Abundant,* defense widespread over local beaches. One hundred to two hundred and fifty individuals defending territories.

former return to their territories after the immediate danger has passed. The Sanderlings at Bodega Bay continue feeding in flocks throughout the day even though the Merlin may appear only once or twice during the day along a given portion of the local beach.

Characteristics of the Sanderling flocks themselves also change during periods when Merlin are resident in ways consistent with models (e.g., Triesman, 1975) for predation benefits of flocking (Myers *et al.,* in preparation). Average flock size increases from 21 ± 8 to 70 ± 36 birds per flock as solitary territorial and nonterritorial birds join flocks and smaller flocks coalesce into larger. Moreover, the within-flock spacing tightens markedly. The relative frequency of loose (interindividual distances averaging > 10 m) vs. tight (distances < 0.1 m) shifts markedly toward the latter.

There was also a marked effect on home range size. Myers *et al.* (in preparation) report that nonterritorial Sanderling home range increased from an average of 1500 m of sandy beach to 3200 m during weeks when the raptor was hunting on the beach. This result may have been due to (1) the simple mechanical effect that repeated spooking might have on flock movements, (2) the increased impact on local Sanderling prey abundance caused by increased flock size, requiring a greater foraging radius, or (3) an adaptive response by Sanderling to lower their predation risk by decreasing the probability they will be at a given location within their home range at any given moment. This last interpretation (the MX hypothesis) is a variation on models interpreting gregariousness as a means of decreasing the likelihood of predator detection (e.g., Triesman, 1975). Its plausibility in this system rests on the hunting style of the Merlin: typically at Bodega Bay this species flies low to the ground either in front or behind the dunes, and burst suddenly out onto a Sanderling foraging area. By expanding their home range size, the Sanderlings decrease the likelihood they will be at the particular foraging site at the moment the Merlin appears. Unfortunately, Myers *et al.* offer no test of any of these hypotheses.

2. Flocking in Relation to Feeding

There have been two questions asked of the relationship between flocking and feeding:

1. How flocking may increase foraging efficiency by decreasing an individual's need for antipredator vigilance or by enhancing the likelihood of finding better feeding sites.
2. How flocking may decrease foraging efficiency through interference.

Thus, (1) considers a selective basis for flocking while (2) focuses on its negative effects.

The Feeding Benefits of Flocking. Several shorebird species decrease vigilance when foraging in flocks: American Avocet *Recurvirostra americana* (Blick, 1980), Ruddy Turnstone *Arenaria interpres* (Fleischer, 1983), European Curlews *Numenius arquata* (Abramson, 1979), Bartailed Godwit *Limosa lapponica* (Smith and Evans, 1973), Sanderling (Silliman *et al.*, 1977; Blick, 1980), Long-billed Dowitcher (Blick, 1980), and Wilson's phalarope *Phalaropus tricolor* (Burger and Howe, 1975). Data for Marbled Godwit *Limosa fedoa* are suggestive but not statistically significant (Blick, 1980).

A series of clever experiments by Blick (1980) shed light on whether his measurements of vigilance actually reflected some tangible change in a bird's chances of detecting a predator. Blick flew a falcon-shaped and -sized glider at specific individual shorebirds of differing species and in flocks of varying sizes, and recorded the proportion of individuals reacting to the stimulus directly, rather than to the flock about them. Solitary birds were far more likely to respond to the stimulus than were individuals within a flock.

Despite the number of species studied and the consistency of results, several factors complicate accepting a relationship between vigilance and flock size as evidence for foraging benefits of flocking. First, the function relating flock size and vigilance in these studies typically reaches an asymptote at relatively low flock sizes: increases in flock size beyong 10–30 birds does not decrease vigilance further, even though flocks may range in size well above several hundred birds.

Second, because of social facilitation (Zajonc, 1965), increased foraging rate may be simply a consequence of being in a flock. While this is not inconsistent with there being foraging benefits, the fact that social facilitation occurs even among sated individuals makes the interpretation somewhat murky (Blick, 1980).

Third, studies of vigilance often examine how foraging rate (e.g., number of foraging efforts per unit time) varies with flock size, rather than consider the actual rate at which prey are ingested (e.g., Burger and Howe, 1975; Silliman et al., 1977). These studies use the former as an indirect indicator of the latter. Blick's (1980) work on American Avocet and Fleischer's (1983) on Ruddy Turnstone showed that decreased vigilance may not result in an increased prey ingestion rate even when foraging rate increases. Unless ingestion rate increases with increasing flock size, it is difficult to argue that the primary benefit of flocking results from an increase in foraging efficiency due to a relaxation of the need for vigilance. It is difficult to control for the large number of relevant variables in this sort of work, and Fleischer's use of multivariate statistics is an important step forward.

Finally, bird density is usually higher in areas of higher prey abundance (e.g., Myers et al., 1979a) and thus flocks are likely to be larger in sites with more food. This means, all other things being equal, that foraging rates should increase with flock size—not because of the effect of flock size on foraging efficiency, but rather because of the fact that larger flocks are feeding in better sites.

These complications leave unresolved the relationship between flock size and foraging benefits derived through decreased vigilance, not only for shorebirds but for any species. Vigilance clearly decreases in many species, but whether the reduced requirement for vigilance in flocks provides real foraging benefits, benefits sufficient to overcome the costs of flock foraging (see below) and to justify flocking as an antipredator strategy, requires further study.

Independent of any effects that flocking may have on vigilance, many hypotheses have been offered on the direct benefits flocking may have for foraging efficiency. These divide into three general classes:

1. Increased efficiency at finding spatially patchy or temporally variable resources through local enhancement (Hinde, 1961; Krebs, 1973; Thompson et al., 1974).

2. Cooperative hunting by flushing (e.g., Bartholomew, 1942; Morse, 1970), by swamping territory holders (Robertson *et al.*, 1976; Myers *et al.*, 1980a), or by permitting capture of prey too large for a single individual (e.g., Caraco and Wolf, 1975).
3. Avoiding duplication of effort when foraging on uniformly distributed food (e.g., Fischer, 1954; Cody, 1971).

For all the noise that has been made about these possible direct benefits of flocking, precious few data exist for shorebirds, and the evidence weighs against accepting them. Much like Kreb's (1974) herons, Drent and Swierstra's (1977) geese, or Caldwell's (1981) egrets, shorebirds can be enticed to land among models of birds simulating a flock (Gerstenberg and Harris, 1976). This is consistent with feeding enhancement, but it is not inconsistent with a purely predation-based interpretation. On the other hand, both Blick (1980) and Fleischer (1983) find that foraging success rate, i.e., actual prey ingestion, does not increase with flock size. Pienkowski (1980a), moreover, reports that Black-bellied Plover *Pluvialis squatarola* prey capture rate decreases with increasing flock size. These results are inconsistent with the local enhancement model.

Too much interference occurs among foraging birds within a flock (e.g., Goss-Custard, 1970) for flushing to be a benefit to flock members. Just the opposite probably prevails, as intertidal invertebrates often withdraw beneath the substrate rather than flush. Thus, flock mates depress local prey availability.

The second subhypothesis within (2) above, that flocking enables nonterritorial birds to swamp territory holders and thereby gain access to protected feeding sites, has only anecdotal support (Myers *et al.*, 1979a). Furthermore, even though many shorebird wintering populations contain a mixture of territorial and nonterritorial individuals (see below, and Myers *et al.*, 1979b), there are also a large number of species without nonbreeding territoriality for which this hypothesis is inapplicable. Thus, even if feeding benefits do accrue in this way, it does not offer a general basis for flocking.

As to the third possibility for (2), there is no indication that cooperative hunting occurs. In fact, the larger the prey item, the longer the handling time (Myers *et al.*, 1980b; Pienkowski, 1980a) and the more likely the food theft will occur (see Section II.B).

No concerted attempt has been made to test (3) for shorebirds, probably because initial observations make it seem so improbable. Flocks within a given area repeatedly cross paths as they move out with the receding tide, while individuals often remain foraging in the same site as the main flock moves on (Blick, 1980).

The Feeding Costs of Flocking. Flocking can also depress feeding rate through foraging interference. This is far better documented than foraging benefits (see review in Goss-Custard, 1980). Interference occurs (1) *directly*, wherein aggression cuts into foraging time and lowers overall foraging rate, or wherein food items are stolen; or (2) *indirectly*, wherein through their physical presence and activity flockmates depress prey availability.

Many species steal prey from flockmates. Eurasian Oystercatchers *Haematopus ostralegus* run an increasing risk of losing prey to conspecifics as nearest-neighbor distance decreases (Vines, 1980). Differences among individuals in the rates at which they lost or stole mussels contributed significantly to variability in Eurasian Oystercatcher intake rates (Goss-Custard *et al.*, 1982). Sanderlings squabble incessantly over sandcrabs that are too large to swallow immediately (personal observation). In Red Knots *Calidris canutus* fighting over food items, prey capture rate was 20% lower during periods when aggressive encounters occurred and encounter rate increased with decreasing nearest-neighbor distance (Goss-Custard, 1977). Thus, the tighter the flock, the greater the feeding interference.

Direct foraging interference without food theft has also been found. Eurasian Oystercatcher density was artificially increased using bird models (Drent, cited in Goss-Custard, 1980). The models could not have stolen food, nor could they have altered prey density or availability. The proportion of Eurasian Oystercatchers feeding at any given time also decreases with increased Eurasian Oystercatcher density (Zwarts and Drent, 1981). Redshank *Tringa totanus* fed more slowly when close together than when more dispersed, even though prey density was not varying nor was food theft occurring (Goss-Custard, 1976). These reductions in foraging rate most likely result from distraction: as bird density increases, birds must devote increasing attention to neighbors and thus less to foraging. This implies that the magnitude of direct foraging interference will vary among differing foraging styles and social contexts, especially when food theft or aggression are possibilities (Goss-Custard, 1970, 1976; Blick, 1980). Red Knot feeding rates were not related to bird density during observations when the knots were feeding by touch on *Macoma* (Goss-Custard, 1977). Strangely, Black-bellied Plover foraging rate increased with decreasing distance to nearest neighbor (Pienkowski, 1980a).

Indirect foraging interference depends upon the birds depressing prey availability, either through depleting a portion of the crop of prey available at a given moment or by eliciting a response on the part of a prey organism that renders it less available (Goss-Custard, 1980). Both are plausible mechanisms, although only the latter has supporting field data: foraging

Redshank induce their prey *Corophium volutator* to withdraw from the surface when they pass by (Goss-Custard, 1970).

3. Within-Flock Density

Within-flock density has clear effects on foraging success by setting the likelihood of foraging interference. What then determines within-flock density? Two mechanistic hypotheses are that interindividual distance is controlled through aggression or through avoidance (Burger *et al.*, 1979; Stinson, 1977; Vines, 1980). The weight of the evidence, discussed in Section II.B, favors the second mechanism.

However interindividual distance is mediated, within-flock densities in shorebirds are known to vary in relation to food density and to predation risk. Rands and Barkham (1981) show that within-flock density in Eurasian Oystercatcher and Dunlin *Calidris alpina* correlated directly with the density of preferred prey. For Dunlin, several prey were involved, including *Nereis, Hydrobia*, and *Macoma*. While apparently linear within each prey type, the slope of the relationship differed from one prey to another. Dunlin within-flock density was most sensitive to *Nereis* density. For Eurasian Oystercatcher the preferred prey were second-year *Cerastoderma*, and the relationship of prey with bird density was linear over the entire range of prey density sampled. Comparing their results, with average densities of 0.3 to 1.9 Eurasian Oystercatchers/m, to those of Vines (1980), whose modal interindividual attack distance for Eurasian Oystercatcher feeding on *Mytilus* corresponds to a density of 0.03 birds/m, indicates that the relationship for Eurasian Oystercatcher also varies among prey species, especially as Rands and Barkham interpret the linearity of their relationships to imply that their within-flock densities are below levels at which feeding interference becomes important.

4. The Spatiotemporal Cohesiveness of Flocks

To this point I have considered variability in flock size and tightness. Shorebird flocks also differ in their cohesiveness through time and space, although little work has focused on the parameters or significance of this flock characteristic.

Myers (1983a) presents the only detailed study on the persistence of association among individuals within flocks. Sanderlings feeding along an outer coast sandy beach in California change flock membership so rapidly that flock composition at any given time approaches a random sample of the local population. This result obtains even though most individuals feed in what appear at face value to be cohesive foraging flocks comparable to Sanderling flocks throughout the species' winter distribution.

The openness of Sanderling flocks implies that benefits for flocking rest solely on being in a flock per se, independent of the identity of flockmates. This leaves little room in this species for reciprocal altruism, or kin or group selection, and it implies that dominance relationships based on individual identity are unlikely (see also Owens and Goss-Custard, 1976).

Group openness varies widely among other taxonomic groups [e.g., passerines (Grieg-Smith, 1983)], and there are hints that the same will be found in shorebirds. Redshank roosts appear to be organized assemblages with persistent substructuring through time (Furness and Galbraith, 1980). Unfortunately, it is unknown whether the substructuring was due to the site fidelity of individuals within the roost, or to associations among particular birds. Color-dyed Red Knots in a migration stopover clustered nonrandomly both within and between flocks (Harrington and Leddy, 1982).

5. Flocking: A Summary

The weight of the evidence on shorebird flocking points to predation benefits and foraging costs. Flocks reduce the predation risk per flock member but they heighten foraging interference. To date there are no compelling analyses that empirically demonstrate a direct foraging advantage for flocking in shorebirds. If there are foraging benefits related to the decreased need for vigilance, these too derive ultimately from antipredation benefits. Thus, Goss-Custard's (1970) model has aged well. Major questions remain, nonetheless:

1. Why are flocks as large as they are? Models and data indicate that benefits asymptote at relatively small flock sizes. Do costs also reach an asymptote?
2. Is better predator detection the major benefit of flocking or predator avoidance after detection? Does the relative importance of these benefits differ among foraging styles or for different predator tactics?
3. Is the reduction in vigilance an essential foraging benefit for flock formation?
4. Does foraging interference have a significant impact on net energy intake rate? How does the importance of foraging interference, both direct and indirect, vary among different foraging styles and different prey species?

5. Does within-flock foraging enhancement occur?
6. To what extent do individuals derive foraging benefits from flocking because of a flock's swamping effect on a territory holder?
7. Are all shorebird flocks as open as those of Sanderling, and if not, what are the ecological/economic factors underlying that variability?

B. Aggressiveness

Aggressiveness in shorebirds varies widely among species and ecological circumstances. It ranges in intensity from minor jostling within flocks to drawn-out, knock-down battles. In its most intense form it is usually linked to territorial defense, considered separately below. I separate aggressiveness per se from territoriality because they are fundamentally different quantities, not points along a single continuum. Territoriality is a different level of behavioral organization, one component of which is aggressiveness. Treating them as a unitary phenomenon simply confuses their ecological bases.

1. Aggressiveness in Relation to Density

The intensity and per capita frequency of aggressive acts rise with increasing shorebird density. This holds intraspecifically, over short-time periods, and within similar ecological settings for a diverse array of species: Eurasian Oystercatchers (Vines, 1980), Red Knots (Goss-Custard, 1977; Burger et al., 1977), Semipalmated Sandpipers *Calidris pusilla* (Recher and Recher, 1969), and Short-billed Dowitchers (Burger et al., 1979). The same shorebird density, however, does not necessarily yield the same aggression level when compared across habitats, through time, or among species (Burger et al., 1979).

The fact that the changes involve per capita alterations in frequency is important. It means that the positive correlation between aggressiveness and density is not simply an artifact of watching a greater number of individuals. On the other hand, this positive correlation leaves unresolved a more detailed question about controlling mechanism. Is the enhanced frequency of aggressive acts in high density due only to the increased rate at which individuals encounter appropriate stimuli, that is, to smaller interindividual distances? Or is it due to a changing response threshold, to an increased responsiveness to identical stimuli?

The usual assumption is that aggressiveness controls spacing. For example, "aggression functions to maintain individual distance" (Burger *et al.*, 1979). As attractive as this may seem, supportive data on wader behavior are lacking. Showing a correlation between density and aggressiveness is insufficient. For one, that begs the question of cause and effect. Moreover, the critical relationship is not that of aggression vs. density, but rather the effect of a *change* in aggression on density. In fact, one could readily interpret the usual positive relationship between density and aggression as showing that aggression has no effect whatsoever on spacing.

In this light, work by Puttick (1981) becomes highly relevant. She found a per capita decrease in the frequency of aggressive interactions with increased Curlew Sandpiper *Calidris ferruginea* density. By her interpretation, aggression was suppressed at high population densities. This echoes an earlier assertion by Recher and Recher (1969). But this negative relationship actually could be reinterpreted as demonstrating that increased aggression decreases density, i.e., the usual assumption noted above. The irony is that support for this assumption is usually derived from just the opposite correlation (as in Burger *et al.*, 1979). We need experimental or statistical analyses that compare actual densities with what they would have been had aggressiveness been different.

Puttick's measurements are also relevant to the question raised above about changing response thresholds. If her interpretation is correct (i.e., suppression of aggression at high densities), then a change in threshold must occur.

Vines' (1980) work on Eurasian Oystercatchers indicates that in this species, avoidance behavior rather than aggressive interactions controls spacing. Aggressive interactions figure in food theft but they do not affect flock dispersion. The distinction Vines makes here is not trivial. The threat of aggressive food theft may provide a motive for avoiding other individuals, and thereby offer an advantage for avoidance, but aggression itself does not maintain individual distances in a proximate sense. Vines study supports Goss-Custard's early (1970) interpretation of the importance of avoidance for several wader species, and is consistent also with Stinson (1977).

2. Aggressiveness in Relation to Prey Type, Dispersion, and Availability

A theme common to much literature on shorebird spacing is that the nature of prey—prey type or prey dispersion or prey availability—influences the level and characteristics of shorebird aggression. This parallels numerous studies in other organisms (see Wolf, 1978). Recher and Recher (1969) threw up a series of hypotheses and interpretations for shorebirds, but unfortunately neither there nor in most subsequent papers have they been substantiated.

The logic of these predictions requires some discussion:

1. Prey type. Prediction: An individual should be more aggressive when feeding on a prey type whose availability is decreased by increased bird density or decreased interbird distance (Goss-Custard, 1970, 1980).

This prediction assumes that aggression controls spacing (see above). Alternatively, if aggression occurs solely as a part of food or site theft (Vines, 1980), then an individual should become more aggressive when feeding on prey types that are more "stealable." How "stealable" a prey item is will be particularly sensitive to handling time because longer handling times will provide greater opportunities for food theft. Variables such as prey hardness and size should thus predict aggression levels (all other things being equal). Prey size should also influence the value of theft.

2. Prey dispersion. Prediction: Individual aggressiveness should rise in areas with patchy food distributions (Recher and Recher, 1969).

Mallory and Schneider (1979) offer a rationale for this prediction by arguing that patchy food leads to an increased local, among-bird variance in foraging rate. Their interpretation requires that prey depletion occur and that aggression slows the rate of depletion.

3. Prey availability. Prediction: Aggressiveness should increase as resources become less available (Recher and Recher, 1969).

This prediction has two important, implicit assumptions, that aggressiveness has a positive effect on foraging efficiency and that is has appreciable costs (either energetic or risk or both). For aggression per se, independent of territoriality, the logic of this prediction has not been developed explicitly. It rests on the idea that an animal ought to be willing to risk more or to spend more in aggressing as food becomes harder to get (but see Caraco, 1979). More will be said on this below.

None of these predictions has been tested with shorebirds, although they are all readily supported by anecdotal information and in a few cases by quantitative study. With respect to prey type, Goss-Custard (1977) asserts that fighting over food items "occurs in waders that feed on prey large enough to be worth contesting." Fleischer (1983) interprets a change in aggressiveness in Ruddy Turnstones during the tidal cycle as a response to a change in prey type. Sanderlings feeding on sandcrabs *Emerita analoga* in California, Peru, and Chile pass a threshold prey size (roughly 15 mm long) above which they can no longer swallow prey immediately. Fighting over sandcrabs that cannot be swallowed instantly occurs at high frequency, while fights do not occur if swallowing is not prolonged (personal observation).

Mallory and Schneider's (1979) observations on prey dispersion in relation to aggression by Short-billed Dowitchers show that aggressive birds were using a patchily distributed food, egg masses of horseshoe crabs *Limulus polyphemus*. Their prey differed, however, in other respects (density, size), rendering the interpretation ambiguous.

Virtually no published work on shorebird behavior examines individual aggressiveness in relation to resource availability. This is an appalling gap. Studies have been made on passerines with contradictory results: low food availability may increase aggression as subordinates approach dominants more frequently when food is scarce (Marler, 1956), or it may decrease aggression because individuals must devote more time to foraging (Caraco, 1979). Goss-Custard's (1977) measurements of the numbers of encounters between Red Knots in winter vs. spring may provide indirect evidence that for shorebirds, aggression increases during periods of food stress: at a given distance between nearest neighbors, aggression is more frequent in winter. The contribution of other variables to this difference is wholly unknown.

Basic to predictions on the relationships between aggressiveness and prey type, dispersion, and availability is the assumption that aggression improves foraging efficiency. Goss-Custard *et al.* (1982) test this assumption directly, finding that the most aggressive Eurasian Oystercatchers in a local foraging area increase their intake through food stealing and suffer food theft least. Several other investigations, notably Goss-Custard (1977), touch on the issue, but more with a view toward whether foraging rates decrease in high shorebird densities as a result of increased encounter rates, than with concern for the consequences of the aggressive act from the point of view of the aggressor.

3. Other Factors Related to Aggression

Age and sex differences in aggressiveness have been documented in wintering shorebirds. Juvenile Semipalmated Sandpipers aggressed more frequently than adults feeding in the same local area (Harrington and Groves, 1977). Adults and juveniles were more likely to be the object of juvenile aggression, and juveniles dominated adults more than adults dominated juveniles. In contrast, adult Ruddy Turnstones in flocks with juveniles aggressed selectively toward juveniles and dominance interactions between ages invariably favored the older bird (Groves, 1978). Adult Eurasian Oystercatchers are more aggressive than juveniles (Goss-Custard *et al.*, 1982).

The only study describing sexual differences in nonbreeding aggression is Puttick's (1981) on Curlew Sandpipers. She found that intrasexual aggression was more frequent than intersexual aggression, and that males were more aggressive than females.

None of the studies above, however, controlled adequately for confounding variables. Thus, while age and sex differences do exist, whether they result from the difference in age and sex per se, or indirectly from age- and sex-related differences in prey type, prey availability, flock density, etc., cannot be determined.

4. Aggression: A Summary

Considerable literature exists on the phenomenology of aggression within wintering shorebirds, but the data are too sparse and unfocused to test a comprehensive theory on its ecological bases. Many fundamental questions remain:

1. What is cause and what is effect in the correlation between aggression and interindividual distance? Is aggression suppressed at high densities, or does the lack of aggression permit small interindividual distances?
2. Does aggression provide an energetic net benefit to the aggressor and/or a net cost to the aggressee? How do the magnitudes of cost and benefit vary among feeding style, prey type of dispersion, and social context? The work of Goss-Custard et al. (1982) is only a beginning. We need detailed analyses comparable to Caldwell's (1980, 1981) work with herons and egrets.
3. Does the likelihood of aggression increase or decrease with resource level? Both theory and data from other organisms are contradictory on this point (see above).
4. Does high predation risk suppress aggression?
5. What factors control the likelihood of success in aggression? Can dominance be predicted on the basis of size or sex? Why do juveniles dominate adults in some situations? Do dominance heirarchies based on individual identity exist in shorebird flocks?

C. Territoriality

Many shorebirds defend individual territories during the nonbreeding season, both at migration stopovers and on wintering grounds (Myers et al., 1979b, and references therein; Ens, 1979; Townshend, 1979; Myers,

1980a,b; Myers *et al.*, 1980a; Johnson *et al.*, 1981; Mallory, 1982; Myers and McCaffery, 1984; Townshend *et al.*, 1983). In a great majority of cases their behavior is classically territorial, with strenuous defense of area (*not* merely individual distance), well-defined boundaries, displays, long-term (months +) occupancy, and effective exclusive use of resources within the defended area (Myers *et al.*, 1979b). But all manner of variation can also be found: ephemeral defense lasting a few hours, boundaries defined by a moving resource, incomplete control over resources, etc. (Myers *et al.*, 1979b).

The frequency of territoriality fluctuates broadly among species, habitats, locations, and time or date of observation. Within a local shorebird population, territorial individuals mix spatially with nonterritorial birds: defended sites intersperse with undefended areas, and even the defended sites are regularly penetrated by marauding groups in flocks of varying size.

Only occasionally does the majority of an entire local population defend territories. In only a few species, such as Spotted Sandpiper, Solitary Sandpiper, and Wandering Tattler, is territoriality the most common spacing behavior (Myers *et al.*, 1979b; personal observation).

Four main issues arise with respect to this behavior:

1. What benefits does an individual derive from territorial defense, and what are its costs?
2. What ecological factors affect these costs and benefits and thereby presumably, set the likelihood for territoriality?
3. How large an area should be defended?
4. Does territorial behavior influence local population size?

1. Benefits and Costs of Nonbreeding Territoriality

Myers *et al.* (1979b) considered four hypotheses for possible benefits of nonbreeding shorebird territoriality: (1) net energetic gain, (2) reduction in predation risk, (3) "practice" for breeding territorial defense, and (4) nonadaptive carryover from the breeding season. The weight of the evidence reviewed was inconsistent with the latter three. Briefly, (2) shorebirds abandon territories when predation risk increases; (3) both sexes defend even in those species in which the mating system only involves male defense; and (4) juveniles and females that did not defend during the prior breeding season defend on the wintering ground. Evidence published since Myers and colleagues' review has supported these points, although Johnson *et al.* (1981) note a strong adult male bias in the probability of territoriality of Lesser Golden Plovers *Pluvialis dominica* wintering in Hawaii.

Direct tests of the first hypothesis, however, were—and still are—lacking. One preliminary study suggests that territorial Eurasian Curlews are heavier than nonterritorial individuals of the same sex (Ens, 1979). Comparable data comparing territorial and nonterritorial Sanderling reveal no such difference (Myers, unpublished). Beyond this, no data are available on the energetic consequences of territorial defense in wintering shorebirds. The energetic benefits hypothesis thus remains an untested interpretation of shorebird nonbreeding territoriality.

Efforts to test this hypothesis have confronted important empirical roadblocks. Measuring caloric ingestion rates with multiple prey species each varying in size and with the quality of observation changing to unknown degrees has presented severe problems, particularly given the likely level of precision necessary for a rigorous test. For example, the size of prey captured, if recorded at all, is usually taken as prey length. Observations without prey size at best are worthless while at worst are misleading. But even if prey size is recorded, caloric value varies roughly the square to the cube of prey length. Thus, a 4-mm *Ermita* contains some 10 calories while a 10-mm one contains almost 300 calories (Myers, unpublished data). Getting an accurate estimate of prey length with prey handling times of less than 1 sec is difficult to achieve consistently. This problem is compounded by the fact that when testing the benefits of territoriality, one must measure ingestion rates in both territorial and nonterritorial birds, which by the nature of their spacing behavior may be focusing on different resource bases.

Measuring defense costs is also difficult. While offering a start, time-activity budgets have severe limitations (Weathers and Nagy, 1980), especially when comparing across tactics as distinct as territorial vs. nonterritorial behavior.

2. Limited Space as a Cause of Territoriality

Competition for space has been invoked repeatedly as a basis for territoriality (e.g., Recher and Recher, 1969). This interpretation is a naive application of concepts from the community ecology of sessile organisms (e.g., Connell, 1961) and begs the issue of why defending an area is beneficial. The notion that space per se is limited in a winter shorebird population has never been tested and seems rather unlikely, given the number of shorebird bodies that can be crammed into a very small area. One might argue that through territoriality, space could *become* limited, but this is a different issue (see on regulation of population size, below).

3. Ecological Control of Territoriality

Three main patterns have been noted in the occurrence of shorebird nonbreeding territoriality. First, within similar habitat, territories are defended at intermediate resource levels (Myers *et al.*, 1979a, 1980a; Myers, 1980b). Territorial defense disappears when food is scarce and also when it is abundant. This resembles patterns seen in several other food-based territorial systems (e.g., Gill and Wolf, 1975; Carpenter and MacMillen, 1976; Bradbury and Vehrencamp, 1977).

Economic cost/benefit interpretations have been argued for the disappearance at both upper and lower ends of resource availability (reviewed by Myers *et al.*, 1980a): below a lower threshold, the resource base does not support existence plus defense costs, however the sum of these costs might compare to using the same site nonterritorially. At high resource levels, defense is not economical because (1) resources are superabundant (Carpenter and MacMillen, 1976), (2) the asymptotic relationship between food abundance and intake rate decreases the energetic return on defense at high resource levels (Gill and Wolf, 1975), or (3) increased competition in areas of high resource density so increases the costs of defense that territoriality is uneconomical (Myers *et al.*, 1980a). None of these hypotheses has been tested, nor need they be mutually exclusive.

The second pattern in the occurrence of territoriality is that defense is less likely with heightened predation risk (Myers 1980a,b; see discussion of Section IIA and Table III). The appearance of a predator can have both short- and long-term effects on the likelihood of territorial defense in a local population, probably because of the antipredator benefits gained through tight flocking. Thus, heightened predation risk must be considered one of the costs of territorial defense.

The third pattern is that, when comparing the frequency of territoriality within species but across habitats, defense is more common in nontidal areas (Myers *et al.*, 1979b), or in tidal sites with relatively small movements in water line position such as sloughs or deep creeks (Myers *et al.*, 1979b; Dugan, 1981a; Townshend *et al.*, 1983). Myers *et al.* (1979b) interpret this as a result of increased habitat stability: compared to an open tidal flat where a small vertical movement of the water line results in a large horizontal displacement of foraging position, foraging sites in sloughs, creeks, and nontidal areas can be used for much longer continuous time periods, making an energetic return on an initial defense investment more likely. Simply put, the idea here is that establishing a territory has setup costs that must be recouped through increased foraging efficiency, and this will take time. In sites where the best foraging position shifts rapidly and over a distance too large for defense, the initial investment will not be recoverable.

Townshend *et al.* (1983) interpret the same pattern as due to the protected nature of deep creeks, which enable better foraging during winds. Neither interpretation has been tested critically.

4. Territory Size

Shorebird territory size varies considerably both among and within species. Larger species tend to defend larger areas (Myers *et al.*, 1979b), and intraspecifically territories are smaller in areas of higher prey abundance (Myers *et al.*, 1979b, 1980a; Mallory, 1982).

In Sanderling, the relationship to prey abundance reflects increased competition for territories at good feeding sites, rather than a direct adjustment of territory size to energetic need (Myers *et al.*, 1979a). Myers *et al.* (1980a) interpret their results to indicate that Sanderling defend as large an area as local competition for sites permits. They further argue that by defending as large an area as possible, Sanderling buffer themselves against catastrophic reductions in prey density caused by beach erosion.

5. Regulation of Local Population Size

A perennial issue associated with territoriality is whether it regulates population size (e.g., Brown, 1969). This has received scant attention within the literature on winter shorebird spacing behavior. Given the fact that all sites are usually not defended—and that among the undefended sites are those with highest prey density—the effects of competitor density on territory size, and the swamping effect that flocks have on territory holders (see above), it seems unlikely that territoriality could wield a strong influence.

Townshend *et al.* (1983) nevertheless argue that a combination of territories plus some mechanism of density regulation in undefended areas could "encourage" some individuals to leave an area. Their argument rests more on foraging interference in general than on territoriality in particular. It is an intriguing notion, especially when considered in light of two additional facts and one theory: The facts are (1) that nonbreeding territoriality commences early in the fall, often before juveniles begin to arrive, and (2) that most birds return year after year to the site where they spent their first winter.

The theory is Pulliam and Parker's (1979), that a bird's choice of a wintering ground reflects its foraging success and the level of aggression it encounters as it samples different sites in migration southward. While simplistically ignoring Perdeck's (1958, 1964) and subsequent studies (e.g., Berthold, 1973) on endogenous control of migration distance, Pulliam and Parker's model, in combination with the facts outlined above, suggests a means by which early fall territoriality could strongly influence local population size: preventing juveniles from settling in their first fall migration would lower long-term recruitment to the wintering population. Work on this would be highly desirable.

6. Territoriality: A Summary

Nonbreeding territoriality is widespread in shorebirds and it appears to be resource-based, although convincing tests of this hypothesis have not been made. Territories are most common at intermediate food densities and in areas or at times of low predation risk. Territory size varies with the intensity of competition for feeding sites, leading to an inverse correlation between prey abundance and territory size.

We are thus left with some empirical ability to predict when and where winter territories ought to be found, but uncertain as to their selective benefit:

1. Do territorial and nonterritorial birds differ in their winter survivorship, in their condition prior to migration, or in their breeding success?
2. If these territories are resource-based, what is the time scale of reward? Are territorial birds winning short-term foraging benefits by moment-to-moment exclusion of conspecifics as in nectivore systems (Gill, 1978) or Davies' (1977) wagtails? Or are benefits built over long intervals—weeks or months—as territorial residents prevent prey depletion?
3. If this behavior is resource-based, why is interspecific nonbreeding territoriality so rare?
4. Does resource stability per se influence the likelihood of territoriality?
5. Are sexual biases in nonbreeding territoriality commonplace, and if so, how do they relate to breeding systems?
6. How are trade-offs made between economic benefits and predation risks? Does the likelihood of territorial defense in the presence of a predator fluctuate with resource level, with the degree of economic benefit for defense, or with the hunting style of the predator?

7. In species with stable prey density, is territory size adjusted directly to resource level or does it fluctuate in response to competitive pressure?

III. THE LOCAL LEVEL

The "local level" of spacing behavior refers to processes occurring within the daily home range of a foraging shorebird: its movements from roosts to foraging sites and between different foraging areas, and the pattern of roost dispersion. Far less attention has focused on this spatial scale than on the individual level.

Virtually all shorebirds move cyclically during the nonbreeding season among roosting and foraging sites. Even individuals defending territories do not remain continuously in their foraging areas. Two distinct questions are raised by these movements: why go to roosts, and why switch foraging sites? As with behaviors at the individual level, the answers proposed revolve around foraging and predation.

A. Roosting

In interior wetlands, shorebird movement cycles follow a diurnal rhythm, with most if not all of the day spent in foraging areas (Swinebroad, 1964; Myers and Myers, 1979). Upland species behave similarly, although their diurnal pattern is often broken in midday or afternoon by trips for water (Myers, 1980a). In tidal areas, shorebirds often congregate in roosts during high tide and at night (Heppleston, 1971; Prater, 1972; Hartwick and Blaylock, 1979; Kelly and Cogswell, 1979) For example, small groups of Black Oystercatchers during midwinter in British Columbia fly each morning from their roost toward the feeding area on a nearby mudflat (Hartwick and Blaylock, 1979). The timing of their arrival is linked to tides: if the high is high early in the day, then their arrival is delayed until late morning or early afternoon. They forage throughout the remainder of the day and then depart before dark.

The pattern is not always this simple. Many species spend little time in roosts during midwinter, allocating more time to feeding at night, at high tide, or during storms (Heppleston, 1971; Prater, 1972; Goss-Custard et al., 1977; Kelly and Cogswell, 1979; Maron and Myers, unpublished). This appears to be a response to decreased daylength and seasonal reductions in prey availability. A second complication is that, within a given season, prey are not always most available at low tide (Puttick, 1980; Connors *et al.*, 1981). For example, when *Emerita* are plentiful along the outer beaches of Bodega Bay, Sanderling often feed on them at intermediate tide levels and roost at low tide (Maron, personal communication). The Black Oystercatchers mentioned above often roost at or near the time of low tide, particularly if the birds have been feeding on the outgoing tide (Hartwick and Blaylock, 1979).

Moreover, shorebird roosting in tidal areas is not always tidally controlled. Sanderlings feeding intertidally on Pacific Coast beaches in northern Chile forage early in the morning and again late in the day, mostly roosting at other times independent of the tidal cycle (Myers *et al.*, 1984).

Shorebird roosts vary in size from a few birds to tens of thousands. In some cases all members of a local population coalesce into the same roosting site, whereas in many they divide into several separate roosts (e.g., Kelly and Cogswell, 1979; Blick, 1980). Some roosts are composed wholly of one species, while others are mixtures, usually imperfectly segregated by species.

Roost placement differs considerably, especially among species. Unfortunately, almost no published data are available on this issue. In coastal Buenos Aires Province, Argentina, most shorebird species gather at night from distances of 5 + km in large multispecific roosts at the edge of lagoons and swamps (Myers and Myers, 1979). Species with very different foraging habitats roost together while a few species roost in monospecific groups away from the main gathering. Buff-breasted Sandpipers, a grassland foraging species, roost up to 4 to 6 km from their main foraging sites in monospecific flocks numbering several hundred birds. Lesser Golden Plovers, which feed in the same fields as Buff-breasted Sandpipers, sleep in the main multispecific roosts at comparable distances from the feeding fields. Two-banded Plovers, muddy stream foragers, roost immediately adjacent to their foraging habitat in monospecific groups of 4 to 10 individuals. These small groups are scattered at irregular intervals along the length of muddy stream banks. During foraging hours White-rumped Sandpipers are the most common shorebird foraging with them, even though White-rumps sleep in the main multispecific roosts several kilometers distant.

Similar variability is found in tidal habitats. In coastal California, most shorebirds gather just above the high tide line in large multispecific groups, usually in salt marsh vegetation or on raised sand or mud spits. They roost on beaches only during the highest tides, when their usual roosting sites are covered. A few species roost consistently apart from the main group. During the daytime around high tide, Sanderling roosts containing a few to several hundred individuals are scattered along sandy beaches above the high tide line, often just out of the reach of waves but frequently several tens of meters away from the wave-washed zone. At night the same birds all gather into one or two very large roosts on the open beach, usually within 2 or 3 meters of the highest waves. During the daytime, territorial Sanderlings usually remain on their territories and roost alone, although not invariably, while at night they join the main roost. Snowy Plovers also roost by day and night on sandy beaches. Their roosts are smaller and internally far less compact than those of Sanderling.

Roost sites often are used for many consecutive years. Frequently this is clearly due to unique habitat configurations coupled with the absence of disturbance. Most Marbled Godwits and Willets at Bodega Bay, for example, have roosted for at least the last 10 years in one of two salt marsh points, switching between them depending upon the height of the tide. At times, however, the persistent attachment to particular roosting sites has no clear basis. Sanderlings at Bodega Bay will settle onto the same spot along a 4-km open beach for weeks on end, despite repeated nighttime disturbance while they are netted for research. The site may appear to be indistinguishable from 50 other locations on the beach. In fact, because of wave action, the topography of the roost location can change through time so much that it is less similar to itself, over time, than it is to other unused sites.

1. Why Go to a Roost?

At first glance, roosting appears simply to be a default activity, time spent not feeding. That is not to say it is time spent completely idle, as undoubtedly birds must allocate time to maintenance during some portion of their daily cycle. At high tide, birds are waiting for foraging areas to emerge with the receding tide. Foraging profitability on intertidal flats usually decreases at higher tidal levels (Prater, 1972; Connors et al., 1981; Fleischer, 1983; but see Puttick, 1980) and during the highest tides foraging sites are physically unavailable to wading birds. At night, foraging may be less efficient (Heppleston, 1971; Hulscher, 1974; Dugan, 1981b) or it may entail greater predation risk (Townshend et al., 1983).

But at the same time, stopping to feed does not require an individual to fly several kilometers or more, nor to congregate in a large flock. This decided nonrandomness to roosting behavior has prompted much speculation about its adaptive significance, for shorebirds as well as for other groups (Ward and Zahavi, 1973; Blick, 1980; Myers, 1980a). One set of hypotheses focuses on the antipredator function of roosts: that the chance of predator detection or escape will be enhanced, similar to the antipredator benefits of flocking while feeding (see above). The other set posits feeding benefits: that information about good feeding sites will be transferred among roostmates. No convincing tests of these ideas have been carried out for shorebirds.

With respect to predation benefits, Blick (1980) showed that vigilance of individuals decreases with roosting flock size, but that large roosts are more likely to spot approaching predators than birds roosting alone. This second result was derived mathematically from the percentage of time that birds spent roosting with their eyes open. It assumed that once a roosting bird's eyes were open, it was equally likely to spot a predator whether solitary or in a flock. No data on predator success as a function of roost size are available.

The data on feeding advantages are virtually nonexistent. Myers (1980a) argued that feeding benefits from information transfer are unlikely because many of the Buff-breasted Sandpipers in the roosts he observed defended feeding territories during the day, and were thus unlikely to switch feeding sites on the basis of information transferred within the roost. The multispecific roosts in his study area, moreover, were composed of species whose foraging habitat did not overlap, while several species that shared foraging habitat roosted in separate sites (see above). Blick (1980) found no support for feeding advantages either.

2. Roosts: A Summary

Surprisingly little work has been done on shorebird roosts. Even at the descriptive level, little can be said, other than to outline the dimensions of variability in roosting behavior and to highlight certain basic patterns. Whether some design underlies interspecific differences in roost dispersion, site, and timing of occupancy awaits quantitative study of several issues:

1. What environmental factors predict roost size and dispersion? Why do some species roost in many small groups, whereas others coalesce into a few large ones, and still others fluctuate between large and small roosts?
2. What controls the choice of roost sites? Physical characteristics, minimizing flight distances to multiple foraging areas, avoiding predators or human disturbance?
3. Is foraging information transferred within roosts?
4. Does predator success vary with roost size?

B. Movements among Local Feeding Areas

In addition to moving between roosting and foraging sites, individual shorebirds regularly switch feeding sites on a daily schedule. These movements are most pronounced in tidal areas.

For example, during midwinter in England, Redshank (Goss-Custard, 1969), Eurasian Oystercatcher (Heppleston, 1971) and Red Knot (Prater, 1972) often leave estuarine habitats at high tide to forage in nearby fields. The situation is more complicated in Eurasian Curlew (Townshend, 1981): some birds (mostly females) almost never switch, whereas others move between flats at low water and fields at high. Still other individuals (mostly male) specialize on field feeding, remaining there even during low tide. The sexual difference in habitat preference appears to be related to bill length dimorphism: females, with longer bills than males, have better access to mudflat prey (chiefly *Nereis*), especially as cumulative prey depletion and lower temperatures on the flats reduce prey availability near the surface of the substrate (Townshend, 1981).

Many species in California switch regularly among habitats. Page *et al.* (1979) report that in winter most Black-bellied Plovers, Dunlins, and some Least Sandpipers alternate between estuarine foraging sites at low tides and pastures at high tides. During rains Willets and Marbled Godwits abandon their normal roosts in San Francisco Bay to forage on upland feeding sites (Kelly and Cogswell, 1979). Snowy Plovers and Sanderlings spend low-tide foraging hours on a broad sandflat and then move to outer coast sandy beaches as the tide rises (Connors *et al.*, 1981). A suite of species in New Jersey including Black-bellied Plover, Ruddy Turnstone, Semipalmated Sandpiper, among others, shift between foraging micro-habitats on a tidal schedule (Burger *et al.*, 1977).

Connors *et al.* (1981) analyze habitat switching as a tactic to enhance foraging profitability. For the Sanderlings they studied, caloric density of prey is higher on the sandflat during the lower portions of the tidal cycle, but as the tide rises not only does prey availability decrease on the sand-flat, but prey availability increases on the beach as the tide line passes over new invertebrate zones. Sanderlings track this change in relative profitability by switching habitat. Connors *et al.* show that the timing of the switching varies seasonally, occurring at much lower tide levels in the spring at a time when, because of the annual cycle of *Emerita*, foraging is more profitable on the beach at intermediate tide levels.

Most of these examples entail use of alternative habitats during high tides, especially in midwinter when foraging time is shortened by decreased daylength, when food availability on the flats has declined, or when storms make foraging difficult along the coast. Some species such as Sanderlings (Connors *et al.*, 1981) and European Curlews (Townshend, 1981) carry this farther by making habitat switching a standard routine. All cases, nonetheless, are consistent with the hypothesis that these movements occur in response to tide-related changes in foraging profitability.

As ever, we are left with several questions:

1. What fraction of daily caloric intakes comes from different stops within these cyclic movements? Studies from Britain and elsewhere clearly show that birds move among sites, but leave unresolved whether the different foraging areas are truly essential to balancing the daily energy budget or whether they are merely dessert. The answer to this has important conservation implications: if they are essential, then land management plans for wader conservation must be far more encompassing and include areas well outside the estuary proper.

2. Related to the last question, does the presence of alternative feeding sites affect local population size?

3. Do age and sex classes differ in their tendency to move among foraging sites? If Townshend's (1981) interpretation is correct, then sexually based differences should be greater in species with greater sexual size dimorphism in feeding morphology. How does the dominance status of an individual affect its likelihood for habitat switching?

4. What are the consequences of habitat switching for individuals defending a territory in one or more of the habitats used?

5. Is habitat switching absent in nontidal areas?

6. Does habitat switching affect predation risk, either by heightened exposure in transit or by lowering vulnerability via the MX hypothesis (see Section II.A)?

IV. AT THE REGIONAL LEVEL

Within the last few years an additional level of organization has been revealed in shorebird winter spacing behavior, one intermediate between their daily foraging movements and their annual migrations. This level involves movements among estuaries and different wintering sites, distances from tens to hundreds of kilometers. The basic observation is that individuals disappear for weeks on end from one estuary, and reappear weeks to months thereafter, even though they may have been faithful residents up until their disappearance. This simple observation raises a suite of questions that are only now receiving focused attention (Myers, 1980b; Evans, 1981).

Regional Movements

At the onset, it will be useful to distinguish between two types of site faithfulness, *philopatry* vs. *site fidelity*. For our purposes here, philopatry is the pattern of annual return to the same site, whether breeding, wintering, or migration stopover. In contrast, site fidelity involves remaining within a locale through time. Often no distinction is made between these because most discussions of site faithfulness focus on breeding birds, where nesting ensures site fidelity (but not philopatry—e.g., Pitelka *et al.*, 1974; Andersson, 1980; Mikkonen, 1983).

Individual shorebirds of many species show both patterns of site faithfulness. Marked birds often appear year after year as passage migrants or winter residents at the same locale (Brainbridge and Minton, 1978; Johnson *et al.*, 1981; Kelly and Cogswell, 1979; Lank, 1979; Myers and Myers, 1979; Myers *et al.*, 1984). Individuals also frequently remain within local wintering areas for extended periods (Kelly and Cogswell, 1979; Evans *et al.*, 1980; Johnson *et al.*, 1981; Page, 1974; Pienkowski and Clark, 1979; Myers, 1980b). Both these results, but especially the latter, contradict views on the nature of site faithfulness in wintering shorebirds published only 18 years ago (Recher, 1966).

Until recently, studies of shorebird site faithfulness in winter concentrated upon documenting its occurrence. This is not surprising. Both philopatry and site fidelity appear rather implausible given the global scale of shorebird migrations and their seemingly vagile life-style during winter. But in establishing site faithfulness there is a tendency to ignore negative results. With respect to philopatry, it is difficult to separate mortality from nonreturn. With site fidelity, without considerable effort or chance sightings in other areas, one cannot tell whether a bird's absence signals its departure or merely inadequate observation.

Work with shorebirds in two geographic areas now suggests that site fidelity may vary considerably both among and within species. In England, the Durham research group has compared movements by several shorebird species among coastal estuaries (Pienkowski and Clark, 1979; Pienkowski, 1980b; Evans *et al.*, 1980; Evans, 1981). Red Knots flew among feeding sites as distant as 30 km over short periods, and moved over 200 km within a winter (Pienkowski and Clark, 1979). Dunlin, in contrast, rarely moved from their home estuary. Unfortunately, these birds carried only color-dye markings so that identification of individuals was not possible. Hence, the frequency of transit among these estuaries is unknown.

Sanderlings in England also may range widely during winter (Evans *et al.*, 1980). Evans (1981) speculated that individuals differ with respect to site fidelity, and proposed two types of Sanderling, one resident (site faithful), the other transient (nomadic). Unfortunately, his data were too scanty to document this simple dichotomy. He went on, nevertheless, to suggest that the ecological basis for the two different behaviors lay in the unstable nature of sandy beach sediments and the high spatiotemporal variability in Sanderling food resources. He further proposed that differences in resource stability underlay interspecific variability in site fidelity. This paralleled similar reasoning by Myers and McCaffery (1980; see below), and seemed reasonable given previous arguments on the importance of resource variability for alternative spacing behaviors in wintering Sanderlings (Myers *et al.*, 1979b; Myers 1980a) and for the determination of Sanderling territory size (Myers *et al.*, 1980a). Evan's argument also followed a suggestion by Tree (1979) that the vagility of Greenshank *Tringa nebularia* wintering populations in the interior of Africa during the winter exceeds that of coastal populations because of the irregularity of inland rainfall.

We have gathered considerable data on site fidelity of Sanderlings around Bodega Bay during the winters 1979–1983. Our basic procedure is to individually color-band up to 200 Sanderlings each winter, roughly 25–30% of the resident population, and then search intensely for them at Bodega Bay and in surrounding estuaries up to 30 km distant through the winter. A network of 40 volunteers regularly searches beaches for marked Sanderling throughout California, Oregon, and Washington. Observations of marked birds are pooled into observation periods, usually 2 weeks long, giving us a catalog of the birds present at Bodega during each 2-week period from July through May.

The catalog of marked birds present provides the basis for two calculations: a vagility index (VI) for each bird i and the population turnover rate (T) for a given time period m:

$$VI = \frac{100\ O}{Q}$$

O is the observed number of changes in status made by a bird i. A change in status occurs when a bird disappears from Bodega Bay for at least one 2-week period, or when it returns after having been absent for at least one period. First arrival and final departure do not contribute to O. Q is the total number of changes bird i could have made had it changed status every 2-week period following its arrival and before its last departure. Thus, high values of VI reflect high vagility, low values indicate site fidelity. Only birds present at least three 2-week periods were included.

For the turnover calculation,

$$T = \frac{100\,(E + I)}{P_m + P_{m+1}}$$

E is the number of banded emigrants, those present in period m and not present in $m + 1$. I is the number of banded immigrants, the opposite of E. P_m is the total pool of banded birds present during period m. This calculation of turnover rate is adapted from studies of island biogeography and the turnover rate of species (e.g., Jones and Diamond, 1976).

From one to five people were involved in searches within any 2-week period. When fewer people searched, more time was spent by each observer. Between 40 and 80 observer hours were spent during most 2-week periods searching for birds. Searches were continued until it was judged we had seen virtually all birds present. We were aided in locating all birds by three factors: (1) Sanderlings use habitats (sandy beach and sandflats) that are completely accessible to humans, and one of which is one dimensional, allowing very thorough searches for marked birds; (2) Sanderlings at Bodega Bay are forced onto the beaches at high tide, facilitating searches further; and (3) the Bodega Bay sandy beach/estuary complex is small (7 km of beach plus approx. 200 ha of sandflat) and discretely bounded at both ends by several kilometers of rocky intertidal. The study area is described in Myers *et al.* (1979a) and Connors *et al.* (1981).

For turnover calculations the consequences of not seeing a few birds were small, given that the sum of banded birds within consecutive 2-week periods (i.e., the denominator for calculating turnover rate) usually exceeded 100 marked birds, Moreover, for most calculations daily observations were pooled into 2-week observation periods. This decreased our temporal resolution but also reduced the consequences of missing a bird on any given day.

Sanderlings at Bodega Bay vary markedly in site fidelity. First, there is considerable interindividual variation in their tendency to wander away from the home estuary: some never leave while others move in and out of the area regularly (Fig. 1). First-winter birds are more vagile than adults ($p < 0.01$). There is no sign, however, of bimodality in the VI distribution (the gap between 0 and 20 occurs because of the small number of 2-week periods within a winter). Thus, Evan's (1981) hypothesized dichotomy (see above) in behavior is too simple.

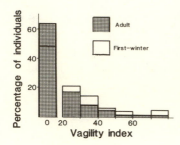

Fig. 1. The distribution of vagility index values for adult and first-winter Sanderlings at Bodega Bay, California. See text for calculation. Values between 0 and 20 were not possible because of the method of calculation of the index.

Second, there are seasonal changes in population stability. Population turnover is high in early fall, during storms in winter, and again in spring (Fig. 2). During the winter period when turnover rates are relatively constant (November–February), the adult population has a lower turnover rate than the first-winter birds ($p < 0.01$, sign test, $n = 9$).

During the 1982–83 winter we intensified searches in order to examine daily turnover rate. On consecutive days within each 2-week period, three observers searched flocks in all Bodega Bay foraging areas at both high, intermediate, and low tides. The daily turnover rate averaged 16% through the year, and in early fall was as high as 25% (Fig. 3). As before, these are not transients, but rather winter resident birds moving in, out, and back again around the Bodega Bay area. Comparing turnover

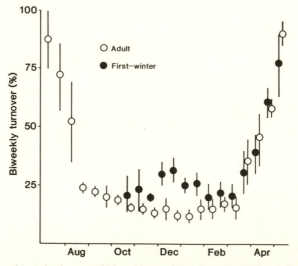

Fig. 2. Mean and standard error of biweekly turnover rates for adult vs. first-winter Sanderlings at Bodega Bay during the nonbreeding seasons of 1979–80 through 1982–83. See text for calculation.

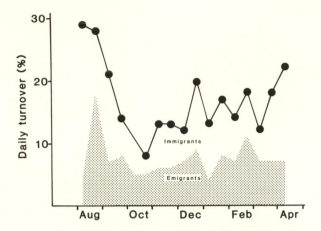

Fig. 3. Daily turnover rate during the 1982–83 nonbreeding season for individually color-marked Sanderlings at Bodega Bay. See text for calculation. The total turnover rate for any particular data (circles) is the sum of immigrants plus emigrants.

rates over successively longer periods (from day 1 to days 2, 4, 8, and 15) showed that turnover reached an asymptote near 20% at 2 weeks (Fig. 4). This implies considerable local churning within the population—birds moving in and out daily—but also a cumulative turnover of birds moving and staying away for at least two weeks.

One consequence of turnover is that the pool of birds using Bodega Bay is actually far greater than that revealed by counts at any given time. We measured the difference between instantaneous counts and the total

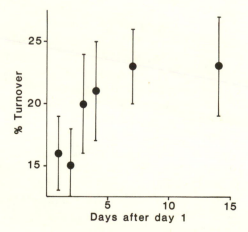

Fig. 4. Turnover rate of Sanderlings at Bodega Bay calculated over successively longer time intervals. See text for calculation.

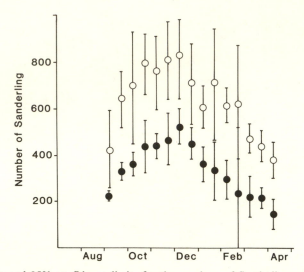

Fig. 5. Means and 95% confidence limits for the numbers of Sanderlings at Bodega Bay during the 1982–83 season. Closed circles are direct, instantaneous counts. Open circles are the cumulative number of individuals using Bodega Bay during a given 2-week period, estimated by a visual Lincoln index method using color-marked individuals (see text).

pool by (1) counting birds directly with three observers spread over the three main feeding areas at high tide, and (2) obtaining an estimate of the total pool, N, through sight–resight methods based on the Lincoln index, where

$$N = \frac{Bn}{b}$$

B is the total number of banded birds seen at Bodega during a given 2-week period; b is the number of banded birds within a given flock, and n is the number of birds (banded and unbanded) within that same flock. Small flocks totaling fewer than 50 birds were added together and treated as a single flock in the analysis.

This approach revealed that at any given time during the 1982–83 winter, roughly 50% of the total pool of birds using Bodega Bay was out of our censused area (Fig. 5).

The high fall and spring values in daily and biweekly turnover rates are more than simply migration to and from the wintering ground. This is a key point. Bodega Bay birds during these periods are wandering broadly around the central California coastal area, especially to and from Point Reyes, an estuary system located 20–40 km south of Bodega Bay

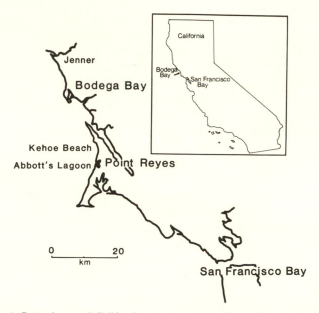

Fig. 6. Central coastal California showing the location of the study area.

(Fig. 6). Many individuals banded in previous years are first seen at Bodega Bay after their return from the Arctic in late July or early August. They stay for a day or two, or up to several weeks, and then leave, appearing next at Point Reyes, where they remain for up to 2 months. By late October, virtually all these fall-wanderers return to Bodega Bay.

Wandering renews in spring, again largely between Bodega Bay and Point Reyes. For example, in March 1980 Abbott's Lagoon (Fig. 6) broke open to the sea after a heavy storm. Within 5 days, 1300 Sanderlings converged on feeding areas newly exposed as the lagoon drained. Among these were 37 banded birds from Bodega Bay. One of these birds moved the 20 km between Bodega Bay and Abbott's Lagoon at least four times in 12 days. In April 1981 almost one-third of the banded Sanderlings at Bodega Bay moved to Kehoe Beach (Fig. 6). Over 80% of the birds moving during these two episodes ultimately returned to Bodega Bay, either during the same spring or the next fall.

Several hypotheses, some more plausible than others, can be invoked to explain the seasonal patterns at Bodega Bay:

1. Ecological Adaptations

1. Sanderlings increase nomadism when spatiotemporal stability of resources decreases.

2. Early fall nomàdism allows "shopping" for the best wintering area. This would predict that at least some birds marked in Bodega Bay during one winter should select another site during a subsequent winter, which almost never happens. In fact, one of the most striking aspects of the seasonal pattern is that the marked birds coalesce into the local Bodega Bay area by midfall.

3. Early fall nomadism allows birds to reacquaint themselves with regional feeding areas in anticipation of winter feeding conditions. Fall wandering is thus a form of map renewal that permits birds to be more efficient at finding good feeding spots in difficult times in winter. Contrary to this hypothesis, many of the birds that wander broadly in fall are among the most site-faithful in winter.

4. Sanderlings change their feeding goals seasonally. In fall and spring, molt and migration force them to maximize their energy intake. In winter, they need merely to meet basic energy requirements and to avoid predation. The changes in site fidelity represent changes in the relative importance of being at the best feeding site.

2. Nonadaptive

1. Food is relatively abundant in early fall and late spring. Increased nomadism during these periods reflects the relative cheapness of energy then compared to midwinter.

2. Seasonal changes in nomadism are a field expression of migratory restlessness.

We are not able yet to test any of these hypotheses definitively.

Regional Movements: A Summary

The data from various sources to data are sketchy but sufficient to show that shorebirds in winter vary in site fidelity to a local estuary. Some species wander more than others, and within species there are individual, seasonal, as well as age-class differences. No clean dichotomy exists between residents and nomads, but rather a spectrum in degree of nomadism. This spectrum raises problem sets at two distinct levels in the ecology of wintering shorebirds. One set involves nomadism as a behavioral tactic. Wherein lies its evolutionary basis and ecological control? The other involves questions of winter population structure: what are the spatial and temporal limits of a winter shorebird population?

Clearly, far more empirical work is needed:

1. Does within-winter survivorship differ between nomads and residents? If so, what mortality agents figure in those differences?

2. Why remain within a single local estuary through the winter? Do residents use resources more efficiently or avoid predators more effectively? Within an estuary do resident individuals switch habitats more than nomads?
3. What proximate factors trigger an individual's departure from a local estuary?
4. How is nomadism integrated into the annual cycles of migration and molt?
5. What maintains variability in nomadic tendencies within a local population? Are there among-year differences in the benefits or costs of nomadism vs. site fidelity?
6. What travel costs come with nomadism? With what frequency and to what distance do individuals wander, and what are the contributions of these flights to winter energetic existence costs? Do sexual size differences introduce sexual differences in travel costs?
7. Is nomadism more likely in species or individuals using temporally unstable resources? If so, does nomadism merely reflect the need to shift among feeding sites, or is it sampling behavior?
8. Do nomads and residents differ in social status? Are nomads bigger or smaller, more or less dominant than residents? Are nomads less likely to defend winter territories than residents?

V. CONCLUSIONS

This chapter began by asking whether variability in shorebird spacing behavior represents predictable responses by individuals to ecological and evolutionary processes. The material reviewed suggest that, at present, the answer has three parts. First, ecological correlates of some spacing behaviors permit prediction as to when or where birds will behave in a particular way; second, these correlations can be strung together in logical stories about their adaptive significance; but third, insofar as empirical tests of theory, we are a long way from definitive evolutionary statements.

The areas best understood lie at the lower end of the hierarchy in Table I:

1. The reduction in predation risk caused by flocking (Page and Whitacre, 1975; Kus et al., 1984) and the empirical correlates of predation risk with various spacing behaviors (Myers 1980a,b, in preparation).

2. The feeding interference costs of flocking (Goss-Custard, 1970, 1976; Vines, 1980; Goss-Custard *et al.*, 1982).
3. The empirical correlates of territoriality, particularly the likelihood of territoriality in relation to prey density and territory size as affected by prey abundance and competitor density (Myers *et al.*, 1980a).
4. Habitat switching in relation to seasonal variation in feeding constraints and foraging profitability (Heppleston, 1971; Prater, 1972; Connors *et al.*, 1981; Townshend, 1981).

At the same time, the lack of progress in other areas is sobering:

1. Separation of cause and effect in the relationship between aggression and density.
2. Relationship of aggression to resource level.
3. Benefits for nonbreeding territoriality.
4. Feeding benefits for flocking.
5. Basic descriptive ecology and adaptive consequences of roosting, site fidelity, and nomadism.
6. Flock structure, including cohesiveness and dominance.

With variable success, ecological conditions do predict behavioral patterns. A central issue therefore becomes the genesis of variability per se in spacing behavior within shorebird populations. Why, for example, do individuals of the same species within a 50-m stretch of beach often behave in radically different ways? The materials reviewed in this chapter suggest there may be at least three possible sources of this variability, all mutually compatible but not necessarily equally important: extrinsic ecological conditions, intrinsic characteristics of individuals, and the frequency-dependency of a behavior's costs or benefits.

By the first hypothesis, the variability arises because each individual experiences different ecological conditions. This seems unlikely in the strictest, instantaneous sense, as conspecifics employing highly different behavioral tactics often use the same local resource (Harrington and Groves, 1977; Mallory and Schneider, 1979; Myers *et al.*, 1979a). Recent work by Caraco *et al.* (1980) and Rubenstein (1982) indicates that the way an animal uses a resource depends upon the potential risks of success or failure, and these will depend upon the current condition of the animal. Risk-seeking tactics, those with a high variance in reward, may actually be preferable to risk-averse tactics for an individual in poor energetic condition. Several of the behavioral axes in Table I run from risk-averse, conservative tactics at one end to risk-seeking, opportunistic tactics at the other (e.g., territorial–nonterritorial, site faithful–nomadic).

By the second hypothesis, the variability is due to intrinsic differences among the individuals. A number of studies showed age- or sex-related differences in spacing behavior, but equally important, many indicated that individuals switched from one tactic to another as conditions changed. Thus, the variability cannot rest solely on this hypothesis.

The final possibility involves frequency-dependent changes in the costs or benefits of behavioral alternatives. It is improbable that these costs or benefits remain constant as the frequency of a given tactic fluctuates within the local population. For example, consider the choices available to a Sanderling in early fall when the local array of territories is being established: competition for a defended site, defense of a hitherto undefended site, solitary nonterritoriality, flocking nonterritoriality, etc. Supposing that sites are filled in approximate rank order of quality, and that competition for sites is proportional to quality, then as sites are filled (1) costs of defense may increase as fewer undefended sites are available, (2) benefits for defense of undefended sites should decrease as the best sites are claimed, (3) costs of solitary nonterritoriality may increase and benefits decrease as more and more feeding locations are claimed by territorial birds, and (4) benefits for flock foraging may increase as it becomes necessary to swamp a territory holder with flockmates to gain access to a feeding site. The fact that multiple options exist, and that their costs and benefits are frequency dependent, must contribute significantly to local behavioral variability.

Stepping back from this detail, several themes—food, predation, individual selection, and optimization—run implicitly throughout the studies reviewed herein. It is important that these themes be recognized, for they channel the type of investigation undertaken and circumscribe potential results. Alternative possibilities often are not even raised.

The first two are ultimate factors upon which most empirical and theoretical treatments have focused. The emphasis on food and predation has been reasonable given what must be the main objectives of a shorebird in winter: eat and avoid being eaten. There are alternatives, nevertheless, even if they seem unlikely at present to be major factors driving variability in winter spacing behavior. Neither parasitism nor disease has been considered and the possibility that winter behavior has breeding benefits beyond reaching the nesting ground in good energetic condition—e.g., practice of territorial defense, obtaining a mate, etc.—scarcely has been mentioned (Myers et al., 1979b). More complicated cross-seasonal interactions akin to the relationship between sandpiper mating system and migration distance (Myers, 1981) have not been comtemplated either.

The second two themes, individual selection and optimization, involve fundamental assumptions about the focus and operational rules for selective processes shaping shorebird winter spacing behavior. No results to date even suggest anything but individual selection in the strictest sense, i.e., no hints of nepotism, kin selection, or group selection. In fact, such a possibility has been considered only twice and rejected each time (Owens and Goss-Custard, 1976; Myers, 1983a). The global scale of migration, coupled with age- and sex-staggered migratory timing, makes anything but simple individual selection highly implausible for most species. As yet, more complicated selective processes cannot be ruled out of hand for nonmigratory species, especially those with prolonged parental care [e.g., Eurasian Oystercatchers (Heppleston, 1971)] or hints of cooperative breeding [e.g., Southern Lapwings *Vanellus chilensis* (Walters and Walters, 1980)].

Few studies in shorebird winter behavior deal explicitly with optimization, either formally (Goss-Custard, 1976) or informally (Myers *et al.*, 1980a; Connors *et al.*, 1981). The implicit influence is there throughout, nevertheless, and will grow because optimization theory offers tractable quantitative modeling approaches and because its basic premise, maximized return of some ecological currency, fits the prevailing current logic of how natural selection ought to work.

Three points are worth bearing in mind. First, although optimization theory has met with considerable success (see review in Krebs *et al.*, 1983), lingering quantitative discrepancies between prediction and data characterize virtually every published result. These can be rationalized within the framework of optimality, but doing so scarcely offers a test of underlying assumption, and certainly never tests alternatives to that assumption, for example, satisficing (Simon, 1956) or some other form of suboptimality (Janetos and Cole, 1981; Myers, 1983b).

Second, to date none of the work on behavioral compromises demonstrate or even test for joint optimization of competing demands. Instead, they show that predation risk alters a foraging behavior that had approached some optimum (e.g., Milinski and Heller, 1978; Sih, 1980). The rules governing those compromises may have involved joint optimization, or they may not.

Third, accepting hypotheses or assumptions not because of empirical findings but rather because the alternatives do not fit our perceptions of the logic of natural selection seems a shaky way to build scientific understanding.

Frankly, I suspect that shorebird winter spacing behavior rarely approaches an optimum. Instead it must be ridden with traditions in which shorebirds drift through routines that meet each requirement but optimize none. This alternative to optimization implies a behavioral inertia: do something as long as it works. Perhaps, as changing ecological conditions finally result in the current tactic no longer working, some form of optimization plays a role in choosing among new alternatives. But for the most part, shorebirds remain far from an instantaneous optimum.

In summary, we are tantalizing close on several fronts to definitive statements about the ecological control and ecological significance of spacing behavior in nonbreeding shorebirds. Overall the results make sense. They fit together within a coherent framework that is internally consistent and also consistent with theory and data from other organisms. Danger lies in that very consistency and the complacency it might produce, because the gaps in our understanding concentrate around basic assumptions.

ACKNOWLEDGMENTS

My work on winter shorebird spacing has been supported by the Thomas J. Watson Foundation, the National Science Foundation, the U.S. Department of Energy, the Committee for Afternoon Projects, and currently by The World Wildlife Fund—US and Penn-Jersey Subaru. Lois Myers, Frank Pitelka, and Peter Connors have been constant sources of advice and insight. The Bodega Marine Laboratory has tolerated unreasonable demands on space and resources, as well as considerable disarray, and I especially thank Paul Siri and Cadet Hand for their support. The people who have assisted me in the field now are too numerous to list, but John Maron, Stephanie Williams, Brian McCaffery, Terry Schick, and Jeff Walters made singular contributions. Lois Myers, Frank Gill, John Maron, Barbara Kus, Joanna Burger, Terry Schick, and Bori Olla commented on versions of this manuscript.

REFERENCES

Abramson, M., 1979, Vigilance as a factor influencing flock formation among curlews *Numenius arquata*, *Ibis* **121**:213–216.

Andersson, M., 1980, Nomadism and site-tenacity as alternative reproductive tactics in birds, *J. Anim. Ecol.* **49**:175–184.

Bainbridge, I. P., and Minton, C. D., 1978, The migration and mortality of the curlew in Britain and Ireland, *Bird Study* **25**:39–50.

Bartholomew, G. A., 1942, The fishing activity of Double-crested Cormorants on San Francisco Bay, *Condor* **44**:13–21.

Berthold, P., 1973, Relationships between migratory restlessness and migration distance in six *Sylvia* species, *Ibis* **115**:594–599.

Blick, D.J., 1980, Advantages of flocking in some wintering birds, Unpublished Ph.D. thesis, University of Michigan.

Bradbury, J. W., and Vehrencamp, S. L., 1977, Social organization and foraging in Emballonurid bats. III. Mating systems, *Behav. Ecol. Sociobiol.* **2**:1–17.

Brown, J. L., 1969, Territorial behavior and population regulation in birds: A review and reevaluation, *Wilson Bull.* **81**:293–329.

Brown, J. L., and Orians, G. H., 1970, Spacing patterns in mobile animals, *Annu. Rev. Ecol. Syst.* **1**:239–622.

Burger, J., and Howe, M., 1975, Notes on winter feeding behavior and molt in Wilson's Phalarope, *Auk* **92**:442–451.

Burger, J., Howe, M. A., Hahn, D. C., and Chase, J., 1977, Effects of tide cycles on habitat selection and habitat partitioning by migrating shorebirds, *Auk* **94**:743–758.

Burger, J., Hahn, D. C., and Chase, J., 1979, Aggressive interactions in mixed species flocks of migrating shorebirds, *Anim. Behav.* **27**:459–469.

Caldwell, G. S., 1980, Underlying benefits of foraging aggression in egrets, *Ecology* **61**:996.

Caldwell, G. S., 1981, Attraction to tropical mixed-species heron flocks: Proximate mechanism and consequences, *Behav. Ecol. Sociobiol.* **8**:99–103.

Caraco, T., 1979, Time budgeting and group size: A test of theory, *Ecology* **60**:618–627.

Caraco, T., and Wolf, L. L., 1975, Ecological determinants of group sizes of foraging lions, *Am. Nat.* **109**:343–352.

Caraco, T., Martindale, S., and Pulliam, H. R., 1980, Avian flocking in the presence of a predator, *Nature (London)* **285**:400–401.

Caraco, T., Martindale, S., and Whittam, T. S., 1980, An empirical demonstration of risk-sensitive foraging preferences, *Anim. Behav.* **28**:820–830.

Carpenter, F. L., and MacMillen, R. E., 1976, Threshold model of feeding territoriality and test with a Hawaiian Honeycreeper, *Science* **194**:639–642.

Cody, M. L., 1971, Finch flocks in the Mohave Desert, *Theor. Pop. Biol.* **2**:142–158.

Connell, J. H., 1961, Effects of competition, predation by *Thais lapillus*, and other factors on natural populations of the barnacle *Balanus balanoides*, *Ecol. Monogr.* **31**:61–104.

Connors, P. G., Myers, J. P., Connors, C. S. W., and Pitelka, F. A., 1981, Interhabitat movements by Sanderlings in relation to foraging profitability and the tidal cycle, *Auk* **98**:49–64.

Davies, N. B., 1977, Prey selection and social behaviour in wagtails (Aves: Motacillidae), *J. Anim. Ecol.* **46**:37–57.

Drent, R., and Swierstra, P., 1977, Goose flocks and food finding: Field experiments with Barnacle Geese in winter, *Wildfowl* **28**:15–20.

Dugan, P. J., 1981a, Seasonal movements of shorebirds in relation to spacing behavior and prey availability, Unpublished Ph.D. thesis, University of Durham, U.K

Dugan, P. J., 1981b, The importance of nocturnal foraging in shorebirds: A consequence of increased invertebrate prey activity, in: *Feeding and Survival Strategies in Estuarine Organisms* (N. V. Jones and W. J. Wolff, eds.), pp. 251–260, Plenum Press, New York.

Ens, B., 1979, Territoriality in curlews *Numenius arquata*, *Wader Study Group Bull.* **26**:28–29.

Evans, P. R., 1976, Energy balance and optimal foraging strategies in shorebirds: Some implications for their distributions and movements in the non-breeding season, *Ardea* **64**:117–139.

Evans, P. R., 1981, Migration and dispersal of shorebirds as a survival strategy, in: *Feeding and Survival Strategies in Estuarine Organisms* (N. V. Jones and W. J. Wolff, eds.), pp. 275–290, Plenum Press, New York.

Evans, P. R., Brearey, D. M., and Goodyer, L. R., 1980, Studies on Sanderling at Tees-mouth, NE England, *Wader Study Group Bull.* **30**:18–20.

Fischer, J., 1954, Evolution and bird sociality, in: *Evolution as a Process* (J. S. Huxley, A. C. Hardy, and E. B. Ford, eds.), Allen & Unwin, London.

Fleischer, R. C., 1983, Relationships between tidal oscillations and Ruddy Turnstone flocking, foraging and vigilance behavior, *Condor* **85**:22–29.

Furness, R. W., and Galbraith, H., 1980, Nonrandom distribution in roosting flocks of waders marked in cannon net catch, *Wader Study Group Bull.* **29**:22–23.

Gerstenberg, R. H., and Harris, S. W., 1976, Trapping and marking of shorebirds at Humboldt Bay, California, *Bird-Banding* **47**:1–7.

Gill, F. B., 1978, Proximate costs of competition for nectar, *Am. Zool.* **18**:753–763.

Gill, F. B., and Wolf, L. L., 1975, Economics of feeding territoriality in the Golden-winged Sunbird, *Ecology* **56**:333–345.

Gochfeld, M., and Burger, J., 1980, Opportunistic scavenging by shorebirds: Feeding behavior and aggression, *J. Field Ornithol.* **51**:373–375.

Goss-Custard, J. D., 1969, The winter feeding ecology of the Redshank *Tringa totanus*, *Ibis* **111**:338–356.

Goss-Custard, J. D., 1970, Feeding dispersion in some overwintering wading birds, in: *Social Behaviour in Birds and Mammals* (J. H. Crook, ed.), pp. 3–35, Academic Press, New York.

Goss-Custard, J. D., 1976, Variation in the dispersion of Redshank *Tringa totanus* on their winter feeding grounds, *Ibis* **118**:257–263.

Goss-Custard, J. D., 1977, The ecology of the Wash. III. Density-behavior and the possible effects of a loss of feeding grounds on wading birds (Charadrii), *J. Appl. Ecol.* **14**:721–739.

Goss-Custard, J. D., 1980, Competition for food and interference among waders, *Ardea* **68**:31–52.

Goss-Custard, J. D., Jones, R. E., and Newberry, P. E., 1977, The ecology of the Wash. I. Distribution and diet of wading birds (Charadrii), *J. Appl. Ecol.* **14**:681–700.

Goss-Custard, J. D., Durell, S. E. A., and Ens, B. J., 1982, Individual differences in aggressiveness and food stealing among wintering oystercatchers, *Haematopus ostralegus* L. *Anim. Behav.* **30**:917–928.

Greenberg, R. S., 1980, Demographic aspects of long-distance migrations, in: *Migrant Birds in the Neotropics: Ecology, Behavior, Distribution and Conservation* (A. Keast and E. S. Morton, eds.), pp. 493–504 Smithsonian Institution Press, Washington, D.C.

Grieg-Smith, P. E., 1982, Distress calling by woodland birds, *Anim. Behav.* **30**:299–301.

Groves, S., 1978, Age related differences in Ruddy Turnstone foraging and aggressive behavior, *Auk* **95**:95–103.

Hamilton, W. D., 1971, Geometry for the selfish herd, *J. Theor. Biol.* **31**:295–311.

Hamilton, w. J., 1959, Aggressive behavior in migrant pectoral sandpipers. *Condor* **61**:161–179.

Harrington, B. A., and Groves, S., 1977, Aggression in foraging migrant Semipalmated Sandpipers, *Wilson Bull.* **89**:336–338.

Harrington, B. A., and Leddy, L. E., 1982, Are wader flocks random groupings?—A knotty problem, *Wader Study Group Bull.* **36**:21–22.

Harrington, B. A., and Morrison, R. I. G., 1979, Semipalmated Sandpiper migration in North America, in: *Studies in Avian Biology No. 2* (F. A. Pitelka, ed.), pp. 83–99, Cooper Ornithological Society, Allen Press, Lawrence, Kans.

Hartwick, E. B., and Blaylock, W., 1979, Winter ecology of a Black Oystercatcher population, in: *Studies in Avian Biology No. 2* (F. A. Pitelka, ed.) pp. 207–215, Cooper Ornithological Society, Allen Press, Lawrence, Kans.

Heppleston, P. B., 1971, The comparative breeding ecology of oystercatchers (*Haematopus ostralegus* in The Netherlands inland and coastal habitats, *J. Anim. Ecol.* **41**:23–51.

Hinde, R. A., 1961, Behavior, in: *Biology and Comparative Physiology of Birds* (A. J. Marshall, ed.), Academic Press, New York.

Hulscher, J. B., 1974, An experimental study of the food intake of the oystercatcher *Haematopus ostralegus* L. in captivity during the summer, *Ardea* **62**:155–171.

Johnson, O. W., Johnson, P. M., and Bruner, P. L., 1981, Wintering behavior and site-faithfulness of Golden Plovers on Oahu, *Elepaio* **41**:123–130.

Jones, H. L., and Diamond, J. M., 1976, Short-time-base studies of turnover in breeding bird populations on the California Channel Islands, *Condor* **78**:526–549.

Kelly, P. R., and Cogswell, H. L., 1979, Movements and habitat use by wintering populations of Willets and Marbled Godwits, in: *Studies in Avian Biology No.* (F. A. Pitelka, ed.), pp. 69–82, Cooper Ornithological Society, Allen Press, Lawrence, Kans.

Kelty, M. P., and Lustic, S. I., 1977, Energetics of the starling (*Sterna vulgaris*) in a pine woods, *Ecology* **58**:1181–1185.

Kenward, R. E., 1978, Hawks and doves: Attack success and selection in goshawk flights at woodpigeons, *J. Anim. Ecol.* **47**:449–460.

Krebs, J. R., 1973, Social learning and the significance of mixed-species flocks of chickadees (*Parus* spp.), *Can. J. Zool.* **51**:1275–1288.

Krebs, J. R., 1974, Colonial nesting and social feeding as strategies for exploiting food resources in the Great Blue Heron (*Ardea herodias*), *Behaviour* **52**:99–131.

Krebs, J. R., Stephens, S., and Sutherland, M. L., 1983, Optimal foraging theory, in: *Perspectives in Ornithology* (A. R. Brush and G. A. Clark, eds.), pp. 165–218, Cambridge University Press, London.

Kruuk, H., 1964, Predators and anti-predator behaviour of the Black-headed Gull, *Larus ridibundus*, *Behaviour* (Suppl.) **11**:1–129.

Kus, B. E., 1980, The adaptive significance of flocking among wintering shorebirds, Unpublished M.S. thesis, University of California, Davis.

Kus, B. E., Ashman, P., Page, G. W., and Stenzel, L. E., 1984, Age-related mortality in a wintering population of Dunlin, *Auk*: in press.

Lack, D., 1968, *Ecological Adaptations for Breeding in Birds*, Metheun, London.

Lank, D., 1979, Dispersal and predation rates of wing-tagged Semipalmated Sandpipers *Calidris pusilla* and an evaluation of the technique, *Wader Study Group Bull.* **27**:41–46.

Mallory, E. P., 1981, Ecological, behavioral, and morphological adaptations of a shorebird (the Whimbrel *Numenius phaeopus hudsonicus*) to its different migratory environments, Unpublished Ph.D. thesis. Dartmouth College, Hanover, N.H.

Mallory, E. P., 1982, Territoriality of Whimbrels *Numenius phaeopus* wintering in Panama, *Wader Study Group Bull.* **34**:37–39.

Mallory, E. P., and Schneider, D. C., 1979, Agonistic behavior in Short-billed Dowitchers feeding on a patchy resource, *Wilson Bull.* **91**:271–278.

Marler, P., 1956, Studies of fighting in Chaffinches. (3) Proximity as a cause of aggression, *Br. J. Anim. Behav.* **5**:29–37.

Mikkonen, A. V., 1983, Breeding site tenacity of the Chaffinch *Fringilla coelebs* and the Brambling *F. montifringilla* in northern Finland, *Ornis Scand* **14**:36–47.

Milinski, M., 1977, Do all members of a swarm suffer the same predation? *Z. Tierpsychol.* **45**:373–388.

Milinski, M., and Heller, R., 1978, Influence of a predator on the optimal foraging behavior of sticklebacks *Gasterosteus aculeatus, Nature (London)* **275**:642–644.

Morse, D., 1970, Ecological aspects of some mixed-species flocking in birds, *Ecol. Monogr.* **40**:119–168.

Myers, J. P., 1980a, Territoriality and flocking by Buff-breasted Sandpipers: Variations in non-breeding dispersion, *Condor* **82**:241–250.

Myers, J. P., 1980b, Sanderlings *Calidris alba* at Bodega Bay: Facts, inferences and shameless speculations, *Wader Study Group Bull.* **30**:26–32.

Myers, J. P., 1981, Cross-seasonal interactions in the evolution of sandpiper social systems, *Behav. Ecol. Sociobiol.* **8**:195–202.

Myers, J. P., 1983a, Space, time and the pattern of individual association in a group-living species: Sanderlings have no friends, *Behav. Ecol. Sociobiol.* **12**:129–134.

Myers, J. P., 1983b, Commentary, in: *Perspectives in Ornithology* (A. R. Brush and G. A. Clark, eds.), pp. 216–221, Cambridge University Press, London.

Myers, J. P., and McCaffery, B. J., 1980, Opportunism and site-faithfulness in wintering Sanderlings, *Wader Study Group Bull.* **26**:43.

Myers, J. P., and McCaffery, B. J., 1984, Paracas revisited: Do shorebirds compete on their wintering ground?, *Auk* **101**:197–199.

Myers, J. P., and Myers, L. P., 1979, Shorebirds of coastal Buenos Aires Province, Argentina, *Ibis* **121**:186–200.

Myers, J. P., Connors, P. G., and Pitelka, F. A., 1979a, Territory size in wintering Sanderlings: The effects of prey abundance and intruder density, *Auk* **96**:551–561.

Myers, J. P., Connors, P. G., and Pitelka, F. A., 1979b, Territoriality in nonbreeding shorebirds, in: *Studies in Avian Biology No. 2* (F. A. Pitelka, ed.), pp. 231–245, Cooper Ornithological Society, Allen Press, Lawrence, Kans.

Myers, J. P., Connors, P. G., and Pitelka, F. A., 1980a, Optimal territory size and the Sanderling: Compromises in a variable environment, in: *Mechanisms of Foraging Behavior* (A. C. Kamil and T. D. Sargent, eds.), pp. 135–158, Garland Press, New York.

Myers, J. P., Williams, S. L., and Pitelka, F. A., 1980b, An experimental analysis of prey availability for Sanderlings *Calidris alba* Pallas feeding on sandy beach crustaceans, *Can. J. Zool.* **58**:1564–1574.

Myers, J. P., Maron, J. L., and Sallaberry, M., 1984, Going to extremes: Why do Sanderlings migrate to the Neotropics, in: *Neotropical Ornithology,* (P. A. Buckley, E. S. Morton, N. G. Smith, and R. S. Ridgely, eds.) A.O.U. Monographs, Allen Press, Lawrence, Kans.

Owens, N. W., and Goss-Custard, J. D., 1976, The adaptive significance of alarm calls given by shorebirds on their winter feeding grounds, *Evolution* **30**:397–398.

Page, G., 1974, Age, sex, molt and migration of Dunlins at Bolinas Lagoon, *West. Birds* **5**:1–12.

Page, G., and Whitacre, D. F., 1975, Raptor predation on wintering shorebirds, *Condor* **77**:73–83.

Page, G. W., Stenzel, L. E., and Wolfe, C. M., 1979, Aspects of the occurrence of shorebirds on a central California estuary, in: *Studies in Avian Biology No. 2* (F. A. Pitelka, ed.), pp. 15–32, Cooper Ornithological Society, Allen Press, Lawrence, Kans.

Perdeck, A. C., 1958, Two types of orientation in migrating starlings, *Sturnus vulgaris*, and Chaffinches, *Fringilla coelebs*, as revealed by displacement experiments, *Ardea* **46**:1–37.

Perdeck, A. C., 1964, An experiment on the ending of autumn migration in starlings, *Ardea* **52**:9–15.

Pienkowski, M. W., 1980a, Aspects of the ecology and behaviour of Ringed and Grey Plovers *Charadrius hiaticula* and *Pluvialis squatarola*, Ph.D. thesis, University of Durham, U.K.

Pienkowski, M. W., 1980b, WSG cooperative project on movements of wader populations in western Europe, *Wader Study Group Bull.* **29**:7.

Pienkowski, M. W., and Clark, H., 1979, Preliminary results of winter dye marking on the Firth of Forth, Scotland, *Wader Study Group Bull.* **27**:16–18.

Pitelka, F. A., Holmes, R. T., and MacLean, S. F., Jr., 1974, Ecology and evolution of social organization in Arctic sandpipers, *Am. Zool.* **14**:185–204.

Powell, G. V. N., 1974, Experimental analysis of the social value of flocking by starlings (*Sturnus vulgaris*) in relation to predation and foraging, *Anim. Behav.* **22**:501–505.

Pulliam, H. R., 1973, On the advantages of flocking, *J. Theor. Biol.* **38**:419–422.

Pulliam, H. R., and Parker, T. A., 1979, Population regulation of sparrows, *Fortschr. Zool.* **25**:137–147.

Puttick, G. M., 1980, Foraging behaviour and activity budgets of Curlew Sandpiper, *Ardea* **67**:111–122.

Puttick, G. M., 1981, Sex-related differences in foraging behaviour of Curlew Sandpipers, *Ornis Scand.* **12**:13–17.

Rands, M. W. R., and Barkham, J. P., 1981, Factors controlling within-flock feeding densities in 3 species of wading bird, *Ornis Scand.* **12**:28–36.

Recher, H. F., 1966, Some aspects of the ecology of migrant shorebirds, *Ecology* **47**:393–407.

Recher, H. F., and Recher, J. A., 1969, Some aspects of the ecology of migrant shorebirds. II. Aggression, *Wilson Bull.* **81**:140–154.

Robertson, D. R., Sweatman, H. P., Fletcher, E. A., and Cleland, M. G., 1976, Schooling as a mechanism for circumventing the territoriality of competitors, *Ecology* **57**:1208–1220.

Rubenstein, D. I., 1982, Risk, uncertainty, and evolutionary strategies, in: *Current Problems in Sociobiology* (King College Sociobiology Group, eds.), pp. 91–111, Cambridge University Press, London.

Sih, A., 1980, Optimal behavior: Can foragers balance two conflicting demands? *Science* **210**:1040–1043.

Silliman, J., Mills, G. S., and Alden, S., 1977, Effect of flock size on foraging activity in wintering Sanderlings, *Wilson Bull.* **89**:434–438.

Simon, H., 1956, Rational choice and the structure of the environment, *Psychology Rev.* **63**:129–130.

Smith, P. C., and Evans, P. R., 1973, Studies of shorebirds at Lindisfarne, Northumberland. I. Feeding ecology and behaviour of the Bar-tailed Godwit, *Wildfowl* **24**:135–139.

Stinson, C. H., 1977, The spatial distribution of wintering Black-bellied Plovers, *Wilson Bull.* **89**:470–472.

Stinson, C. H., 1980, Flocking and predator avoidance: Models of flocking and observations on the spatial dispersion of foraging wintering shorebirds (Charadrii), *Oikos* **34**:35–43.

Swinebroad, J., 1964, Nocturnal roosts of migrating shorebirds, *Wilson Bull.* **76**:155–159.

Thompson, W. A., Vertinsky, I., and Krebs, J. R., 1974, The survival value of flocking in birds: A simulation model, *J. Anim. Ecol.* **43**:785–808.

Tinbergen, N., 1951, *The Study of Instinct*, Oxford University Press, (Clarendon), London.

Townshend, D. J., 1979, The use of space by individual Grey Plovers *Pluvialis squatarola* and curlews *Numenius arquata* on their winter feeding grounds, *Wader Study Group Bull.* **26**:29.

Townshend, D. J., 1981, The importance of field feeding to the survival of wintering male and female curlews *Numenius arquata* on the Tees Estuary, in: *Feeding and Survival Strategies of Estuarine Organisms* (N. V. Jones and W. J. Wolff, eds.), pp. 261–273, Plenum Press, New York.

Townshend, D. J., Dugan, P. J., and Pienkowski, M. W., 1983, The unsociable plover— Use of space by Grey Plover *Pluvialis squatarola*, in: *Coastal Waders and Wildfowl in Winter* (R. H. Drent, P. R. Evans, J. D. Goss-Custard, and W. G. Hale, eds.), Cambridge University Press, London.

Tree, A. J., 1979, Biology of the greenshank in southern Africa, *Ostrich* **50**:240–241.

Triesman, M., 1975, Predation and the evolution of gregariousness. I. Models for concealment and evasion, *Anim. Behav.* **23**:779–900.

Vine, I., 1971, Risk of visual detection and pursuit by a predator and the selective advantage of flocking behavior, *J. Theor. Biol.* **30**:405–422.

Vine, I., 1973, Detection of prey flocks by predators, *J. Theor. Biol.* **40**:207–210.

Vines, G., 1980, Spatial consequences of aggressive behavior in flocks of oystercatcher *Haematopus ostralegus* L., *Anim. Behav.* **28**:1175–1183.

Walters, J. R., and Walters, B. F., 1980, Co-operative breeding by Southern Lapwings *Vanellus chilensis*, *Ibis* **122**:505–509.

Ward, P., and Zahavi, A., 1973, The importance of certain assemblages of birds as "information centers" for food-finding, *Ibis* **115**:517–534.

Weathers, W., and Nagy, K. A., 1980, Simultaneous doubly labeled water and time-budget estimates of daily energy expenditures in *Phainopepla nitens*, *Auk* **97**:861–867.

Wolf, L. L., 1978, Aggressive social organization in nectarivorous birds, *Am. Zool.* **18**:765–778.

Yasukawa, K., 1979, Territory establishment in Red-winged Blackbirds: Importance of aggressive behavior and experience, *Condor* **81**:258–264.

Zajonc, R. B., 1965, Social facilitation, *Science* **149**:269–274.

Zwarts, L., 1981, Intra- and interspecific competition for space in estuarine bird species in a one prey situation, *Proc. XCII Int. Ornithol. Congr.* (*Berlin*) pp. 1045–1050.

Zwarts, L., and Drent, R. H., 1981, Prey depletion and the regulation of predator density: Oystercatchers (*Haematopus ostralegus*) feeding on mussels (*Mytilus edulis*), in: *Feeding and Survival Strategies of Estuarine Organisms* (N. V. Jones and W. J. Wolff, eds), pp. 193–216, Plenum Press, New York.

SPECIES INDEX

323

SUBJECT INDEX